Intelligent Diagnosis and Prognosis of Industrial Networked Systems

AUTOMATION AND CONTROL ENGINEERING
A Series of Reference Books and Textbooks

Series Editors

FRANK L. LEWIS, Ph.D., Fellow IEEE, Fellow IFAC
Professor
Automation and Robotics Research Institute
The University of Texas at Arlington

SHUZHI SAM GE, Ph.D., Fellow IEEE
Professor
Interactive Digital Media Institute
The National University of Singapore

Classical Feedback Control: With MATLAB® and Simulink®, Second Edition, *Boris J. Lurie and Paul J. Enright*

Synchronization and Control of Multiagent Systems, *Dong Sun*

Subspace Learning of Neural Networks, *Jian Cheng Lv, Zhang Yi, and Jiliu Zhou*

Reliable Control and Filtering of Linear Systems with Adaptive Mechanisms, *Guang-Hong Yang and Dan Ye*

Reinforcement Learning and Dynamic Programming Using Function Approximators, *Lucian Buşoniu, Robert Babuška, Bart De Schutter, and Damien Ernst*

Modeling and Control of Vibration in Mechanical Systems, *Chunling Du and Lihua Xie*

Analysis and Synthesis of Fuzzy Control Systems: A Model-Based Approach, *Gang Feng*

Lyapunov-Based Control of Robotic Systems, *Aman Behal, Warren Dixon, Darren M. Dawson, and Bin Xian*

System Modeling and Control with Resource-Oriented Petri Nets, *Naiqi Wu and MengChu Zhou*

Sliding Mode Control in Electro-Mechanical Systems, Second Edition, *Vadim Utkin, Jürgen Guldner, and Jingxin Shi*

Optimal Control: Weakly Coupled Systems and Applications, *Zoran Gajić, Myo-Taeg Lim, Dobrila Skatarić, Wu-Chung Su, and Vojislav Kecman*

Intelligent Systems: Modeling, Optimization, and Control, *Yung C. Shin and Chengying Xu*

Optimal and Robust Estimation: With an Introduction to Stochastic Control Theory, Second Edition, *Frank L. Lewis, Lihua Xie, and Dan Popa*

Feedback Control of Dynamic Bipedal Robot Locomotion, *Eric R. Westervelt, Jessy W. Grizzle, Christine Chevallereau, Jun Ho Choi, and Benjamin Morris*

Intelligent Freight Transportation, *edited by Petros A. Ioannou*

Modeling and Control of Complex Systems, *edited by Petros A. Ioannou and Andreas Pitsillides*

Wireless Ad Hoc and Sensor Networks: Protocols, Performance, and Control, *Jagannathan Sarangapani*

Stochastic Hybrid Systems, *edited by Christos G. Cassandras and John Lygeros*

Hard Disk Drive: Mechatronics and Control, *Abdullah Al Mamun, Guo Xiao Guo, and Chao Bi*

Autonomous Mobile Robots: Sensing, Control, Decision Making and Applications, *edited by Shuzhi Sam Ge and Frank L. Lewis*

Automation and Control Engineering Series

Intelligent Diagnosis and Prognosis of Industrial Networked Systems

Chee Khiang Pang
National University of Singapore,
Singapore

Frank L. Lewis
The University of Texas at Arlington,
USA

Tong Heng Lee
National University of Singapore,
Singapore

Zhao Yang Dong
The Hong Kong Polytechnic University,
Hong Kong SAR

CRC Press
Taylor & Francis Group
Boca Raton London New York

CRC Press is an imprint of the
Taylor & Francis Group, an **informa** business

CRC Press
Taylor & Francis Group
6000 Broken Sound Parkway NW, Suite 300
Boca Raton, FL 33487-2742

International Standard Book Number: 978-1-4398-3933-1 (Hardback)

Visit the Taylor & Francis Web site at
http://www.taylorandfrancis.com

and the CRC Press Web site at
http://www.crcpress.com

Dedication

To those I love, and those who love me.
C. K. Pang

To Galina.
F. L. Lewis

Contents

Preface

In an era of intensive competition when asset usage and plant operating efficiencies must be maximized, unexpected downtime due to machinery failure has become more costly and unacceptable than before. To cut operating costs and increase revenues, industries have an urgent need for prediction of fault progression and remaining lifespan of industrial machines, processes, and systems. As such, predictive maintenance has been actively pursued in the manufacturing industries in recent years where equipment outages are forecasted, and maintenance is carried out only when necessary. Prediction leads to improved management and hence effective usage of equipment, and multifaceted guarantees are increasingly being given for industrial machines, processes, products, and services, etc. To ensure successful condition-based maintenance, it is necessary to detect, identify, and classify different kinds of failure modes in the manufacturing processes as early as possible.

With the pushing need for increased longevity in machine lifetime and its early process fault detection, intelligent diagnosis and prognosis have become an important field of interest in engineering. For example, an engineer who mounts an acoustic sensor onto a spindle motor would like to know when the ball bearings will be worn out and need to be changed without having to halt the ongoing milling processes, which decreases the industrial yield. Or a scientist working on sensor networks would like to know which sensors are redundant during process monitoring and can be pruned off to save operational and computational overheads. These realistic scenarios illustrate the need for new or unified perspectives for challenges in system analysis and design for engineering applications.

Currently, most works on Condition-Based Monitoring (CBM), Fault Detection and Isolation (FDI), or even Structural Health Monitoring (SHM) consider solely the integrity of independent modules, even when the complex integrated industrial processes consist of several mutually interacting components interwoven together. Most literature on diagnosis and prognosis is also mathematically involved, which makes it hard for potential readers not working in this field to follow and appreciate the state-of-art technologies. As such, a good intelligent diagnosis and prognosis architecture should consider crosstalk to facilitate actions and decisions among the synergetic integration of composite systems simultaneously, while maintaining overall stability at the same time. This "big-picture" approach will also limit the inherent intrinsic uncertainties and variabilities within the interacting components, while suppressing any possible extrinsic socio-techno intrusion and uncertainties from the human interface layer.

Adding to the current literature available in this research arena, this book provides an overview of linear systems theory and the corresponding matrix operations required for intelligent diagnosis and prognosis of industrial networked systems. With the essential theoretical fundamentals covered, automated mathematical machineries are developed and applied to targeted realistic engineering systems. Our results show the effectiveness of these tool sets for many *time-triggered* and *event-triggered* industrial applications, which include forecasting machine tool wear in industrial cutting machines, sensors and features reduction for industrial FDI, identification of critical resonant modes in mechatronic systems for systems design of research and development (R&D), probabilistic small signal stability in large-scale interconnected power systems, discrete event command and control for military applications, etc., just to name a few. It should be noted that these developed tool sets are highly portable, and can be readily adopted and applied to many other engineering applications.

Outline

This book is intended primarily as a bridge between academics in universities, practicing engineers in industries, and also scientists working in research institutes. The book is carefully organized into chapters, each providing an introductory section tailored to cover the essential background materials, followed by specific industrial applications to realistic engineering systems and processes. To reach out to a wider audience, linear matrix operators and indices are used to formulate mathematical machineries and provide formal decision software tools that can be readily appreciated and applied. The book is carefully crafted into seven chapters with the following contents:

- Chapter 1: *Introduction*
  Intelligent diagnosis and prognosis using model-based and non-model-based methods in current existing literature are discussed. The various application domains in realistic industrial networked systems are also introduced.
- Chapter 2: *Vectors, Matrices, and Linear Systems*
  Fundamental concepts of linear algebra and linear systems are reviewed along with eigenvalue and singular value decompositions. The usage of both real and binary matrices for diagnosis and prognosis applications are also discussed.
- Chapter 3: *Modal Parametric Identification (MPI)*
  Proposes a Modal Parametric Identification (MPI) algorithm for fast identification of critical modal parameters in R&D of mechatronic systems. A systems design approach with enhanced MPI is proposed for mechatronic systems and verified with frequency responses of dual-stage actuators in commercial hard disk drives (HDDs).
- Chapter 4: *Dominant Feature Identification (DFI)*
  Proposes a Dominant Feature Identification (DFI) software framework for advanced feature selection when using inferential sensing in online monitoring of industrial systems and processes. A mathematical tool set which guarantees minimized least squares error in feature reduction and clustering is developed. The proposed techniques are verified with experiments on tool wear prediction in industrial high speed milling machines and fault detection in a machine fault simulator.
- Chapter 5: *Probabilistic Small-Signal Stability Assessment*
  Proposes analytical and numerical methods to obtain eigenvalue sensitivities with respect to non-deterministic system parameters and load models

for large-scale interconnected power systems. A probabilistic small-signal stability assessment method is proposed, and verified with extensive simulations on the New England 39-Bus Test System.

- Chapter 6: *Discrete Event Command and Control*
  Proposes the use of binary matrices and algebra for command and control of discrete event–triggered systems. A mathematically justified framework is provided for distributed networked teams on multiple missions. This is verified with simulations and experiments on a wireless sensor network (WSN), as well as simulation on a military ambush attack mission.
- Chapter 7: *Future Challenges*
  Provides conclusion and future work directions for intelligent diagnosis and prognosis in areas of energy-efficient manufacturing, life cycle assessment, and systems of systems architecture.

Learning Outcomes

The developed tools allow for higher level decision making and command in synergetic integration between several industrial processes and stages, thereby achieving shorter time in failure and fault analysis in the entire industrial production life cycle. This shortens production time while reducing failure through early identification and detection of the key factors that can lead to potential faults. As such, engineers and managers are empowered with the knowledge and know-how to make important decisions and policies. They can also be used to educate fellow researchers and the public about the advantages of various technologies.

Potential readers not working in the relevant fields can also appreciate the literature therein even without prior knowledge and exposure, and are still be able to apply the tool sets proposed therein to address industrial problems arising from evolving or even emerging behavior in networked systems or processes, e.g., sensor fusion, pattern recognition, and reliability studies, etc. The mathematical machineries proposed aim to analyze methodologies to make autonomous decisions that meet present and uncertain future needs quantitatively, without compromising the ad-hoc "add-on" flexibility of network-centered operations.

Many universities also have established programs and courses in this new field, with cross-faculty and inter-discipline research going on in this arena as well. As such, this book can also serve as a textbook for an intermediate to advanced module as part of control engineering, systems reliability, diagnosis and prognosis, etc. We also hope that the book is concise enough to be used for self-study, or as a recommended text, for a single advanced undergraduate or postgraduate module on intelligent diagnosis and prognosis, FDI, CBM, or SHM, etc.

Acknowledgments

Last but not least, we would like to acknowledge our loved ones for their love, understanding, and encouragement throughout the entire course of preparing this research book. This book was also made possible with the help of our colleagues, collaborators, as well as students and members in our research teams. This work was supported in part by Singapore MOE AcRF Tier 1 Grant R-263-000-564-133, NSF Grant ECCS-0801330, ARO Grant W91NF-05-1-0314, AFOSR Grant FA9550-09-1-0278, and Hong Kong Polytechnic University Grant #ZV3E.

Chee Khiang Pang
Frank L. Lewis
Tong Heng Lee
Zhao Yang Dong

Nomenclature

ADFI	Augmented Dominant Feature Identification
AE	Acoustic Emission
ARFIMA	AutoRegressive Fractionally Integrated Moving Average
ARIMA	AutoRegressive Integrated Moving Average
ARMAX	Auto-Regressive Moving-Average with eXogenous input/s
AVR	Automatic Voltage Regulator
BEP	Best Efficiency Point
BIBO	Bounded-Input-Bounded-Output
BU	Business Unit
C2	Command and Control
CAD	Computer-Aided Design
CAPEX	CAPital EXpenditure
CBM	Condition-Based Monitoring
CDF	Cumulative Distribution Function
CF	Characteristic Function
CNC	Computer Numerical Control
dRAM	disjunctive-input Resource Assignment Matrix
DAE	Differential and Algebraic Equation
DDFI	Decentralized Dominant Feature Identification
DEC	Discrete Event Control
DFI	Dominant Feature Identification
DSA	Dynamic Signal Analyzer
DSP	Digital Signal Processing
EA	Evolutionary Algorithm
ELS	Extended Least Squares
EPRI	Electric Power Research Institute
FACTS	Flexible Alternating Current Transmission Systems
FCS	Future Combat System
FDI	Fault Detection and Isolation
FEA	Finite Element Analysis
FEM	Finite Element Modeling
FFBD	Functional Flow Block Diagram
FFT	Fast Fourier Transform
GA	Genetic Algorithm
GHG	Green House Gas
HDD	Hard Disk Drive
HHT	Hilbert–Huang Transform
HMM	Hidden Markov Model
HTN	Hierarchical Task Network
HVDC	High-Voltage Direct Current
IM	Induction Motor
IPP	Independent Power Producer
ISO	Independent System Operator
JAUGS	Joint Architecture for Unmanned Ground System
KLT	Karhunen–Loève Transform

LCA	Life Cycle Assessment
LDV	Laser Doppler Vibrometer
LS	Least Squares
LSE	Least Square Error
LITP	Linear-In-The-Parameter
LTI	Linear Time-Invariant
MIMO	Multi-Input-Multi-Output
MPI	Modal Parametric Identification
MRE	Mean Relative Error
MRM	Multiple Regression Model
MSE	Mean Square Error
MTBF	Mean Time Between Failure
NN	Neural Network
OODA	Observe, Orient, Decide, and Act
OPEX	OPerations EXpense
OS	Overall Sensitivity
O&S	Operation and Support
PCA	Principal Component Analysis
PDA	Personal Digital Assistant
PDF	Probability Density Function
PFA	Principal Feature Analysis
PHM	Prognostic Health Management
PN	Petri Net
PSS	Power System Stabilizer
PZT	Lead-Zirconate-Titanate (Pb-Zr-Ti)
R&D	Research & Development
RAM	Resource Assignment Matrix
RBF	Radial Basis Function
RBS	Rule-Based System
RDM	Resource Dependency Matrix
RLS	Recursive Least Squares
RMS	Root Mean Square
RTO	Regional Transmission Organization
SD	Standard Deviation
SHM	Structural Health Monitoring
SISO	Single-Input-Single-Output
SNR	Signal-to-Noise Ratio
SoS	System-of-Systems
SVD	Singular Value Decomposition
TCM	Tool Condition Monitoring
TIA	Totally Integrated Automation
TOC	Total Ownership Cost
TPM	Technical Performance Metric
TRADOC	TRAining and DOCtrine command
TSM	Task Sequencing Matrix
UAV	Unmanned Aerial Vehicle
UGS	Unattended Ground Sensor
UGV	Unmanned Ground Vehicle
VCM	Voice Coil Motor
WSN	Wireless Sensor Network
ZIP	Constant impedance (Z), current (I), and power (P)

List of Figures

List of Tables

1 Introduction

Traditionally, *diagnosis* and *prognosis* are terms commonly used in the medical domain; diagnosis being the identification of diseases through careful observation of the patients' symptoms and results from various in-depth examinations, while prognosis is the prediction of the various outcomes of the illness and corresponding remedies, including the anticipation of recovery from the expected course if no contingencies arise. With the introduction of these concepts to the field of engineering, diagnosis now becomes the art of identification of engineering systems' failure through observation of the sensory signals from the machines and equipment, while prognosis is the prediction of failure and provision of corresponding engineering solutions to achieve expected desired outcomes. Intelligent diagnosis and prognosis is extremely important as we are in an era of intensive manufacturing competition, and the new challenges faced by industrial manufacturing processes include maximizing productivity, ensuring high product quality, and reducing the production time while minimizing the production cost simultaneously. Unanticipated and unresolved failures result in machine and equipment downtime, and increases OPerations EXpenditure (OPEX) and CAPital EXpenditure (CAPEX) from halting production and intervention of engineers, respectively. This decreases revenue and is unacceptable in modern competitive manufacturing and production systems.

The desire and need for various accurate diagnostic tools with predictive prognostic capabilities have been around since human beings invented and operated complex and expensive machineries after the Industrial Revolution. With the amount of technological advancement, the area of intelligent diagnosis and prognosis is of extreme importance in today's industrial networked systems ranging from complex manufacturing, large-scale interconnected power systems, aerospace vehicles, military and merchant ships, and automotive industries, etc. While the application domains might be different, the goals are identical to maximize equipment up time and to minimize maintenance and operating costs. As manning levels are reduced and equipment becomes more complex, intelligent maintenance schemes must replace the old prescheduled and labor intensive planned maintenance systems to ensure that equipment continues to function [1].

It is thus obvious that the modern engineering systems require intelligent machine and system fault diagnosis and prognosis, since the positive impacts of employing these capabilities reduces Operation and Support (O&S) costs and Total Ownership Costs (TOCs). In this chapter, we will give a survey of the diagnosis and prognosis methods provided in current literature, as well as those of technical proximity, eg., Conditioned-Based Maintenance (CBM), Prognostic Health Management (PHM), Fault Detection and Isolation (FDI), Structural Health Monitoring (SHM), etc. We will then discuss the novelty of our proposed methodologies and our application to realistic industrial networked systems.

1.1 DIAGNOSIS AND PROGNOSIS

The intelligence of diagnosis and prognosis methodologies has increased tremendously over the past few decades. After a positive diagnosis of impending fault or failure in the engineering system or process, prognosis takes place and provides ample time for rectification before total breakdown or instability. While this time window and corresponding corrective measurements depend on application to application, the main functions include but are not limited to detection, isolation, quantification, prediction, anticipation, and correction, etc. The vast range of applications now range from manual, semi-automated, or fully automated, and have been applied to manufacturing, commercial, and defense systems, etc. Generally, these applications can be classified mainly into *parametric-based* or *non-parametric-based* diagnosis and prognosis.

1.1.1 PARAMETRIC BASED

Parametric-based methods for diagnosis and prognosis are attractive because the underlying physics or statistics can be captured and understood. In this section, we discuss some of these research efforts for realistic engineering systems that appear in current literature.

The Markov Process is a mathematical model used to represent random evolution of a memoryless system, i.e., a model for which the likelihood of a given future state at any given moment depends only on its present state and not on any past states. Such a process is analogous to "threshold-detection." Over the years, Markov Processes have been successfully used by researchers to model past failure states or predict future states of processes or machines in aerospace, automotive, and defense industries, etc.

A Hidden Markov Model (HMM) has only one discrete hidden state variable and a set of discrete or continuous observation nodes. As tool wear degradation is one of the main failure modes in large scale industrial cutting and milling machines, tool wear monitoring and prediction of useful remaining life is a very important and practical consideration. The cutting tool wear monitoring and prediction of useful life was modeled using HMMs in [2][3], via self organizing maps and dynamic HMMs in [4][5], and continuous HMMs in [6]. Similarly, monitoring of bearings' health in rotary machines is also studied in current literature. A strategy to optimize bearing maintenance schedules was proposed with the application of condition-based monitoring techniques in [7]. A robust condition based maintenance algorithm was developed for Gearbox of Westland and SH-60 helicopters and remaining useful life prediction using HMM techniques in [8][9]. Diagnosis of pump systems was also studied using a new autoregressive hidden semi-Markov Model in [10].

The maintenance strategy has also evolved over the years. A real-time health prognosis and dynamic preventive maintenance policy was developed for equipment under aging Markovian deterioration in [11]. Markov analysis technique was used to calculate reliability measures for safety systems that involving several typical reliability related factors in [12]. The structural parallel system with multiple failures was studied using Markov chain-line sampling method in [13].

Besides Markovian methods, time series models are also used to forecast future events based on known past events. Time series analyses are often divided into two classes; namely *frequency-domain* methods and *time-domain* methods. The former is based on frequency information from spectral analysis or wavelet analysis, while time series analyses are based mainly on examination of autocorrelation and cross-correlation information. Autocorrelation analysis examines *serial* dependence, and spectral analysis examine *cyclic* behavior which might not be related to seasonality.

In the Box-Jenkins methodology (named after the statisticians George Box and Gwilym Jenkins), AutoRegressive Moving Average (ARMA) models are used to find the best fit of a time series to past values of this time series for forecasts. Models for time series data can have many forms and represent different stochastic processes. When modeling variations in the level of a process, three broad classes of practical importance are the AR models, Integrated (I) models, and MA models. These three classes depend linearly on previous data points. Combinations of these ideas produce ARMA and AutoRegressive Integrated Moving Average (ARIMA) models. The AutoRegressive Fractionally Integrated Moving Average (ARFIMA) model generalizes the former three. The ARMA model is used to successfully to monitor and forecast past, present, and future conditions of the machine by various researches in automotive and aeronautical fields. Power consumption of active suspension of automotive system is predicted [14] using a novel pseudo-linear method for the estimation of fractionally integrated ARMA. The authors in [15] used AR and ARMA models to predict automobile body shop assembly process variation. The steam turbine rotor failure forecasted by [16] using vibration signals and applying the ARMA model. The condition based maintenance forecast for aerospace engine performance parameters analyzed using the ARMA model [17]. A simple and fast soft-wired tool wear state at every wear condition monitoring is developed using the ARMA model [18]. A method to predict the future conditions of machines based on one-step-ahead prediction of time-series forecasting techniques and regression trees is proposed in [19].

1.1.2 NON-PARAMETRIC BASED

For non-parametric-based approaches, heuristic methods involving Evolutionary Algorithms (EAs), Genetic Algorithms (GAs), Neural Networks (NNs), fuzzy logic, etc., and their combinations have been used for improved intelligence. The GA uses are inspired by the evolutionary nature of biological systems, and is a search technique consisting of mutation, selection, reproduction (inheritance), and recombination stages. GAs are also commonly used for complex optimization, and are used in financial, industrial design using parametrization, time series prediction, and signal processing applications, etc.

In current literature, GAs are used in a layered approach based on Q-learning (a reinforcement learning technique) to determine the best weighting for optimal control and design problems in [20]. The authors in [21] presented an algorithm for determining an optimal loading of elements in series-parallel systems. An early warning method was proposed based on a discrete event simulation that evaluates cost-effects of an arbitrary allocation of failure risks for a simulated 3-machine production system in [22]. The authors in [23] proposed an optimization procedure based on GA to search for the most cost-effective maintenance schedule, considering both production gains and maintenance expenses. A multivariate Bayesian process mean-control problem for a finite production run (under the assumption that the observations are values of independent normally-distributed vectors of random variables) is presented in [24]. The author in [25] modeled repairable systems using hierarchical Bayesian model that are a compromise between the "bad-as-old" and "good-as-new." The Aircraft rotor dynamic system is studied using an applied continuous wavelet transform that utilizes harmonic forcing satisfying combination resonance. The authors in [26] performed health monitoring and evaluation of dynamic characteristics in gear sets using wavelet transform methods. A condition monitoring system was set up for a turbine using vibration data based on state-space whose associated recursive algorithms (Kalman filter and fixed interval smoothing) provide the basis for probability of failure estimation in [27]. Kalman filters were also used to track changes in features like vibration levels, mode frequencies, or other waveform signature features. Prognostic utility for the signature features are determined by transitional failure experiments in [28].

For the case of NNs, information processing algorithms are employed to mimick a human brain, i.e., activation of functions via "firing" of neurons. A typical NN consists of a large number of highly interconnected processing elements (neurons) through proper activation, which work in unison to solve specific problems like pattern recognition, time series prediction, or data classification, etc. In an artificial NN, other intelligent techniques are merged with conventional NN to complete even more difficult or higher hierarchical tasks. In the literature, two popular artificial intelligent techniques, i.e., artificial NNs and expert systems are commonly used for machine diagnosis, which include fuzzy-logic systems, fuzzy-neural networks, and neural-fuzzy systems, etc.

The authors in [29] developed a new hydraulic valve fluid field model based on non-dimensional artificial NNs to provide an accurate and numerically efficient tool in an automatic transmission hydraulic control system design. Their results show better performance than the conventional computational fluid dynamics technique, which is numerically inefficient and time consuming. The turbine, compressor, and gear wear statuses were studied in [30], and an NN model was proposed which predicts temperature faster in comparison with the original models for Honeywell turbine and compressor components. The authors in [31] evaluated the performance of recurrent neural networks and neuro-fuzzy systems predictors using two-benchmark gear wear data sets. Through comparison, it is found that if a neuro-fuzzy system is properly trained, it performs better than recurrent neural networks in both forecasting accuracy and training efficiency. Artificial intelligence techniques have been applied to machine diagnosis and have shown improved performance over conventional approaches.

As such, it is obvious that novel techniques in intelligent diagnosis and prognosis have become an important field of interest in engineering science with the pushing need for increased longevity in machine lifetime and its early fault detection. In this book, we develop the essential theories for tackling issues pertaining to application issues faced during complex decision making in a multi-attribute space. This is done through use of linear matrix operators and indices to formulate mathematical machineries and provide formal decision software tools which can be readily appreciated and used by engineers from industries as well as researchers from research institutes and academia. The developed tools allow for higher level decision making and command in synergetic integration between several industrial processes and stages, for shorter time in failure and fault analysis in the entire industrial production life cycle. The enhanced automated systems level toolsets also address industrial problems arising from evolving and emerging behavior in networked systems or processes, and analyze methodologies to make autonomous decisions that meet present and uncertain future needs quantitatively without compromising the ad hoc "ad-on" flexibility of network-centered operations. This shortens production time while reducing failure by early identification and detection of the key factors which lead to potential faults. As such, engineers and managers are empowered with the knowledge and know-how to make important decisions and policies, as well as educate fellow researchers and public about the advantages of various technologies.

1.2 APPLICATIONS IN INDUSTRIAL NETWORKED SYSTEMS

In this section, we detail some of the realistic engineering domains of industrial networked systems where intelligent diagnosis and prognosis are needed. With the complexity and scale of problems encountered, it is essential that linear operations with closed-form solutions are provided for quick and easy analyses. The identified applications are modeled using linear systems theory, and diagnosis and prognosis using proposed solutions are carried out with advanced matrix manipulations.

1.2.1 MODAL PARAMETRIC IDENTIFICATION (MPI)

Mechatronic systems—the integration of mechanical, electrical, computational, and control systems—have pervaded products ranging from larger scale anti-braking systems in cars to small scale microsystems in portable mobile phones. To sustain market shares in the highly competitive consumer electronics industries, continual improvements in servo evaluation and performance of products are essential, in particular for portable devices requiring ultra-high data capacities and ultra-strong disturbance rejection capabilities. Moreover, pressures to minimize time-to-market of new products imply a necessarily high schedule compression of the end-to-end servomechanical design and evaluation cycle, i.e., from mechanical structural designing, prototyping, to servo control system design to meet target specifications.

Furthering our earlier research to improve yield through rapid Modal Parametric Identification (MPI) of critical resonant modes for prognosis with mechanical actuator redesign using Least-Squares (LS) of frequency response data packed in ***matrices*** [32], we enhance the proposed MPI using Forsythe's method of complex orthogonal polynomial transformation [33]. The new MPI avoids ill-conditioned numerical solutions during computation of matrices [34], and the calculated modal parameters are stored in a central repository for technical sharing. This immediately follows from the proposed integrated systems design methodology in managing complex R&D processes for high-performance mechatronic industries [35].

1.2.2 DOMINANT FEATURE IDENTIFICATION (DFI)

In an era of intensive competition, the new challenges faced by industrial manufacturing processes include maximizing productivity, ensuring high product quality, and reducing the production time while minimizing the production cost simultaneously. As such, it is crucial that asset usage and plant operating efficiency be maximized, as unexpected downtime due to machinery failure has become more costly than before. Predictive maintenance is thus actively pursued in the manufacturing industry in recent years, where equipment outages are predicted and maintenance is carried out only when necessary.

To ensure successful condition based maintenance, it is necessary to detect, identify, and classify different kinds of failure modes in the manufacturing process. One of the causes of delay in manufacturing processes is machine downtime or failure of the machining tools. Failure of machine tools can also affect the production rate and quality of products, and detection of the tool state becomes a vital role in manufacturing [36].

In an industrial cutting machine, it is not possible to estimate and predict the state of the tool directly. However, it is easy to obtain information that is highly correlated to tool through inferential sensing, which can be achieved by signal processing of the collected signals obtained from the embedded sensors. In [37], we developed a Dominant Feature Identification (DFI) algorithm to identify the dominant features from a dynamometer force sensor which affects tool wear based on the features vs. time data concatenated in a ***matrix***. A new DFI approach is proposed in [38] to predict tool wear dynamically from the dominant features selected from an acoustic sensor. With the celebrated success of the effectiveness of DFI, it is also extended in [39] to greatly reduce the number of sensors and features required for industrial fault detection.

1.2.3 PROBABILISTIC SMALL SIGNAL STABILITY ASSESSMENT

Due to the deregulation of power industry in many countries, the traditionally vertically integrated power systems have been experiencing dramatic changes leading to competitive electricity markets. Power system planning in such an environment is now facing increasing requirements and challenges because of the deregulation. In particular, it introduces a variety of uncertainties to system planning. The traditional deterministic power system analysis techniques have been found in many cases to have limited capability to reveal the increasing uncertainties in today's power systems. The power system operation and planning demonstrate probabilistic characteristics which requires emphasis on probabilistic techniques.

To study the probabilistic small signal stability on large-scale interconnected power systems, it is imperative to obtain a linearized model of the entire power system under consideration in state-space form, where the key generation and control parameters are coupled with the state transition ***matrix***. In this chapter, we investigate power system state matrix sensitivity characteristics with respect to system parameter uncertainties with analytical and numerical approaches, and identify those parameters that have great impact on system eigenvalues, therefore, the system stability properties [40]. A key probabilistic power system analysis technique is the probabilistic power system small signal stability assessment technique. With the many factors such as demand uncertainty, market price elasticity, and unexpected system congestions, it is more appropriate to have probabilistic power system stability assessment results rather than a deterministic one, especially for the sake of risk management in a competitive electricity market [41]. We present a framework of probabilistic power system small signal stability assessment technique, fully supported with detailed probabilistic analysis and case studies. The results can be used as a valuable reference for utility power system small signal stability assessment, probabilistically and reliably, and can be used to help Regional Transmission Organizations (RTOs) and Independent System Operators (ISOs) perform planning studies under the open access environment [42].

On the other hand, load modeling also plays an important role in power system dynamic stability assessment. One of the widely used methods in assessing load model impact on system dynamic response is parametric sensitivity analysis. A composite load model-based load sensitivity analysis framework is proposed. It enables comprehensive investigation into load modeling impacts on system stability considering the dynamic interactions between load and system dynamics. The effect of the location of individual as well as patches of composite loads in the vicinity on the sensitivity of the oscillatory modes are also investigated. The impact of load composition on the overall sensitivity of the load is also discussed [43].

1.2.4 DISCRETE EVENT COMMAND AND CONTROL

Traditional design methodologies for discrete event workflow systems that employ trial-and-error approaches for evaluating designs followed by implementations are time consuming and costly. If tasks are not scheduled to be activated correctly, serious problems might occur in the workflow, including blocking and deadlock phenomena, which will halt the entire discrete event system. As such, it is imperative to perform an integrated inquiry of how task structures impact discrete-event task scheduling efficiencies. This is instrumental in the presence of competitive due-date targets and time-sensitive operations in many workflow systems, where managers typically seek process re-engineering solutions using concurrent process engineering, etc.

In military systems, the TRADOC Pamphlet 525-66 Battle Command and Battle Space Awareness capabilities prescribe expectations that networked teams will perform in a reliable manner under changing mission requirements, varying resource availability and reliability, and resource faults, etc., during mission execution in military applications. In [44], a Command and Control (C2) structure is presented that allows for computer-aided execution of the networked team decision-making process, control of force resources, shared resource dispatching, and adaptability to change based on battlefield conditions. A mathematically justified networked computing environment called the Discrete Event Control (DEC) Framework based on ***binary matrices*** is provided. DEC has the ability to provide the logical connectivity among all team participants including mission planners, field commanders, war-fighters, and robotic platforms, etc. The proposed data management tools are developed and demonstrated in a simulation and implementation study using a distributed Wireless Sensor Network (WSN). A simulation example on a battlefield with networked Future Combat System (FCS) teams deploying ambush attack tactics is also presented [45]. The results show that the tasks of multiple missions are correctly sequenced in real-time, and that shared resources are suitably assigned to competing tasks under dynamically changing conditions without conflicts and bottlenecks.

In this chapter, we review the parametric-based and non-parametric-based diagnosis and prognosis methods that exist in current literature. Some examples are given of realistic engineering applications in industrial networked systems where intelligent diagnosis and prognosis are also identified, and linear matrix operations are proposed for fast closed-form analysis.

With these covered, we will introduce the essential theoretical fundamentals of vectors, matrices, and linear systems in the following chapter. Advanced matrix operations like eigenvalue decomposition and Singular Value Decomposition (SVD) are also presented, along with binary matrices, which are useful for modeling and control of event-triggered systems.

2 Vectors, Matrices, and Linear Systems

Linear algebra has a wide range of applications in various domains for diagnosis and prognosis of engineering systems. In this chapter, we first review the fundamental concepts of linear algebra, including domain, range, transformation, and null spaces, etc. This is followed by the introduction of linear systems, which covers linearization of non-linear systems. Finally, we investigate a special type of matrix called *Boolean* matrices, and its usage in system modeling of graphs and Discrete Event Control (DEC) of *event-triggered* discrete event systems.

2.1 FUNDAMENTAL CONCEPTS

In this section, we introduce the basic definitions of vectors and matrices. We also review the concepts of norms and spaces, as well as some useful matrix operators which are commonly encountered in different fields of engineering.

2.1.1 VECTORS

A vector is an n-tuple of numbers arranged either in a column or row [46]. This column or row arrangement hence gives a vector magnitude and direction.

For example,

$$\mathbf{x} = \begin{bmatrix} x_1 \\ x_2 \\ \vdots \\ x_n \end{bmatrix}, \tag{2.1}$$

where $\mathbf{x}$ is a *column* vector of length n, while

$$\mathbf{y} = \begin{bmatrix} y_1 & y_2 & \cdots & y_m \end{bmatrix} \tag{2.2}$$

is a *row* vector of length m. The elements in $\mathbf{x}$ and $\mathbf{y}$ can be real or complex depending on application domain.

Norms are generally functions which map a vector onto a scalar. Using $\mathbf{x}$ in (2.1) as an example, the *Euclidean norm* $||\mathbf{x}||_2$ of $\mathbf{x}$ is

$$\begin{aligned} ||\mathbf{x}||_2 &= \sqrt{x_1^2 + x_2^2 + \cdots + x_n^2} \\ &= \sqrt{\mathbf{x}^T\mathbf{x}}, \end{aligned} \tag{2.3}$$

where the superscript ^{T} denotes the transpose operation. The Euclidean norm is also called the *Euclidean length*, $\mathfrak{L}_2$ *distance*, or $\mathfrak{L}^2$ *norm*. As such, the difference in

Euclidean norms of two vectors is in essence the inverse of the physical proximity between them, i.e., the greater the difference in Euclidean norms, the "farther" or "disalike" the two vectors are.

Other interesting vector norms include the *Taxicab norm* or *Manhattan norm*

$$||\mathbf{x}||_1 = \sum_{i=1}^{n} |x_i|, \tag{2.4}$$

which is the $\mathfrak{L}_1$ *distance* or $\mathfrak{L}^1$ *norm.*

The *infinite norm* or *supremum* is

$$||\mathbf{x}||_\infty = \max\big(|x_1|, |x_2|, \cdots, |x_n|\big), \tag{2.5}$$

where max(.) denotes the maximum value of the elements therein.

Now given a set of k vectors $\mathbf{x_1}$, $\mathbf{x_2}$, $\cdots$, $\mathbf{x_k}$ of the same length of n, the *inner product* of $\mathbf{x_1}$ and $\mathbf{x_2}$ is denoted as $\mathbf{x_1} \cdot \mathbf{x_2}$ or $\langle \mathbf{x_1}, \mathbf{x_2} \rangle$, where

$$\begin{aligned} \langle \mathbf{x_1}, \mathbf{x_2} \rangle &= \mathbf{x_1}^T \mathbf{x_2} \\ &= \mathbf{x_2}^T \mathbf{x_1} \\ &= \sum_{i=1}^{n} x_{1,i} x_{2,i} \\ &= ||\mathbf{x_1}|| \, ||\mathbf{x_2}|| \cos\theta, \end{aligned} \tag{2.6}$$

where θ is the angle between vectors $\mathbf{x_1}$ and $\mathbf{x_2}$. The *outer product* of $\mathbf{x_1}$ and $\mathbf{x_2}$ is $\rangle \mathbf{x_1}, \mathbf{x_2} \langle$, where

$$\begin{aligned} \rangle \mathbf{x_1}, \mathbf{x_2} \langle &= \mathbf{x_1} \mathbf{x_2}^T \\ &= \mathbf{x_2} \mathbf{x_1}^T. \end{aligned} \tag{2.7}$$

Note that the inner product produces a scalar, while the outer product produces a matrix of size $n \times n$.

Relating the concept of inner product back to norms, the norms $||\mathbf{x_1}||$ and $||\mathbf{x_2}||$ of $\mathbf{x_1}$ and $\mathbf{x_2}$ satisfy

$$||\mathbf{x_1} + \mathbf{x_2}|| \leq ||\mathbf{x_1}|| + ||\mathbf{x_2}||, \tag{2.8}$$

or the *triangle inequality*, and

$$|\langle \mathbf{x_1}, \mathbf{x_2} \rangle| \leq ||\mathbf{x_1}|| \cdot ||\mathbf{x_2}||, \tag{2.9}$$

or the *Cauchy-Schwarz inequality.*

The *cross product* between $\mathbf{x_1}$ and $\mathbf{x_2}$ is defined as $\mathbf{x_1} \times \mathbf{x_2}$ and produces a vector *normal* to both $\mathbf{x_1}$ and $\mathbf{x_2}$ of magnitude

$$||\mathbf{x_1} \times \mathbf{x_2}|| = ||\mathbf{x_1}|| \, ||\mathbf{x_2}|| \sin\theta. \tag{2.10}$$

A *linear combination* of these vectors is defined as the vector

$$c_1\mathbf{x_1} + c_2\mathbf{x_2} + \cdots + c_k\mathbf{x_k}, \tag{2.11}$$

where $c_1, c_2, \cdots, c_k$ are scalars. The vectors are said to be *linearly dependent* if there exist k scalars $c_1, c_2, \cdots, c_k$ being non-zero and satisfying

$$c_1\mathbf{x_1} + c_2\mathbf{x_2} + \cdots + c_k\mathbf{x_k} = 0, \tag{2.12}$$

but excluding the trivial solution. As such, the vectors are *linearly independent* if

$$c_1\mathbf{x_1} + c_2\mathbf{x_2} + \cdots + c_k\mathbf{x_k} = 0 \tag{2.13}$$

has only the trivial solution, i.e., $c_1 = c_2 = \cdots = c_k = 0$.

Now, given a set of k vectors $\mathbf{x_1}, \mathbf{x_2}, \cdots, \mathbf{x_k}$ which have the same length of n. The set $\mathbb{V}$ containing $\mathbf{x_1}, \mathbf{x_2}, \cdots, \mathbf{x_k}$ and their linear combinations is said to form a *vector space*. Alternatively, the vectors $\mathbf{x_1}, \mathbf{x_2}, \cdots, \mathbf{x_k}$ are said to *span* the vector space $\mathbb{V}$.

Some properties of a vector space include

- Addition: If $\mathbf{x_1} \in \mathbb{V}$ and $\mathbf{x_2} \in \mathbb{V}$, then $\mathbf{x_1} + \mathbf{x_2} \in \mathbb{V}$, and
- Scalar Multiplication: If $\mathbf{x_1} \in \mathbb{V}$, then $c_1\mathbf{x_1} \in \mathbb{V}$ for all arbitrary scalars c_1.

It is clear that the *null vector* $\mathbf{0}$ is a vector in all vector spaces $\mathbb{V}$.

For any arbitrary vector $\mathbf{x} \in \mathbb{V}$, we can express as

$$\mathbf{x} = c_1\mathbf{x}_1 + c_2\mathbf{x}_2 + \cdots + c_k\mathbf{x}_n. \tag{2.14}$$

These vectors are said to constitute a *basis* for the vector space $\mathbb{V}$. The largest number of linearly independent vectors in $\mathbb{V}$ is d, where d is the *dimension* of $\mathbb{V}$ or $\dim(\mathbb{V})$. Note that although the choice of a basis is non-unique, d is fixed for a given vector space $\mathbb{V}$.

2.1.2 MATRICES

A *matrix* $\mathbf{A}$ is a rectangular array of numbers where

$$\mathbf{A} = \begin{bmatrix} a_{11} & a_{12} & \cdots & a_{1n} \\ a_{21} & a_{22} & \cdots & a_{2n} \\ \vdots & \vdots & a_{ij} & \vdots \\ a_{m1} & a_{m2} & \cdots & a_{mn} \end{bmatrix}, \tag{2.15}$$

and $\mathbf{A}$ is said to be a $m \times n$ matrix of m *rows* and n *columns*. The entry or element of the matrix in the ith row and j^{th} column is a_{ij}. As such, $\mathbf{A}$ can be seen as a rectangular arrangement of mn-tuple numbers, or a concatenation of vectors of the same length. A matrix with one column or one row is hence, a vector.

The 1-*norm* for $\mathbf{A}$ is

$$||\mathbf{A}||_1 = \max_{1 \leq j \leq n} \sum_{i=1}^{m} |a_{ij}|, \tag{2.16}$$

which is simply the maximum absolute column sum of the $\mathbf{A}$. The ∞-*norm* of $\mathbf{A}$ is defined as

$$||\mathbf{A}||_\infty = \max_{1\le i\le m} \sum_{j=1}^{n} |a_{ij}|, \tag{2.17}$$

or the maximum absolute row sum of $\mathbf{A}$. The 2-*norm* of $\mathbf{A}$ is

$$\begin{aligned} ||\mathbf{A}||_2 &= \sqrt{\lambda_{\max}(\mathbf{A}^H\mathbf{A})} \\ &= \sigma_{\max}(\mathbf{A}), \end{aligned} \tag{2.18}$$

where $\lambda_{\max}(\cdot)$ denotes the maximum *eigenvalue* of $\cdot$, and $\sigma_{\max}(\cdot)$ denotes the maximum *singular value* of $\cdot$. The superscript ^{H} is the *Hermitian* or conjugate transpose operator.

Another interesting matrix norm is the *Frobenius norm* $||\mathbf{A}||_F$ of $\mathbf{A}$ where

$$\begin{aligned} ||\mathbf{A}||_F &= \sqrt{\mathrm{tr}\{\mathbf{A}^H\mathbf{A}\}} \\ &= \sqrt{\sum_{i=1}^{m}\sum_{j=1}^{n} |a_{ij}|^2} \\ &= \sqrt{\sum_{i=1}^{\min\{m,n\}} \sigma_i^2}, \end{aligned} \tag{2.19}$$

where $\mathrm{tr}\{\cdot\}$ is the *trace* operation, or sum of diagonal elements of the square matrix. σ_i are the singular values of $\mathbf{A}$. More details on eigenvalues and singular values will be provided in future sections.

Now given an $m \times n$ matrix $\mathbf{A}$, the vector spaces spanned by the set of linearly independent row vectors and column vectors of $\mathbf{A}$ are called the *row space* and *column space* of $\mathbf{A}$, respectively. The vector space consisting of all possible solutions of $\mathbf{Ax} = \mathbf{0}$ is called the *null space* of $\mathbf{A}$, *kernel* of $\mathbf{A}$, or simply $\mathrm{Ker}(\mathbf{A})$. This is because if $\mathbf{A}$ is viewed as a *linear transformation*, the null space of a matrix is precisely the set of vectors that is mapped to the null vector.

The maximum number of linearly independent vectors of a matrix $\mathbf{A}$ (or the dimension of the row space of $\mathbf{A}$) is called the *row rank* or simply the *rank* of $\mathbf{A}$. *Nullity* of a matrix $\mathbf{A}$ is defined as the dimension of the null space of $\mathbf{A}$. For an $m \times n$ matrix $\mathbf{A}$, the following theorems hold [47].

Theorem 2.1

The number of linearly independent rows of $\mathbf{A}$ is the number of linearly independent columns of $\mathbf{A}$. In other words, the row rank of $\mathbf{A}$ is also the column rank of $\mathbf{A}$ and the rank of $\mathbf{A}$. ■

Theorem 2.2

For a given $m \times n$ matrix $\mathbf{A}$, rank($\mathbf{A}$)+ nullity($\mathbf{A}$) $= n$. ■

The theorems are self-explanatory and the proofs are omitted here for brevity.

In linear algebra, linear transformations can be represented by matrices. If T is a *linear transformation* mapping $\mathbb{R}^n \to \mathbb{R}^m$ where $\mathbf{x} \in \mathbb{R}^n$ and $\mathbf{y} \in \mathbb{R}^m$ are vectors, then $T :\to \quad \mathbf{y} = \mathbf{A}\mathbf{x}$ for some $m \times n$ matrix $\mathbf{A}$. Here, $\mathbf{A}$ is called the *transformation matrix*.

The *domain* of a function is the set of "input" or argument values for which the function is defined. In other words, the function provides a unique "output" or "value" for each member of the domain [48]. As such, $\mathbf{A}$ can now be viewed as an operator which manipulates $\mathbf{x}$ to produce $\mathbf{y}$. In the same sense as matrix norm is discussed above, we can define an *operator norm* or *induced norm* $||\mathbf{A}||$ of $\mathbf{A}$ as

$$
\begin{aligned}
||\mathbf{A}|| &= \sup_{\mathbf{x} \neq \mathbf{0}} \frac{||\mathbf{A}\mathbf{x}||}{||\mathbf{x}||} \\
&= \sup_{||\mathbf{x}||=1} ||\mathbf{A}\mathbf{x}||.
\end{aligned} \tag{2.20}
$$

For a linear transformation T which is represented by a transformation matrix $\mathbf{A}$ where $\mathbf{A} \in \mathbb{R}^{m \times n}$ that maps $\mathbb{R}^n$ to $\mathbb{R}^m$, we define the *domain space* of $\mathbf{A}$ as all possible "input" vectors $\mathbf{x}$ where $\mathbf{x} \in \mathbb{R}^n$. If the linear transformation is unconstrained, one easily sees that its domain is exactly $\mathbb{R}^n$.

Conversely, given a transformation matrix $\mathbf{A} \in \mathbb{R}^{m \times n}$ that maps $\mathbb{R}^n \to \mathbb{R}^m$, the *range space* of $\mathbf{A}$ is defined as the set of all possible "output" vectors $\mathbf{y} \in \mathbb{R}^m$ [47]. As such, it is easy to see that the range space of $\mathbf{A}$ is exactly the column space of itself or *column space*. This is also the *image* of the corresponding matrix transformation, as the mapping $\mathbf{A}\mathbf{x} = \mathbf{y}$ is actually taking all possible linear combinations of column vectors of $\mathbf{A}$. Therefore, for a transformation matrix $\mathbf{A} \in \mathbb{R}^{m \times n}$, we denote $\mathrm{R}(\mathbf{A})$ as the range space of $\mathbf{A}$.

Let $\mathbf{A}$ be an $m \times n$ matrix. The product of $\mathbf{A}$ and the n-dimensional vector $\mathbf{x}$ can be written in terms of the *inner product* of vectors as follows

$$
\mathbf{A}\mathbf{x} = \begin{bmatrix} \mathbf{r_1 x} \\ \mathbf{r_2 x} \\ \vdots \\ \mathbf{r_m x} \end{bmatrix}, \tag{2.21}
$$

where $\mathbf{r_1}, \mathbf{r_2}, \cdots, \mathbf{r_m}$ denote the rows of the matrix $\mathbf{A}$. It follows that $\mathbf{x}$ is in the null space of $\mathbf{A}$ if and only if $\mathbf{x}$ is orthogonal (or perpendicular) to each of the row vectors of $\mathbf{A}$. This is because the inner product of two vectors is zero if they are orthogonal.

The row space of a matrix $\mathbf{A}$ is the span of the row vectors of $\mathbf{A}$. Using a similar reasoning as above, the null space of $\mathbf{A}$ is the orthogonal complement to the row space. That is, a vector $\mathbf{x}$ lies in the null space of $\mathbf{A}$ **if and only if** it is perpendicular to every vector in the row space of $\mathbf{A}$.

The left null space of $\mathbf{A} \in \mathbb{R}^{m\times n}$ is the set of all vectors $\mathbf{x} \in \mathbb{R}^m$ such that $\mathbf{x}^T\mathbf{A} = \mathbf{0}$. It is the same as the null space of the transpose of $\mathbf{A}$, or $\mathbf{A}^T$. The left null space is the orthogonal complement to the column space of $\mathbf{A}$. This can be seen by writing the product of the matrix $\mathbf{A}^T$ and the vector $\mathbf{x}$ in terms of the inner product of vectors

$$\mathbf{A}^T\mathbf{x} = \begin{bmatrix} \mathbf{c_1 x} \\ \mathbf{c_2 x} \\ \vdots \\ \mathbf{c_n x} \end{bmatrix}, \tag{2.22}$$

where $\mathbf{c_1}, \mathbf{c_2}, \cdots, \mathbf{c_n}$ are the column vectors of $\mathbf{A}$. Thus, $\mathbf{A}^T\mathbf{x} = \mathbf{0}$ if and only if $\mathbf{x}$ is orthogonal (perpendicular) to each of the column vectors of $\mathbf{A}$. It follows that the null space of $\mathbf{A}^T$ is the orthogonal complement to the column space of $\mathbf{A}$. As such for a matrix $\mathbf{A}$, the column space, row space, null space, and left null space are sometimes referred to as the four *fundamental subspaces* [49].

2.2 LINEAR SYSTEMS

After introducing the concepts of vectors and matrices, we will discuss their perusal in representation of linear systems and linearized non-linear systems in this section.

2.2.1 INTRODUCTION TO LINEAR SYSTEMS

A typical dynamic linear system model is shown in Figure 2.1.

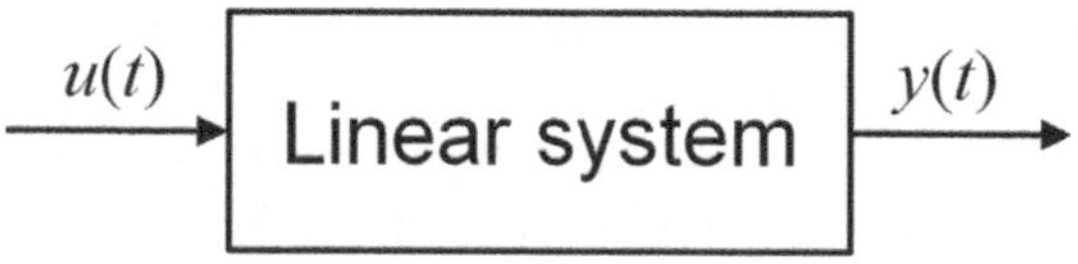

FIGURE 2.1 Dynamic model of linear system.

Let $y_1(t)$ be the output produced by an input signal $u_1(t)$ and $y_2(t)$ be the output produced by another input signal $u_2(t)$. The system is said to be linear if when [50]

1. the input is $\alpha u_1(t)$, the corresponding output is $\alpha y_1(t)$, where α is scalar; and
2. the input is $u_1(t) + u_2(t)$, the corresponding output is $y_1(t) + y_2(t)$.

Or equivalently, when the input is $\alpha u_1(t) + \beta u_2(t)$, the corresponding output is $\alpha y_1(t) + \beta y_2(t)$. As such, a linear system is said to obey the *Principle of Superposition* since it has such a property.

A system is said to be *Linear Time-Invariant* (LTI) if it is linear and for a time-shifted input signal $u(t-t_0)$, the output of the system is also time-shifted $y(t-t_0)$. To see if a system satisfies the LTI criterion, we first find the output $y_1(t)$ that corresponds to the input $u_1(t)$. Next, we let $u_2(t) = u_1(t-t_0)$ and then find the corresponding output $y_2(t)$. If $y_2(t) = y_1(t-t_0)$, the linear system is said to be LTI. In other words, an LTI system satisfies the Principle of Superposition and for the same input signal, the output produced by the system today will be exactly the same as that produced by the system at any other time. This reproducibility and precision allow us to predict and analyze linear systems.

A system is said to have *memory* if the value of $y(t)$ at any particular time t_1 depends on the time from $(-\infty, t_1]$, i.e., depends on the historical profile of $y(t)$. Conversely, a system is said to have no memory if the value of $y(t)$ at any particular time t_1 depends only on the time t_1. On the other hand, a *causal* system is a system where the output $y(t)$ at a particular time t_1 depends solely on its input for $t \leq t_1$. All physical systems are causal by nature. Similarly, a system is said to be *non-causal* if the value of $y(t)$ at a particular time t_1 depends on its input $u(t)$ for some $t > t_1$.

The signal $u(t)$ is said to be *bounded* if $|u(t)| < \beta < \infty \quad \forall t$, where β is a positive, real, and finite scalar. A system is said to be *Bounded-Input-Bounded-Output* (BIBO) stable if the output $y(t)$ produced by any bounded input $u(t)$ is also bounded.

2.2.2 STATE-SPACE REPRESENTATION OF LTI SYSTEMS

LTI systems in general can be represented by *transfer functions* (ratios of orders of Laplace operator "s" in continuous-time domain or z-transform operator "z" in discrete-time domain) in the frequency domain, or *state-space representation* using ***matrices*** in time domain. As such, a state-space representation is a mathematical model of a physical system as a set of inputs, outputs, and state variables, related by first-order differential equations and expressed as ***vectors*** in matrix form. The state space representation provides a systematic and convenient way to represent and analyze systems with multiple inputs and outputs.

Consider a single input n^{th} order LTI differential equation

$$\frac{d^n x(t)}{dt^n} + a_{n-1}\frac{d^{n-1}x(t)}{dt^{n-1}} + \cdots + a_i \frac{d^i x(t)}{dt^i} \cdots + a_0 x(t) = b_0 u(t), \tag{2.23}$$

where $u(t)$ is the input, $x(t)$ is the *state*, and a_i and b_0 are constants. The most straightforward method for choosing n state variables to represent this system is to let the state variables be equal to $x(t)$ and its first $(n-1)$ derivatives. If the state variables are denoted by ξ, we can rewrite (2.23) as

$$\begin{aligned} \xi_1(t) &= x(t), \\ \xi_2(t) &= \frac{dx(t)}{dt}, \\ &\vdots \\ \xi_n(t) &= \frac{d^{n-1}x(t)}{dt^{n-1}}, \end{aligned} \tag{2.24}$$

and the n differential equations resulting from these definitions become

$$\begin{aligned}
\dot{\xi}_1(t) &= \xi_2(t), \\
\dot{\xi}_2(t) &= \xi_3(t), \\
&\vdots \\
\dot{\xi}_{n-1}(t) &= \xi_n(t), \\
\dot{\xi}_n(t) &= -a_0\xi_1(t) - a_1\xi_2(t) - \cdots - a_{n-1}\xi_n + b_0u(t).
\end{aligned} \tag{2.25}$$

Putting all the states into a vector, the above system of equations can be rewritten as

$$\begin{bmatrix} \dot{\xi}_1 \\ \dot{\xi}_2 \\ \vdots \\ \dot{\xi}_{n-1} \\ \dot{\xi}_n \end{bmatrix} = \begin{bmatrix} 0 & 1 & 0 & \cdots & 0 \\ 0 & 0 & 1 & \ddots & 0 \\ 0 & 0 & 0 & \ddots & 0 \\ 0 & 0 & \cdots & 0 & 1 \\ -a_0 & -a_1 & -a_2 & \cdots & -a_{n-1} \end{bmatrix} \begin{bmatrix} \xi_1 \\ \xi_2 \\ \vdots \\ \xi_{n-1} \\ \xi_n \end{bmatrix} + \begin{bmatrix} 0 \\ 0 \\ \vdots \\ 0 \\ b_0 \end{bmatrix} u(t). \tag{2.26}$$

Now if the output of the dynamic system $y(t)$ is one (or a linear combination) of the state variables, e.g., $\xi_1(t)$, we can write $y(t)$ in the *output equation* as

$$y(t) = \begin{bmatrix} 1 & 0 & \cdots & 0 & 0 \end{bmatrix} \begin{bmatrix} \xi_1 \\ \xi_2 \\ \vdots \\ \xi_{n-1} \\ \xi_n \end{bmatrix}. \tag{2.27}$$

Defining $\mathbf{x}(t) = \begin{bmatrix} \xi_1 & \xi_2 & \cdots & \xi_{n-1} & \xi_n \end{bmatrix}^T$, we can rewrite the system model in the form of *state-space representation* as

$$\begin{aligned}
\dot{\mathbf{x}}(t) &= \mathbf{A}\mathbf{x}(t) + \mathbf{B}u(t), \\
y(t) &= \mathbf{C}\mathbf{x}(t) + \mathbf{D}u(t),
\end{aligned} \tag{2.28}$$

where

$$\begin{aligned}
\mathbf{A} &= \begin{bmatrix} 0 & 1 & 0 & \cdots & 0 \\ 0 & 0 & 1 & \ddots & 0 \\ 0 & 0 & 0 & \ddots & 0 \\ 0 & 0 & \cdots & 0 & 1 \\ -a_0 & -a_1 & -a_2 & \cdots & -a_{n-1} \end{bmatrix}, \\
\mathbf{B} &= \begin{bmatrix} 0 \\ 0 \\ \vdots \\ 0 \\ b_0 \end{bmatrix}, \\
\mathbf{C} &= \begin{bmatrix} 1 & 0 & \cdots & 0 & 0 \end{bmatrix}, \\
\mathbf{D} &= 0.
\end{aligned} \tag{2.29}$$

Now, $\mathbf{x} \in \mathbb{R}^n$ is the *state vector*, y is *output vector*, and u is the *input* (or *control*) *vector*. $\mathbf{A} \in \mathbb{R}^{n \times n}$ is the *state matrix*, $\mathbf{B} \in \mathbb{R}^n$ is the *input matrix*, $\mathbf{C} \in \mathbb{R}^{1 \times n}$ is the *output matrix*, and $\mathbf{D} = 0$ is the *feedthrough* (or *feedforward*) *matrix*. In this example, we have chosen a Single-Input-Single-Output (SISO) system for simplicity but without loss of generality, and this framework can be readily applied to systems of higher dimensions with multiple inputs and outputs.

The state equation can now be expressed as a transfer function in frequency domain as [46]

$$\frac{Y(s)}{U(s)} = \mathbf{C}(s\mathbf{I} - \mathbf{A})^{-1}\mathbf{B} + \mathbf{D}, \tag{2.30}$$

where $U(s)$ and $Y(s)$ are the system's input and output, respectively, and s is the Laplace Transform operator.

For general Multi-Input-Multi-Output (MIMO) systems, (2.28) is written as

$$\begin{aligned}
\dot{\mathbf{x}}(t) &= \mathbf{A}\mathbf{x}(t) + \mathbf{B}\mathbf{u}(t), \\
\mathbf{y}(t) &= \mathbf{C}\mathbf{x}(t) + \mathbf{D}\mathbf{u}(t),
\end{aligned} \tag{2.31}$$

where $\mathbf{A} \in \mathbb{R}^{n \times n}$, $\mathbf{B} \in \mathbb{R}^{n \times p}$, $\mathbf{C} \in \mathbb{R}^{q \times n}$, and $\mathbf{D} \in \mathbb{R}^{q \times p}$ for a system with p inputs and q outputs. In this general formulation, all matrices are allowed to be time-variant, i.e., their entries may vary with time. However for LTI systems, these matrices contain constant real numbers.

Depending on the assumptions taken, the state-space model representation can assume the following forms, and variable t can be continuous or discrete. In the latter case, the time variable is usually indicated as sampled instant k. In general, the following representations are typical and most encountered in literature and practice

- Continuous time-invariant
  $\dot{\mathbf{x}}(t) = \mathbf{A}\mathbf{x}(t) + \mathbf{B}\mathbf{u}(t)$,
  $\mathbf{y}(t) = \mathbf{C}\mathbf{x}(t) + \mathbf{D}\mathbf{u}(t)$,

- Continuous time-variant
  $\dot{\mathbf{x}}(t) = \mathbf{A}(t)\mathbf{x}(t) + \mathbf{B}(t)\mathbf{u}(t)$,
  $\mathbf{y}(t) = \mathbf{C}(t)\mathbf{x}(t) + \mathbf{D}(t)\mathbf{u}(t)$,
- Discrete time-invariant
  $\mathbf{x}(k+1) = \mathbf{A}\mathbf{x}(k) + \mathbf{B}\mathbf{u}(k)$,
  $\mathbf{y}(k) = \mathbf{C}\mathbf{x}(k) + \mathbf{D}\mathbf{u}(k)$,
- Discrete time-variant
  $\mathbf{x}(k+1) = \mathbf{A}(k)\mathbf{x}(k) + \mathbf{B}(k)\mathbf{u}(k)$,
  $\mathbf{y}(k) = \mathbf{C}(k)\mathbf{x}(k) + \mathbf{D}(k)\mathbf{u}(k)$

depending on domain of application.

2.2.3 LINEARIZATION OF NON-LINEAR SYSTEMS

However, all physical systems are inherently non-linear by nature. In order to create a linear model from a non-linear system, the non-linear system under consideration is linearized about a particular point based on the *Taylor series* expansion.

We recall that the Taylor series expansion for a general function $f(x)$ is

$$f(x) = \sum_{n=0}^{\infty} \frac{1}{n!} \left. \frac{d^n f(x)}{dx^n} \right|_{x=x_0} (x - x_0)^n, \tag{2.32}$$

and this series is said to be expanded about the point $x = x_0$, commonly referred to as the *operating point* or *equilibrium point*. For functions that are relatively smooth, the magnitudes of the terms in this series decrease as higher order derivatives are introduced, and an approximation of a function can be achieved by selecting only some lower order terms.

Using the above definition, we have

$$\begin{aligned} f(x) &= f(x_0) + f'(x_0)(x - x_0) + \frac{1}{2} f''(x_0)(x - x_0)^2 + \cdots \\ &\approx f'(x_0)x + \left[f(x_0) - f'(x_0)x_0 \right], \end{aligned} \tag{2.33}$$

and a linear relation is obtained if we retain only the first two terms of expansion. This approximation is referred to as the *linearization* of $f(x)$, and it is important that the region of operation be near x_0 for the approximation to be valid.

In a MIMO setting, the function $\mathbf{f}$ depends on multiple variables and each function in $\mathbf{f}$ can be expanded into a Taylor series and thus linearized separately. Alternatively, one can use matrix-vector notation and the linearized version of the non-linear function $\mathbf{f}$ is

$$\begin{aligned} \mathbf{f}(\mathbf{x}) &= \mathbf{f}(\mathbf{x}_0) + \left. \frac{\partial \mathbf{f}(\mathbf{x})}{\partial \mathbf{x}} \right|_{\mathbf{x}_0} (\mathbf{x} - \mathbf{x}_0) + \frac{1}{2} (\mathbf{x} - \mathbf{x}_0)^T \left. \frac{\partial^2 \mathbf{f}(\mathbf{x})}{\partial \mathbf{x}^2} \right|_{\mathbf{x}_0} (\mathbf{x} - \mathbf{x}_0) + \cdots \\ &\approx \mathbf{f}(\mathbf{x}_0) + \left. \frac{\partial \mathbf{f}(\mathbf{x})}{\partial \mathbf{x}} \right|_{\mathbf{x}_0} (\mathbf{x} - \mathbf{x}_0), \end{aligned} \tag{2.34}$$

and the linear relation is obtained if we retain the first two terms as usual. In this expression, the first order derivative of $\mathbf{f}(\mathbf{x})$ is a derivative of an $m \times 1$ vector with respect to an $n \times 1$ vector. This results in an $m \times n$ matrix commonly known as the *Jacobian*, whose $(i, j)^{\text{th}}$ element is $\frac{\partial f_i}{\partial x_j}$.

2.3 EIGENVALUE DECOMPOSITION AND SENSITIVITY

With the fundamental concepts of vector, matrices, and their uses in representation of linear systems and linearized non-linear systems, we will further discuss the concepts eigenvalues, eigenvectors, and their corresponding computational methods in this section. Specific applications to find the sensitivity of eigenvalues to non-deterministic system parameters and link parameters will also be detailed.

2.3.1 EIGENVALUE AND EIGENVECTOR

Given a square matrix $\mathbf{A} \in \mathbb{R}^{n \times n}$. A non-zero vector $\mathbf{x}$ is defined to be an *eigenvector* of $\mathbf{A}$ if it satisfies

$$\mathbf{A}\mathbf{x} = \lambda \mathbf{x} \tag{2.35}$$

for some scalar λ. λ is called an *eigenvalue* of $\mathbf{A}$ corresponding to the eigenvector $\mathbf{x}$. If $\mathbf{A}$ is the state matrix of a linear system, λ is also the *pole* of the system, and the corresponding eigenvector $\mathbf{x}$ is also known as the *mode shape*.

The above equation can also be rewritten as

$$(\mathbf{A} - \lambda \mathbf{I})\mathbf{x} = 0, \tag{2.36}$$

where $\mathbf{I} \in \mathbb{R}^{n \times n}$ is an identity matrix. We can therefore also interpret an eigenvector $\mathbf{x}$ as being a vector from the null space of $(\mathbf{A} - \lambda \mathbf{I})$ corresponding to an eigenvalue λ. Each eigenvector is associated with a specific eigenvalue, but one eigenvalue can be associated with several or even with an infinite number of eigenvectors.

If $\mathbf{A}$ is viewed as a linear transformation, the eigenvector $\mathbf{x}$ has the property that its direction is not changed by the transformation $\mathbf{A}$, but that it is only scaled by a factor of λ. In general, most vectors $\mathbf{x}$ will not satisfy (2.36), since linear transformation $\mathbf{A}$ generally scales, translates, or shears $\mathbf{x}$. Alternatively, if $\mathbf{A}$ is a multiple of the identity matrix, i.e., no vectorial change in directions, then all non-zero vectors are also eigenvectors.

Under the linear transformation $\mathbf{A}$, the eigenvectors experience merely a change in magnitude but no change in direction. If $\lambda = 1$, the vector remains unchanged (unaffected by the transformation). If $\lambda = -1$, the vector flips to the opposite direction. This is defined as a *reflection*.

With these concepts in mind, we have the following lemmas [51].

Lemma 2.1

If $\mathbf{x}$ is an eigenvector of the linear transformation $\mathbf{A}$ with eigenvalue λ, then any scalar multiple $\alpha\mathbf{x}$ is also an eigenvector of $\mathbf{A}$ with the same eigenvalue. Similarly, if more than one eigenvector shares the same eigenvalue λ, any linear combination of these eigenvectors will itself be an eigenvector with eigenvalue λ. ■

Together with the zero vector, the eigenvectors of $\mathbf{A}$ with the same eigenvalue form a linear subspace of the vector space called an *eigenspace*. The set of all eigenvalues is called the *spectrum* of $\mathbf{A}$.

Lemma 2.2

The eigenvectors corresponding to different eigenvalues are linearly independent, in particular, that in an n-dimensional space the linear transformation $\mathbf{A}$ cannot have more than n eigenvectors with different eigenvalues. ■

Conventionally, the word eigenvector formally refers to the *right* eigenvector $\mathbf{v}$ where $\mathbf{A}\mathbf{v} = \lambda\mathbf{v}$. $\mathbf{v}$ is the most commonly used eigenvector. However, there is the less commonly known *left* eigenvector $\mathbf{w}$ as well, and is defined by $\mathbf{w}^T\mathbf{A} = \lambda\mathbf{w}^T$. It is worth noting the eigenvalues are exactly *identical* since the poles of the system remain unchanged under different representations, and that $\mathbf{v}\mathbf{w}^T = \mathbf{w}^T\mathbf{v} = 1$. Both left and right eigenvalues are important for our proposed probabilistic small signal assessment methodology, and more details will be provided in Chapter 5.

In order to compute eigenvectors, one has to first find the eigenvalues depicted earlier in (2.36). As such, it is imperative that $\mathbf{A} - \lambda\mathbf{I}$ has to be *non-singular* or not *invertible*. These conditions are equivalent to solving

$$\det(\mathbf{A} - \lambda\mathbf{I}) = 0, \tag{2.37}$$

where det(.) is the determinant operation.

Equation (2.37) is also called the *characteristic equation* of $\mathbf{A}$, and the left-hand side is called the *characteristic polynomial.* When expanded, this gives a polynomial equation for λ. Neither the eigenvector nor its components are present in the characteristic equation. With the computed eigenvalues λ, we can then substitute them back into (2.36) to obtain the corresponding eigenvectors $\mathbf{x}$ of $\mathbf{A}$. A simple example is provided for illustration.

Example 2.1: A linear transformation on the real plane $\mathbf{A}$ is given by

$$\mathbf{A} = \begin{bmatrix} 1 & 0 \\ 1 & 2 \end{bmatrix}. \tag{2.38}$$

The eigenvalues of **A** can be obtained by solving the characteristic equation

$$\begin{aligned}
\det(\mathbf{A}-\lambda\mathbf{I}) &= \det\left(\begin{bmatrix}1 & 0\\ 1 & 2\end{bmatrix} - \begin{bmatrix}\lambda & 0\\ 0 & \lambda\end{bmatrix}\right)\\
&= \det\begin{bmatrix}1-\lambda & 0\\ 1 & 2-\lambda\end{bmatrix}\\
&= 0.\\
\Rightarrow (1-\lambda)(2-\lambda) &= 0\\
\therefore \lambda &= 1,\ 2, \qquad (2.39)
\end{aligned}$$

and the eigenvalues of **A** are thus $\lambda = 1$ and $\lambda = 2$. With these eigenvalues, we can proceed to find the corresponding eigenvectors.

Consider the eigenvalue $\lambda = 2$ and let $\mathbf{x} = \begin{bmatrix} x_1 & x_2 \end{bmatrix}^T$. Now, the characteristic equation becomes

$$\begin{bmatrix}-1 & 0\\ 1 & 0\end{bmatrix}\begin{bmatrix}x_1\\ x_2\end{bmatrix} = \mathbf{0}. \qquad (2.40)$$

This matrix equation represents a system of two linear equations

$$-x_1 = 0, \qquad (2.41)$$

$$x_1 = 0, \qquad (2.42)$$

which implies that $x_1 = 0$. We are free to choose any real number for x_2 except zero. By selecting $x_2 = 1$, the corresponding eigenvector is $\begin{bmatrix} 0 & 1 \end{bmatrix}^T$.

Similarly, consider the eigenvalue $\lambda = 1$, the characteristic equation is thus

$$\begin{bmatrix}0 & 0\\ 1 & 1\end{bmatrix}\begin{bmatrix}x_1\\ x_2\end{bmatrix} = \mathbf{0}. \qquad (2.43)$$

This matrix equation represents a linear equation

$$x_1 + x_2 = 0. \qquad (2.44)$$

As per above, we are free to choose a any real number for x_1 except zero. By selecting $x_1 = 1$ and $x_2 = -x_1 = -1$, the corresponding eigenvector is $\begin{bmatrix} 1 & -1 \end{bmatrix}^T$.

It is obvious that the complexity of the eigenvalue problem increases rapidly with increasing the degree of the polynomial or the dimension of the vector space. As the dimension increases, no exact solutions exist and numerical methods are used to find the eigenvalues and their corresponding eigenvectors.

2.3.2 EIGENVALUE DECOMPOSITION

A matrix can be factorized into the *canonical form* and represented in terms of its eigenvalues and eigenvectors. This is known as *eigenvalue decomposition*.

Let $\mathbf{A} \in \mathbb{R}^{n \times n}$ with n linearly independent eigenvectors $\mathbf{x_1}, \mathbf{x_2}, \cdots, \mathbf{x_n}$. As such, $\mathbf{A}$ can be factorized as

$$\mathbf{A} = \mathbf{M}\mathbf{\Lambda}\mathbf{M}^{-1}, \tag{2.45}$$

where $\mathbf{M} \in \mathbb{R}^{n \times n}$ whose i^{th} column is the eigenvector $\mathbf{x_i}$ of $\mathbf{A}$ and $\mathbf{\Lambda}$ is a diagonal matrix whose diagonal elements are the corresponding eigenvalues, i.e., $\mathbf{\Lambda}_{ii} = \lambda_i$.

In general, the eigenvectors $\mathbf{x_1}, \mathbf{x_2}, \cdots, \mathbf{x_n}$ are normalized though it is not a necessity, since a normalized eigenvector is also a valid eigenvector itself. A non-normalized set of eigenvectors can also be used as the columns of $\mathbf{M}$. This is always true and can be intuitively understood by noting that the magnitude of the eigenvectors in $\mathbf{M}$ is "cancelled" in the decomposition with $\mathbf{M}^{-1}$.

If $\mathbf{A}$ can be eigenvalue-decomposed and if none of its eigenvalues are zero, then $\mathbf{A}$ is non-singular and its inverse is given by

$$\mathbf{A}^{-1} = \mathbf{M}\mathbf{\Lambda}^{-1}\mathbf{M}^{-1}. \tag{2.46}$$

Since $\mathbf{\Lambda}$ is a diagonal matrix, its inverse is simply given by

$$\mathbf{\Lambda}^{-1} = \text{diag}\left(\lambda_i^{-1}\right) \tag{2.47}$$

and $\mathbf{A}^{-1}$ can be easily obtained.

Eigenvalue decomposition has many practical engineering applications, and is numerically efficient when higher orders of multiplicities (or power series) of $\mathbf{A}$ are required. To compute $\mathbf{A}^n$ where $n \in \mathbb{Z}^+$, it is easy to see that

$$\begin{aligned} \mathbf{A}^n &= \underbrace{\mathbf{M}\mathbf{\Lambda}\mathbf{M}^{-1}\mathbf{M}\mathbf{\Lambda}\mathbf{M}^{-1} \times \cdots \times \mathbf{M}\mathbf{\Lambda}\mathbf{M}^{-1}}_{n} \\ &= \mathbf{M}\mathbf{\Lambda}^n\mathbf{M}^{-1}, \end{aligned} \tag{2.48}$$

and the required calculations are greatly simplified with $\mathbf{\Lambda}^n = \text{diag}\left(\lambda_i^n\right)$.

Furthermore, if $\mathbf{A}$ is symmetrical, i.e., $a_{ij} = a_{ji}$, it will have n linearly independent eigenvectors which can be chosen such that they are orthogonal to each other with unity norm. As such, we can decompose $\mathbf{A}$ as

$$\mathbf{A} = \mathbf{M}\mathbf{\Lambda}\mathbf{M}^T, \tag{2.49}$$

where the superscript ^{T} is the matrix transpose operator. The eigenvectors obtained are real, mutually orthogonal, and provide a basis for $\mathbb{R}^n$. Also, $\mathbf{M}$ is *orthonormal*, i.e., $\mathbf{M}^T = \mathbf{M}^{-1}$, and $\mathbf{\Lambda}$ is also real.

However, if $\mathbf{A} \in \mathbb{C}^{n \times n}$ is normal and has an orthogonal eigenvector basis, it can be decomposed as

$$\mathbf{A} = \mathbf{U}\mathbf{\Lambda}\mathbf{U}^H, \tag{2.50}$$

where the superscript ^{H} is the *Hermitian* operator and $\mathbf{U}$ is an *unitary* matrix. Furthermore if $\mathbf{A}$ is Hermitian, $\mathbf{\Lambda}$ will be real. If $\mathbf{A}$ is unitary, $\mathbf{\Lambda}$ takes all its values on the complex unit circle.

In general, the product of the eigenvalues is equal to the *determinant* of **A**, or

$$\det(\mathbf{A}) = \prod_{i=1}^{n} \lambda_i, \tag{2.51}$$

while the sum of the eigenvalues is equal to the *trace* of **A**

$$\operatorname{tr}(\mathbf{A}) = \sum_{i=1}^{n} \lambda_i. \tag{2.52}$$

The eigenvectors of $\mathbf{A}^{-1}$ are the same as the eigenvectors of **A**.

If $\mathbf{A} \in \mathbb{R}^{n\times n}$ does not have a complete set of n linearly independent eigenvectors, one or more of the eigenvalues are repeated or the same λ_i is a multiple root of the characteristic equation. Due to this multiplicity of repeated eigenvalues, the computation of the corresponding eigenvectors will be more difficult, and eigenvalue decomposition of **A** cannot be performed since we will not have a sufficient set of n eigenvectors to construct the $n \times n$ modal matrix **M**. To resolve this difficulty, the notion of a *generalized eigenvector* [52] will be presented here.

2.3.3 GENERALIZED EIGENVECTORS

A generalized eigenvector of a matrix **A** is a non-zero vector **v**, which has associated with it an eigenvalue λ_i having algebraic multiplicity $m > 1$ and satisfying

$$(\mathbf{A} - \lambda\mathbf{I})^m \mathbf{v} = 0. \tag{2.53}$$

In order to compute the generalized eigenvectors, there are currently several methods available. The two most widely-used algorithms are the *bottom-up* and *top-down* approaches with different pros and cons [46]. In this chapter, we will introduce the top-down algorithm to compute the generalized eigenvectors.

The *index* of a repeated eigenvalue λ_i is denoted η_i, where η_i is the smallest integer η such that

$$\operatorname{rank}(\mathbf{A} - \lambda_i\mathbf{I})^{\eta} = n - m_i, \tag{2.54}$$

where n is the dimension of the space of $(\mathbf{A} - \lambda_i\mathbf{I})$, and m_i is the algebraic multiplicity of λ_i. The top-down algorithm is a procedure depicted by the following steps [52]:

- *Step 1*: For an eigenvalue λ_i with index η_i, find all linearly independent solutions to the simultaneous set of matrix equations give by

$$(\mathbf{A} - \lambda_i\mathbf{I})^{\eta_i}\mathbf{x} = 0, \tag{2.55}$$

$$(\mathbf{A} - \lambda_i\mathbf{I})^{\eta_i - 1}\mathbf{x} \neq 0. \tag{2.56}$$

  Each such solution will start a different "chain" of generalized eigenvectors. Because $\operatorname{rank}(\mathbf{A} - \lambda_i\mathbf{I})^{\eta_i} = n - m$, there will be no more than $n - (n - m_i) = m_i$ solutions, each of which is a generalized eigenvector. We denote these solutions as $\mathbf{v}_1^1, \mathbf{v}_2^1, \cdots, \mathbf{v}_{m_i}^1$.

- *Step 2*: Begin generating further generalized eigenvectors by computing the chain for each $j = 1, 2, \cdots, m_i$ by solving

$$\begin{aligned}(\mathbf{A}-\lambda_i\mathbf{I})\mathbf{v}_j^1 &= \mathbf{v}_j^2,\\ (\mathbf{A}-\lambda_i\mathbf{I})\mathbf{v}_j^2 &= \mathbf{v}_j^3,\\ (\mathbf{A}-\lambda_i\mathbf{I})\mathbf{v}_j^3 &= \mathbf{v}_j^4,\\ &\vdots\end{aligned} \tag{2.57}$$

  until we get

$$(\mathbf{A}-\lambda_i\mathbf{I})\mathbf{v}_j^{\eta_i} = 0, \tag{2.58}$$

  which indicates that $\mathbf{v}_j^{\eta_i}$ is a regular eigenvector.
- *Step 3*: The length of the chains is η_i. There may also may be chains of shorter length. If the chains of length η_i do not produce the full set of generalized eigenvectors, begin the procedure again by finding all solutions to

$$(\mathbf{A}-\lambda_i\mathbf{I})^{\eta_i-1}\mathbf{x} = 0, \tag{2.59}$$
$$(\mathbf{A}-\lambda_i\mathbf{I})^{\eta_i-2}\mathbf{x} \neq 0, \tag{2.60}$$

  and repeating the procedure when necessary. This will produce chains of lengths $\eta_i - 1$. Continue until all generalized eigenvectors have been found.
- *Step 4*: Repeat this procedure for other repeated eigenvalues.

It is now clear that the case of n independent eigenvectors makes construction of the modal matrix much simpler. However, this is generally not true for the case of repeated eigenvalues, and the motivation for computing generalized eigenvectors is that we can use them in the modal matrix when we have an insufficient number of regular eigenvectors. When generalized eigenvectors are included in the modal matrix, the resulting matrix operator in the new basis will be *almost* diagonal, since generalized eigenvectors $\mathbf{x}_i$ are chained to the regular eigenvectors $\mathbf{x}_\eta$ as

$$\begin{aligned}\mathbf{A}\mathbf{x}_\eta &= \lambda\mathbf{x}_\eta,\\ \mathbf{A}\mathbf{x}_{\eta-1} &= \lambda\mathbf{x}_{\eta-1}+\mathbf{x}_\eta,\\ &\vdots\\ \mathbf{A}\mathbf{x}_2 &= \lambda\mathbf{x}_2+\mathbf{x}_3,\end{aligned} \tag{2.61}$$
$$\mathbf{A}\mathbf{x}_1 = \lambda\mathbf{x}_1+\mathbf{x}_2, \tag{2.62}$$

and hence a *Jordan block* as a submatrix with the following structure

$$\begin{bmatrix} \lambda & 1 & 0 & 0 & \cdots & 0 \\ 0 & \lambda & 1 & 0 & \ddots & 0 \\ 0 & 0 & \lambda & \ddots & \ddots & \vdots \\ 0 & 0 & 0 & \ddots & 1 & 0 \\ 0 & 0 & 0 & \ddots & \lambda & 1 \\ 0 & 0 & 0 & \cdots & 0 & \lambda \end{bmatrix} \tag{2.63}$$

can be obtained.

The size of this block will be the length of the chain. This implies that the size of the largest Jordan block will be the size of the longest chain of eigenvectors, which is the index η of the eigenvalues. The modal matrix in Jordan canonical form will be constructed with both regular and generalized eigenvectors by concatenating the regular eigenvectors with the chains of generalized eigenvectors. We illustrate these concepts with an example [46].

Example 2.2: Given a defective or rank-deficient matrix

$$\mathbf{A} = \begin{bmatrix} 3 & -1 & 1 & 1 & 0 & 0 \\ 1 & 1 & -1 & -1 & 0 & 0 \\ 0 & 0 & 2 & 0 & 1 & 1 \\ 0 & 0 & 0 & 2 & -1 & -1 \\ 0 & 0 & 0 & 0 & 1 & 1 \\ 0 & 0 & 0 & 0 & 1 & 1 \end{bmatrix}, \tag{2.64}$$

we have the following characteristic polynomial as

$$|\mathbf{A} - \lambda \mathbf{I}| = (\lambda - 2)^5 \lambda, \tag{2.65}$$

which provide eigenvalues of $\lambda_1 = 0$ and $\lambda_2 = 2$ with $m_2 = 5$. For $\lambda_1 = 0$, we can compute the corresponding eigenvector as $\mathbf{x}_1 = \begin{bmatrix} 0 & 0 & 0 & 0 & 1 & -1 \end{bmatrix}^T$.

However for $\lambda_2 = 2$, we must first compute the index of λ_2. Following the above-mentioned procedure, we have $n - m_2 = 6 - 5 = 1$, and the index η_2 can be found from

$$\begin{aligned} \text{rank}(\mathbf{A} - 2\mathbf{I})^1 &= 4 \neq 1, && (2.66) \\ \text{rank}(\mathbf{A} - 2\mathbf{I})^2 &= 2 \neq 1, && (2.67) \\ \text{rank}(\mathbf{A} - 2\mathbf{I})^3 &= 1, && (2.68) \end{aligned}$$

which gives $\eta_2 = 3$. This implies that there will be the chain of eigenvector of length three. We proceed to find a solution to

$$\begin{aligned} (\mathbf{A} - 2\mathbf{I})^3 \mathbf{x} &= 0, && (2.69) \\ (\mathbf{A} - 2\mathbf{I})^2 \mathbf{x} &\neq 0, && (2.70) \end{aligned}$$

and one such solution is the vector $\mathbf{v}_1^1 = \begin{bmatrix} 0 & 0 & 0 & 1 & 1 & 1 \end{bmatrix}^T$.

Beginning the chain, we can compute

$$\begin{aligned} \mathbf{v}_1^2 &= (\mathbf{A} - 2\mathbf{I})\mathbf{v}_1^1 \\ &= \begin{bmatrix} 1 & -1 & 2 & -2 & 0 & 0 \end{bmatrix}^T, \\ \mathbf{v}_1^3 &= (\mathbf{A} - 2\mathbf{I})\mathbf{v}_1^2 \\ &= \begin{bmatrix} 2 & 2 & 0 & 0 & 0 & 0 \end{bmatrix}^T, \\ \mathbf{v}_1^4 &= (\mathbf{A} - 2\mathbf{I})\mathbf{v}_1^3 \\ &= \begin{bmatrix} 0 & 0 & 0 & 0 & 0 & 0 \end{bmatrix}^T. \end{aligned} \tag{2.71}$$

$\mathbf{v}_1^3$ is a regular eigenvector, chained in order to obtain generalized eigenvectors $\mathbf{v}_1^2$ and $\mathbf{v}_1^1$. However, we still need two more generalized eigenvectors in order to construct the modal matrix. We continue to carry on the algorithm by computing the solution to

$$\begin{aligned} (\mathbf{A} - 2\mathbf{I})^2\mathbf{x} &= 0, \\ (\mathbf{A} - 2\mathbf{I})^1\mathbf{x} &\neq 0, \end{aligned} \tag{2.72}$$

and one such solution is $\mathbf{v}_2^1 = \begin{bmatrix} 0 & 0 & 0 & 0 & 1 & 1 \end{bmatrix}^T$. This produces the following chain of

$$\begin{aligned} \mathbf{v}_2^2 &= (\mathbf{A} - 2\mathbf{I})\mathbf{v}_2^1 \\ &= \begin{bmatrix} 0 & 0 & 2 & -2 & 0 & 0 \end{bmatrix}^T, \\ \mathbf{v}_2^3 &= (\mathbf{A} - 2\mathbf{I})\mathbf{v}_2^2 \\ &= \begin{bmatrix} 0 & 0 & 0 & 0 & 0 & 0 \end{bmatrix}^T, \end{aligned} \tag{2.73}$$

and so the chain consists of regular eigenvector $\mathbf{v}_2^2$ and generalized eigenvector $\mathbf{v}_2^1$.

In the previous section, we showed that the eigenvalue decomposition of a full rank matrix $\mathbf{A}$ is $\mathbf{A} = \mathbf{M}\mathbf{\Lambda}\mathbf{M}^{-1}$, where column vectors of modal matrix $\mathbf{M}$ are the linearly independent eigenvectors of $\mathbf{A}$. In this case of a rank defective matrix, the modal matrix $\mathbf{M}$ is constructed as

$$\begin{aligned} \mathbf{M} &= \begin{bmatrix} \mathbf{x}_1 \mid \mathbf{v}_1^3 \mid \mathbf{v}_1^2 \mid \mathbf{v}_1^1 \mid \mathbf{v}_2^2 \mid \mathbf{v}_2^1 \end{bmatrix} \\ &= \begin{bmatrix} 0 & 2 & 1 & 0 & 0 & 0 \\ 0 & 2 & -1 & 0 & 0 & 0 \\ 0 & 0 & 2 & 0 & 2 & 0 \\ 0 & 0 & -2 & 1 & -2 & 0 \\ 1 & 0 & 0 & 1 & 0 & 1 \\ -1 & 0 & 0 & 1 & 0 & 1 \end{bmatrix}. \end{aligned} \tag{2.74}$$

As such, we obtain $\mathbf{\Lambda}$ as

$$\begin{aligned} \mathbf{\Lambda} &= \mathbf{M}^{-1}\mathbf{A}\mathbf{M} \\ &= \begin{bmatrix} 0 & 0 & 0 & 0 & 0 & 0 \\ 0 & 2 & 1 & 0 & 0 & 0 \\ 0 & 0 & 2 & 1 & 0 & 0 \\ 0 & 0 & 0 & 2 & 0 & 0 \\ 0 & 0 & 0 & 0 & 2 & 1 \\ 0 & 0 & 0 & 0 & 0 & 2 \end{bmatrix}, \end{aligned} \tag{2.75}$$

which is seen to be a Jordan canonical form with a single 1×1 Jordan block corresponding to $\lambda_1 = 0$, with two Jordan blocks of sizes 3×3 and 2×2. This is the order in which the eigenvectors were arranged in modal matrix $\mathbf{M}$.

2.3.4 EIGENVALUE SENSITIVITY TO NON-DETERMINISTIC SYSTEM PARAMETERS

In this section, the eigenvalues' sensitivities to non-deterministic system parameters will be derived from its left eigenvector $\mathbf{w}$ and right eigenvector $\mathbf{v}$. Let $\mathbf{w}_i$ be the ith left eigenvector and $\mathbf{v}_i$ be the ith right eigenvector of $\mathbf{A}$. As such, we have

$$\begin{aligned} \mathbf{A}\mathbf{v}_i &= \lambda_i \mathbf{v}_i, && (2.76) \\ \mathbf{w}_i^T \mathbf{A} &= \lambda_i \mathbf{w}_i^T. && (2.77) \end{aligned}$$

Taking the partial derivative of (2.76) with respect to the jth non-deterministic system parameter $\mathbf{K}_j$ where $\mathbf{K}$ is a vector of critical non-deterministic system parameters of interest, we get

$$\frac{\partial \mathbf{A}}{\partial \mathbf{K}_j}\mathbf{v}_i + \mathbf{A}\frac{\partial \mathbf{v}_i}{\partial \mathbf{K}_j} = \lambda_i \frac{\partial \mathbf{v}_i}{\partial \mathbf{K}_j} + \frac{\partial \lambda_i}{\partial \mathbf{K}_j}\mathbf{v}_i. \tag{2.78}$$

Now taking the inner product of each term with the ith left eigenvector $\mathbf{w}_i$, we have

$$\begin{aligned} \frac{\partial \mathbf{A}}{\partial \mathbf{K}_j}\mathbf{v}_i \cdot \mathbf{w}_i + \mathbf{A}\frac{\partial \mathbf{v}_i}{\partial \mathbf{K}_j} \cdot \mathbf{w}_i &= \lambda_i \frac{\partial \mathbf{v}_i}{\partial \mathbf{K}_j} \cdot \mathbf{w}_i + \frac{\partial \lambda_i}{\partial \mathbf{K}_j}\mathbf{v}_i \cdot \mathbf{w}_i \\ &= \frac{\partial \mathbf{v}_i}{\partial \mathbf{K}_j} \cdot \lambda_i \mathbf{w}_i + \frac{\partial \lambda_i}{\partial \mathbf{K}_j}\mathbf{v}_i \cdot \mathbf{w}_i \\ &= \frac{\partial \mathbf{v}_i}{\partial \mathbf{K}_j} \cdot \mathbf{A}^T \mathbf{w}_i + \frac{\partial \lambda_i}{\partial \mathbf{K}_j}\mathbf{v}_i \cdot \mathbf{w}_i. && (2.79) \end{aligned}$$

Since

$$\mathbf{A}\frac{\partial \mathbf{v}_i}{\partial \mathbf{K}_j} \cdot \mathbf{w}_i = \frac{\partial \mathbf{v}_i}{\partial \mathbf{K}_j} \cdot \mathbf{A}^T \mathbf{w}_i, \tag{2.80}$$

we have

$$\begin{aligned} \frac{\partial \lambda_i}{\partial \mathbf{K}_j} &= \frac{\frac{\partial \mathbf{A}}{\partial \mathbf{K}_j}\mathbf{v}_i \cdot \mathbf{w}_i}{\mathbf{v}_i \cdot \mathbf{w}_i} \\ &= \frac{\mathbf{w}_i \cdot \frac{\partial \mathbf{A}}{\partial \mathbf{K}_j}\mathbf{v}_i}{\mathbf{v}_i \cdot \mathbf{w}_i} \\ &= \frac{\mathbf{w}_i^T \frac{\partial \mathbf{A}}{\partial \mathbf{K}_j}\mathbf{v}_i}{\mathbf{w}_i^T \mathbf{v}_i}, && (2.81) \end{aligned}$$

or in matrix form

$$\Delta \mathbf{\Lambda} = \mathbf{S} \Delta \mathbf{\Gamma}, \tag{2.82}$$

where $\Delta\boldsymbol{\Gamma} = \begin{bmatrix} \Delta\mathbf{K}_1 & \Delta\mathbf{K}_2 & \cdots & \Delta\mathbf{K}_j \end{bmatrix}^T$, $\Delta\boldsymbol{\Lambda} = \begin{bmatrix} \Delta\lambda_i & \Delta\lambda_i & \cdots & \Delta\lambda_i \end{bmatrix}^T$, and $\mathbf{S} \in \mathbb{R}^{i\times j}$ is the *eigenvalue sensitivity matrix*. $\mathbf{S}$ relates how the ith eigenvalue is affected by the jth non-deterministic system parameter $\mathbf{K}_j$, i.e., the rate of change of λ_i with respect to $\mathbf{K}_j$.

From (2.81), it can be seen that the main obstacle in computing $\mathbf{S}$ arises from the term $\frac{\partial\mathbf{A}}{\partial\mathbf{K}_j}$. If $\mathbf{A}$ can be expressed explicitly in terms of $\mathbf{K}_j$, an analytic solution can be obtained [42]. However, if the converse is true, one has to resort to numerical approximations and methods to approximate $\frac{\partial\mathbf{A}}{\partial\mathbf{K}_j}$ [41]. More details on both methods will be covered in future sections.

2.3.5 EIGENVALUE SENSITIVITY TO LINK PARAMETERS

In this section, the eigenvalue sensitivity to *link parameter* a_{kl} in state matrix $\mathbf{A} \in \mathbb{R}^{n\times n}$ for a state equation $\dot{\mathbf{x}} = \mathbf{A}\mathbf{x}$ is detailed. For our discussion, we have summarized some important derivations in [53] here.

Since $\dot{\mathbf{x}} = \mathbf{A}\mathbf{x}$, this homogeneous system can also be rewritten as

$$\mathbf{A} = \frac{\partial\dot{\mathbf{x}}}{\partial\mathbf{x}}, \tag{2.83}$$

or the rate of change of the time derivative of $\mathbf{x}$ with respect to $\mathbf{x}$.

Consider now the solution to the homogeneous system. If we assume that $\mathbf{A}$ has a complete set of n linearly independent right eigenvectors $(\mathbf{v}_1, \mathbf{v}_2, \cdots, \mathbf{v}_n)$ with corresponding eigenvalues $(\lambda_1, \lambda_2, \cdots, \lambda_n)$ for simplicity but without loss of generality, the state vector $\mathbf{x}$ can be expressed as a linear combination of the eigenvectors as

$$\mathbf{x}(t) = b_1(t)\mathbf{v}_1 + b_2(t)\mathbf{v}_2 + \cdots + b_n(t)\mathbf{v}_n, \tag{2.84}$$

where $[b_1(t), b_2(t), \cdots, b_n(t)]$ are scalars in functions of time. This is true since the eigenvectors are linearly independent and span $\mathbb{R}^n$, even when the eigenvalues might not be distinct. The derivative of (2.84) is hence

$$\dot{\mathbf{x}}(t) = \dot{b}_1(t)\mathbf{v}_1 + \dot{b}_2(t)\mathbf{v}_2 + \cdots + \dot{b}_n(t)\mathbf{v}_n. \tag{2.85}$$

Since $\dot{\mathbf{x}} = \mathbf{A}\mathbf{x}$, we combine with (2.84) to get

$$\begin{aligned} \dot{\mathbf{x}} &= \mathbf{A}\mathbf{x} \\ &= b_1(t)\mathbf{A}\mathbf{v}_1 + b_2(t)\mathbf{A}\mathbf{v}_2 + \cdots + b_n(t)\mathbf{A}\mathbf{v}_n \\ &= b_1(t)\lambda_1\mathbf{v}_1 + b_2(t)\lambda_2\mathbf{v}_2 + \cdots + b_n(t)\lambda_n\mathbf{v}_n. \end{aligned} \tag{2.86}$$

Combining (2.85) and (2.86), we have

$$\dot{b}_1(t)\mathbf{v}_1 + \dot{b}_2(t)\mathbf{v}_2 + \cdots + \dot{b}_n(t)\mathbf{v}_n = b_1(t)\lambda_1\mathbf{v}_1 + b_2(t)\lambda_2\mathbf{v}_2 + \cdots + b_n(t)\lambda_n\mathbf{v}_n. \tag{2.87}$$

Since the eigenvectors are assumed to be linearly independent, (2.87) holds only when $\dot{b}_i(t) = b_i(t)\lambda_i$ for $i = 1, 2, \cdots, n$, and the corresponding solution can then be given by

$$b_i(t) = e^{\lambda_i t} b_i(0). \tag{2.88}$$

Finally, we substitute (2.88) into (2.84) to link the state vector **x** as a *trajectory* of its eigenvectors as

$$\mathbf{x}(t) = e^{\lambda_1 t} b_1(0)\mathbf{v}_1 + e^{\lambda_2 t} b_2(0)\mathbf{v}_2 + \cdots + e^{\lambda_n t} b_n(0)\mathbf{v}_n, \tag{2.89}$$

and the behavior of each state $x_i(t)$ in the system can be described by

$$x_i(t) = v_{1i} e^{\lambda_1 t} b_1(0) + v_{2i} e^{\lambda_2 t} b_2(0) + \cdots + v_{ni} e^{\lambda_n t} b_n(0), \tag{2.90}$$

where v_{1i} is the ith component of the first eigenvector, or in matrix form

$$\begin{bmatrix} x_1(t) \\ x_2(t) \\ \vdots \\ x_n(t) \end{bmatrix} = \begin{bmatrix} v_{11} & v_{21} & \cdots & v_{n1} \\ v_{12} & v_{22} & \cdots & v_{n2} \\ \vdots & \vdots & \ddots & \vdots \\ v_{1n} & v_{2n} & \cdots & v_{nn} \end{bmatrix} \begin{bmatrix} e^{\lambda_1 t} b_1(0) \\ e^{\lambda_2 t} b_2(0) \\ \vdots \\ e^{\lambda_n t} b_n(0) \end{bmatrix}. \tag{2.91}$$

This tells us that the overall trajectory of state variable $x_j(t)$ is determined by the linear combination of the product of eigenvector components v_{ji}, behavior mode $e^{\lambda_j t}$ generated by eigenvalue λ_j, and initial condition $b_j(0)$. This is important because we usually analyze the trajectory of a state $x_j(t)$ mainly by finding the critical resonant modes via their poles and natural frequencies, which are specified by a specific eigenvalue λ_j and corresponding eigenvector $\mathbf{v}_j$.

However, a change in the link parameter a_{kl} in state matrix **A** will yield a totally different $\mathbf{A}'$, hence eigenvalues λ'_j and eigenvector components v'_{ji}. Taking partial derivative of the ith state $x_i(t)$ with respect to the changes in link gain a_{kl} using (2.90), the changes in link gain on overall system behavior can be obtained as

$$\frac{\partial x_i(t)}{\partial a_{kl}} = \frac{\partial}{\partial a_{kl}} \left[v_{1i} e^{\lambda_1 t} b_1(0) + v_{2i} e^{\lambda_2 t} b_2(0) + \cdots + v_{ni} e^{\lambda_n t} b_n(0) \right]. \tag{2.92}$$

Now taking the derivative of individual components, we obtain

$$\begin{aligned} \frac{\partial x_1(t)}{\partial a_{kl}} &= \frac{\partial v_{1i}}{\partial a_{kl}} e^{\lambda_1 t} b_1(0) + v_{1i} \frac{\partial e^{\lambda_1 t}}{\partial \lambda_1} \frac{\partial \lambda_1}{\partial a_{kl}} b_1(0) + \cdots \\ & \frac{\partial v_{2i}}{\partial a_{kl}} e^{\lambda_2 t} b_2(0) + v_{2i} \frac{\partial e^{\lambda_2 t}}{\partial \lambda_2} \frac{\partial \lambda_2}{\partial a_{kl}} b_2(0) + \cdots \\ & \cdots + \frac{\partial v_{ni}}{\partial a_{kl}} e^{\lambda_n t} b_n(0) + v_{ni} \frac{\partial e^{\lambda_n t}}{\partial \lambda_n} \frac{\partial \lambda_n}{\partial a_{kl}} b_n(0), \end{aligned} \tag{2.93}$$

or

$$\frac{\partial x_i(t)}{\partial a_{kl}} = \sum_{j=1}^{n} \left(\frac{\partial v_{ji}}{\partial a_{kl}} e^{\lambda_j t} + v_{ji} \frac{\partial e^{\lambda_j t}}{\partial \lambda_j} \frac{\partial \lambda_j}{\partial a_{kl}} \right) b_j(0). \tag{2.94}$$

Since the eigenvalues and eigenvectors in linear systems are constant, the derivative of the exponential of the jth behavior mode $e^{\lambda_j t}$ with respect to its eigenvalue λ_j yields a term that depends on time $te^{\lambda_j t}$. As such, (2.94) can be rewritten as

$$\frac{\partial x_i(t)}{\partial a_{kl}} = \sum_{j=1}^{n} \left(\frac{\partial v_{ji}}{\partial a_{kl}} + v_{ji} \frac{\partial \lambda_j}{\partial a_{kl}} t \right) e^{\lambda_j t} b_j(0), \tag{2.95}$$

which suggests that a change in behavior of state $x_i(t)$ due to a change in link gain a_{kl} will be composed of two terms for each behavior mode $e^{\lambda_j t}$ contributing to the overall behavior trajectory of state variable $x_i(t)$. Each of the terms corresponds to

1. The derivative of eigenvector component v_{ji} with respect to link gain a_{kl}; and
2. The product of eigenvector component v_{ji}, the derivative of eigenvalue λ_i with respect to link gain a_{kl}, and time t.

This suggests that when $t = 0^+$, the behavior mode $e^{\lambda_j t}$ will be mainly influenced by the first term, i.e., the derivative of the eigenvector with respect to the link gain. As the transient dies down and during steady-state or as $t \longrightarrow \infty$, the behavior mode $e^{\lambda_j t}$ will be more influenced by the second term, i.e., the derivative of the eigenvalue with respect to the link gain. As such, this leads to defining the two sensitivities which are important to understanding the overall trajectory of $x_i(t)$ as discussed in [54][55].

Firstly, $\frac{\partial \lambda_j}{\partial a_{kl}}$ is the sensitivity of eigenvalue λ_i with respect to link a_{kl} and allows us to understand the strength of a link a_{kl} on behavior mode $e^{\lambda_j t}$. This is denoted by $S_{\lambda_i kl}$, where

$$S_{\lambda_i kl} = \frac{\partial \lambda_i}{\partial a_{kl}}. \tag{2.96}$$

(2.96) can be normalized by multiplying $S_{\lambda_i kl}$ with the ratio of the magnitude of the link gain $|a_{kl}|$ to the magnitude of the eigenvalue $|\lambda_i|$ (or $||\lambda_i||$ if λ_i is complex), i.e.,

$$E_{ikl} = \frac{\partial \lambda_i}{\partial a_{kl}} \frac{|a_{kl}|}{|\lambda_i|}. \tag{2.97}$$

The normalized sensitivity E_{ikl} is defined as the *eigenvalue elasticity with respect to link gain* or *link gain (eigenvalue) elasticity* [55].

Next, $\frac{\partial v_{ji}}{\partial a_{kl}}$ is the sensitivity of an eigenvector component v_{ji} with respect to a specific link a_{kl} [53]. This is denoted by $S_{r_{ji}kl}$, where

$$S_{v_{ji}kl} = \frac{\partial v_{ji}}{\partial a_{kl}}. \tag{2.98}$$

Similarly, (2.98) can be normalized by multiplying $S_{v_{ji}kl}$ with the ratio of the magnitude of the link gain $|a_{kl}|$ to the magnitude of the eigenvector component $|v_{ji}|$ (or $||v_{ji}||$ if v_{ji} is complex), i.e.,

$$E_{v_{ji}kl} = \frac{\partial v_{ji}}{\partial a_{kl}} \frac{|a_{kl}|}{|v_{ji}|}. \tag{2.99}$$

The normalized sensitivity $E_{v_{ji}kl}$ is defined as the *eigenvector component* v_{ji} *elasticity with respect to link gain* or *link gain eigenvector component elasticity* [53].

As such, (2.95) can be written as

$$\frac{\partial x_i(t)}{\partial a_{kl}} = \sum_{j=1}^{n} \left(S_{\lambda_i kl} + v_{ji} S_{v_{ji}kl} t \right) e^{\lambda_j t} z_j(0), \tag{2.100}$$

and the contribution of the eigenvalue sensitivity to the weight changes with time, becoming the main determinant of weight of behavior mode $e^{\lambda_j t}$ as time elapses.

Similarly, defining *loop gain* g_k as the product of the link gains (a_{kl}) of all links present in the loop, the derivative (2.90) with respect to g_k can be expressed as

$$\begin{aligned} \frac{\partial x_i(t)}{\partial g_k} &= \frac{\partial}{\partial g_k} \left[v_{1i} e^{\lambda_1 t} b_1(0) + v_{2i} e^{\lambda_2 t} b_2(0) + \cdots + v_{ni} e^{\lambda_n t} b_n(0) \right] \\ &= \sum_{j=1}^{n} \left(\frac{\partial v_{ji}}{\partial g_k} + v_{ji} \frac{\partial \lambda_j}{\partial g_k} t \right) e^{\lambda_j t} b_j(0) \end{aligned} \tag{2.101}$$

for linear systems [53].

Equation (2.101) also suggests that a change in behavior of state $x_i(t)$ due to a change in loop gain g_k will be composed of two terms for each behavior mode $e^{\lambda_j t}$ contributing to the overall behavior trajectory of state variable $x_i(t)$. Each of the terms corresponds to

1. the derivative of eigenvector component v_{ji} with respect to loop gain g_k; and
2. the product of eigenvector component v_{ji}, the derivative of eigenvalue λ_i with respect to loop gain g_k, and time t.

These results can also be extended to concepts of *link sensitivity* $S_{\lambda_i k}$ and link elasticity to loop sensitivity and *loop elasticity* E_{ik}, where

$$\begin{aligned} S_{\lambda_i k} &= \frac{\partial \lambda_i}{\partial g_k}, \\ E_{ik} &= \frac{\partial \lambda_i}{\partial g_k} \frac{|g_k|}{|\lambda_i|}, \end{aligned} \tag{2.102}$$

and also to loop *eigenvector component sensitivity* $S_{v_{ij}k}$ and eigenvector component elasticity with respect to loop gain or *loop gain eigenvector component elasticity* $E_{v_{ij}k}$ with [55]

$$\begin{aligned} S_{v_{ij}k} &= \frac{\partial v_{ij}}{\partial g_k}, \\ E_{v_{ij}k} &= \frac{\partial v_{ij}}{\partial g_k} \frac{|g_k|}{|v_{ij}|}. \end{aligned} \tag{2.103}$$

As such, (2.101) can also be written as

$$\frac{\partial x_i(t)}{\partial g_k} = \sum_{j=1}^{n} \left(S_{v_{ij}k} + v_{ji} S_{\lambda_j k} t \right) e^{\lambda_j t} b_j(0), \tag{2.104}$$

and the contribution of the eigenvalue sensitivity to the weight changes with time, becoming the main determinant of weight of behavior mode $e^{\lambda_j t}$ as time grows.

This section covers the concepts of eigenvalues and eigenvectors and their properties, as well as methods to obtain them using the respective decompositions. The *sensitivities* to both non-deterministic system parameters and link parameters are also discussed. In the next section, another common mathematical tool called the Singular Value Decomposition (SVD) will be covered, along with its properties and some practical applications to engineering systems.

2.4 SINGULAR VALUE DECOMPOSITION (SVD) AND APPLICATIONS

Singular Value Decomposition (SVD) is an important factorization of a rectangular real or complex matrix, with many applications in industrial control and signal processing applications. In general, the numerical algorithm for SVD of matrices is extremely efficient and stable as compared to that of eigenvalue decomposition. In this section, we will introduce the theory of SVD and some relevant engineering examples.

2.4.1 SINGULAR VALUE DECOMPOSITION (SVD)

Consider a matrix $\mathbf{A} \in \mathbb{R}^{m\times n}$ of rank $n < m$. The SVD of $\mathbf{A}$ is

$$\mathbf{A} = \mathbf{U}\boldsymbol{\Sigma}\mathbf{V}^T, \tag{2.105}$$

with $\mathbf{U} \in \mathbb{R}^{m\times n}$ and $\mathbf{V} \in \mathbb{R}^{n\times n}$. $\mathbf{U}$ and $\mathbf{V}$ are unitary, i.e., $\mathbf{U}^T\mathbf{U} = \mathbf{V}^T\mathbf{V} = \mathbf{I}_n$, with $\mathbf{I}_n \in \mathbb{R}^{n\times n}$ being an identity matrix of dimension n. $\boldsymbol{\Sigma} \in \mathbb{R}^{n\times n}$ is a diagonal matrix whose elements are corresponding singular values (or principal gains) arranged in descending order, i.e., with $\boldsymbol{\Sigma} = \text{diag}(\sigma_1, \sigma_2, \cdots, \sigma_n)$ and $\sigma_1 \geq \sigma_2 \geq \cdots \geq \sigma_n > 0$. This is illustrated with a simple numerical example.

Example 2.3: Find the SVD of $\mathbf{A}$, where

$$\mathbf{A} = \begin{bmatrix} 1 & -5 \\ 2 & 5 \\ 3 & 6 \\ -1 & 6 \end{bmatrix}. \tag{2.106}$$

Obviously, $\mathbf{A} \in \mathbb{R}^{4\times 2}$ with $m = 4$ and $n = 2$. The SVD of $\mathbf{A}$ yields the following matrices

$$\begin{aligned} \mathbf{U} &= \begin{bmatrix} 0.4289 & -0.4989 \\ -0.4700 & -0.3442 \\ -0.5723 & -0.5816 \\ -0.5174 & 0.5425 \end{bmatrix}, \\ \mathbf{\Sigma} &= \begin{bmatrix} 11.1640 & 0 \\ 0 & 3.5162 \end{bmatrix}, \\ \mathbf{V}^T &= \begin{bmatrix} -0.1532 & -0.9882 \\ -0.9882 & 0.1532 \end{bmatrix}. \end{aligned} \tag{2.107}$$

It can be easily verified that $\mathbf{U} \in \mathbb{R}^{4\times 2}$, $\mathbf{\Sigma} \in \mathbb{R}^{2\times 2}$, and $\mathbf{V} \in \mathbb{R}^{2\times 2}$. $\mathbf{U}^T\mathbf{U} = \mathbf{V}^T\mathbf{V} = \mathbf{I}_2 = \text{diag}(1,1)$, and $\mathbf{\Sigma} = \text{diag}(11.1640, 3.5162)$ with $11.1640 \geq 3.5162 > 0$.

From the above example and previous discussions, it can be seen that the SVD can be applied to any $\mathbf{A} \in \mathbb{R}^{m\times n}$ matrix, while the eigenvalue decomposition can only be applied to certain classes of square matrices. Although the methods used for both decompositions differ, the following equations hold

$$\begin{aligned} \mathbf{A}^T\mathbf{A} &= \mathbf{V}\mathbf{\Sigma}^T\mathbf{U}^T\mathbf{U}\mathbf{\Sigma}\mathbf{V}^T \\ &= \mathbf{V}\mathbf{\Sigma}^T\mathbf{\Sigma}\mathbf{V}^T, \end{aligned} \tag{2.108}$$

$$\begin{aligned} \mathbf{A}\mathbf{A}^T &= \mathbf{U}\mathbf{\Sigma}\mathbf{V}^T\mathbf{V}\mathbf{\Sigma}^T\mathbf{U}^T \\ &= \mathbf{U}\mathbf{\Sigma}\mathbf{\Sigma}^T\mathbf{V}^T \end{aligned} \tag{2.109}$$

since both $\mathbf{U}$ and $\mathbf{V}$ are unitary. With these manipulations, it is easy to verify that the right-hand sides of the above equations are *already* in the eigenvalue-decomposed forms as shown earlier in (2.45).

As such, the following properties hold

- The column vectors of $\mathbf{U}$ are known as *left singular vectors*. They form a set of orthonormal "output" basis vector directions for $\mathbf{A}$, and are the eigenvectors of $\mathbf{A}\mathbf{A}^T$;
- The column vectors of $\mathbf{V}$ are known as *right singular vectors*. They form a set of orthonormal "input" basis vector directions for $\mathbf{A}$, and are the eigenvectors of $\mathbf{A}^T\mathbf{A}$;
- The singular values σ_i in $\mathbf{\Sigma}$ for $i = 1, 2, \cdots, n$, are always non-negative. They can be visualized as scalar "gain" controls by which each corresponding input is multiplied to give a corresponding output;
- The singular values are the positive square roots of the eigenvalues of $\mathbf{A}\mathbf{A}^T$ and $\mathbf{A}^T\mathbf{A}$ that correspond with the same columns in $\mathbf{U}$ and $\mathbf{V}$.

Moreover, the left singular vectors (columns of $\mathbf{U}$) are eigenvectors of $\mathbf{A}\mathbf{A}^T$, and the right singular vectors (columns of $\mathbf{V}$) are eigenvectors of $\mathbf{A}^T\mathbf{A}$.

Applications which employ the SVD include computing the pseudo-inverse, least-squares system identification of data, matrix approximation, as well as determining the rank, range, and null space of a matrix, etc. In this section, we introduce some applications of SVD to realistic engineering applications.

2.4.2 NORMS, RANK, AND CONDITION NUMBER

In general, 2-norms are computed using SVD through the relation

$$\|\mathbf{A}\|_2 = \sigma_1, \tag{2.110}$$

or the largest singular value of **A**. If **A** is square and $\mathbf{A}^{-1}$ exists, then

$$\|\mathbf{A}^{-1}\|_2 = \frac{1}{\sigma_n}, \tag{2.111}$$

or the inverse of the smallest singular value of **A** [46].

The *Frobenius norm* introduced earlier can also be computed using SVD through

$$\|\mathbf{A}\|_F = \sqrt{\sigma_1^2 + \sigma_2^2 + \cdots + \sigma_n^2}. \tag{2.112}$$

SVD is also by far the most common method for determining the rank of a matrix **A**. The rank of **A** is simply the *number of non-zero singular values* after performing SVD on **A**.

The *condition number* of **A** or $\text{cond}(\mathbf{A})$ is a relative measure of how close **A** is to rank deficiency, i.e., how close **A** is to being *singular* or non-invertible, and can be interpreted as a measure of how much numerical error is likely to be introduced by computations involving **A**. The condition number is defined as $\text{cond}(\mathbf{A}) = \frac{\sigma_1}{\sigma_n}$, i.e., the ratio of the largest singular value σ_1 to the smallest singular value σ_n. As such, it is obvious that singular matrices have *at least* one zero singular value, which leads to an infinite or undefined condition number. A matrix **A** with a large $\text{cond}(\mathbf{A})$ is said to be *ill-conditioned* or *stiff*, and the integrity of numerical solutions obtained when handling these matrices should be confronted. On the other hand, a matrix **A** with a small $\text{cond}(\mathbf{A})$ close to unity is desirable, and is said to be *well-conditioned*. Matrices with a small condition number near unity are usually *balanced*, with small variations amongst entries in the matrix or *link gains* as discussed in the previous section.

2.4.3 PSEUDO-INVERSE

SVD can be used for computing the *pseudo-inverse* of a matrix **A**. In fact, the pseudo-inverse of the matrix **A** is $\mathbf{A}^+$, and is obtained using

$$\mathbf{A}^+ = \mathbf{V}\mathbf{\Sigma}^+\mathbf{U}^T, \tag{2.113}$$

where $\mathbf{\Sigma}^+$ is the pseudo-inverse of $\mathbf{\Sigma}$. $\mathbf{A}^+$ can be easily calculated because $\mathbf{\Sigma}^+$ can be readily obtained by simply replacing every non-zero entry with its reciprocal and transposing the resulting matrix. The pseudo-inverse is a numerically efficient way to solve linear Least Squares (LS) problems.

2.4.4 LEAST SQUARES SOLUTION

Let $n \leq m$ and assume **A** has full column rank n for simplicity but without loss of generality. As such, the null space of **A** is $N(\mathbf{A}) = \phi$ or $N^{\perp}(\mathbf{A}) = \mathbb{R}^n$, where $N(\centerdot)$ denotes the nullity. This is an immediate result from Theorem 2.2.

The SVD of $\mathbf{A}$ is

$$\begin{aligned}\mathbf{A} &= \mathbf{U}\boldsymbol{\Sigma}\mathbf{V}^T \\ &= \begin{bmatrix}\mathbf{U}_1 & \mathbf{U}_2\end{bmatrix}\begin{bmatrix}\boldsymbol{\Sigma}_1 \\ \mathbf{0}\end{bmatrix}\mathbf{V}^T,\end{aligned} \tag{2.114}$$

with $\mathbf{U}_1 \in \mathbb{R}^{m\times n}$ of full column rank. $\boldsymbol{\Sigma}_1 \in \mathbb{R}^{n\times n}$ and $\mathbf{V} \in \mathbb{R}^{n\times n}$ are both non-singular.

Now, define a projection $R(\mathbf{A})$ on $\mathbf{X}$ as $\text{proj}\{R(\mathbf{A})\} = \mathbf{U}_1\mathbf{U}_1^T$, and a projection on the range *perpendicular* of $\mathbf{A}$ as $\text{proj}\{R^{\perp}(\mathbf{A})\} = \mathbf{U}_2\mathbf{U}_2^T = \mathbf{I} - \mathbf{U}_1\mathbf{U}_1^T$. As such, a vector $\mathbf{x}$ under the projection of $\mathbf{A}$ (or linear transformation) yields vector $\mathbf{z}$, where

$$\begin{aligned}\mathbf{z} &= \mathbf{A}\mathbf{x} \\ &= \mathbf{U}\boldsymbol{\Sigma}\mathbf{V}^T\mathbf{x} \\ &= \begin{bmatrix}\mathbf{U}_1 & \mathbf{U}_2\end{bmatrix}\begin{bmatrix}\boldsymbol{\Sigma}_1 \\ \mathbf{0}\end{bmatrix}\mathbf{V}^T\mathbf{x} \\ &= \mathbf{U}_1\boldsymbol{\Sigma}_1\mathbf{V}^T\mathbf{x}.\end{aligned} \tag{2.115}$$

It is obvious that

$$\begin{aligned}\mathbf{U}^T\mathbf{z} &= \begin{bmatrix}\mathbf{U}_1^T \\ \mathbf{U}_2^T\end{bmatrix}\mathbf{z} \\ &= \begin{bmatrix}\boldsymbol{\Sigma}_1\mathbf{V}^T\mathbf{x} \\ \mathbf{0}\end{bmatrix},\end{aligned} \tag{2.116}$$

or

$$\mathbf{U}_1^T\mathbf{z} = \boldsymbol{\Sigma}_1\mathbf{V}^T\mathbf{x}, \tag{2.117}$$

$$\mathbf{U}_2^T\mathbf{z} = \mathbf{0}. \tag{2.118}$$

Solving (2.117), we have

$$\hat{\mathbf{x}} = \mathbf{V}\boldsymbol{\Sigma}_1^{-1}\mathbf{U}_1^T\mathbf{z}, \tag{2.119}$$

and the estimate for $\mathbf{z}$ is $\hat{\mathbf{z}}$ and is given by

$$\begin{aligned}\hat{\mathbf{z}} &= \mathbf{A}\hat{\mathbf{x}} \\ &= \begin{bmatrix}\mathbf{U}_1 & \mathbf{U}_2\end{bmatrix}\begin{bmatrix}\boldsymbol{\Sigma}_1 \\ \mathbf{0}\end{bmatrix}\mathbf{V}^T\mathbf{V}\boldsymbol{\Sigma}_1^{-1}\mathbf{U}_1^T\mathbf{z} \\ &= \mathbf{U}_1\mathbf{U}_1^T\mathbf{z} \\ &= \text{proj}\{R(\mathbf{A})\}\mathbf{z},\end{aligned} \tag{2.120}$$

and the estimation error $\tilde{\mathbf{z}}$ is given by

$$\begin{aligned}\tilde{\mathbf{z}} &= \mathbf{z} - \hat{\mathbf{z}} \\ &= \mathbf{z} - \mathbf{U}_1\mathbf{U}_1^T\mathbf{z} \\ &= (\mathbf{I} - \mathbf{U}_1\mathbf{U}_1^T)\mathbf{z} \\ &= \mathbf{U}_2\mathbf{U}_2^T\mathbf{z} \\ &= \text{proj}\{R^{\perp}(\mathbf{A})\}\mathbf{z}.\end{aligned} \tag{2.121}$$

The square of the estimation error is thus

$$\begin{aligned}\tilde{\mathbf{z}}^T\tilde{\mathbf{z}} &= \mathbf{z}^T\mathbf{U}_2\mathbf{U}_2^T\mathbf{z}\\ \therefore \|\tilde{\mathbf{z}}\|^2 &= \tilde{\mathbf{z}}^T\tilde{\mathbf{z}}\\ &= \mathbf{z}^T\mathbf{U}_2\mathbf{U}_2^T\mathbf{z}\\ &\leq \|\mathbf{U}_2\|^2\|\mathbf{z}\|^2. \end{aligned} \tag{2.122}$$

Recall that $\mathbf{U}_2^T\mathbf{U}_2 = \mathbf{I}$. Hence, the 2-norm of $\mathbf{U}_2$ is equal to one, i.e., its maximum singular value. As such, (2.122) gives the *least squares estimation error solution* of a projection by $\mathbf{A}$.

The Frobenius norm $\mathbf{U}_2$ is also the root of the sum of the squares of all the elements given by

$$\begin{aligned}\|\mathbf{U}_2\|_F^2 &= \mathrm{tr}\{\mathbf{U}_2^T\mathbf{U}_2\}\\ &= \mathrm{tr}\{\mathbf{U}_2\mathbf{U}_2^T\}.\end{aligned} \tag{2.123}$$

The Frobenius norm is also compatible with the 2-norm via the *Cauchy-Schwarz Inequality* where $\|\mathbf{Ax}\| \leq \|\mathbf{A}\|_F\|\mathbf{x}\|$. Since $\mathbf{U}_2^T\mathbf{U}_2 = \mathbf{I}$, one has

$$\begin{aligned}\|\mathbf{U}_2\|_F^2 &= \mathrm{tr}\{\mathbf{U}_2^T\mathbf{U}_2\}\\ &= \mathrm{tr}\{\mathbf{I}_{m-n}\}\\ &= m-n\\ &= \dim\{R^{\perp}(\mathbf{A})\}.\end{aligned} \tag{2.124}$$

2.4.5 MINIMUM-NORM SOLUTION USING SVD

Similarly, let $n \leq m$, and assume $\mathbf{A}$ has full row rank m for simplicity but without loss of generality. As such, the *range* of linear transformation $\mathbf{A}$ is $R(\mathbf{A}) = \mathbb{R}^m$ and $R^{\perp}(\mathbf{A}) = \phi$.

The SVD of $\mathbf{A}$ is

$$\mathbf{A} = \mathbf{U}\begin{bmatrix}\boldsymbol{\Sigma}_1 & 0\end{bmatrix}\begin{bmatrix}\mathbf{V}_1^T\\ \mathbf{V}_2^T\end{bmatrix}, \tag{2.125}$$

with $\mathbf{V}_1^T \in \mathbb{R}^{m\times n}$ of full row rank. $\boldsymbol{\Sigma}_1 \in \mathbb{R}^{m\times m}$ and $\mathbf{U} \in R^{m\times m}$ are both non-singular.

Now, one has projection on $N^{\perp}(\mathbf{A})$ as $\mathrm{proj}\{N^{\perp}(\mathbf{A})\} = \mathbf{V}_1\mathbf{V}_1^T$, and projection on $N(\mathbf{A})$ as $\mathrm{proj}\{N(\mathbf{A})\} = \mathbf{V}_2\mathbf{V}_2^T = \mathbf{I} - \mathbf{V}_1\mathbf{V}_1^T$. Then,

$$\begin{aligned}\mathbf{z} &= \mathbf{Ax}\\ &= \mathbf{U}\boldsymbol{\Sigma}\mathbf{V}^T\mathbf{x}\\ &= \mathbf{U}\begin{bmatrix}\boldsymbol{\Sigma}_1 & \mathbf{0}\end{bmatrix}\begin{bmatrix}\mathbf{V}_1^T\\ \mathbf{V}_2^T\end{bmatrix}\mathbf{x}\\ &= \mathbf{U}\boldsymbol{\Sigma}_1\mathbf{V}_1^T\mathbf{x}.\end{aligned} \tag{2.126}$$

We have

$$\begin{aligned} \mathbf{A}\mathbf{A}^T &= \mathbf{U}\begin{bmatrix}\boldsymbol{\Sigma}_1 & \mathbf{0}\end{bmatrix}\begin{bmatrix}\mathbf{V}_1^T \\ \mathbf{V}_2^T\end{bmatrix}\begin{bmatrix}\mathbf{V}_1 & \mathbf{V}_2\end{bmatrix}\begin{bmatrix}\boldsymbol{\Sigma}_1 \\ \mathbf{0}\end{bmatrix}\mathbf{U}^T \\ &= \mathbf{U}\boldsymbol{\Sigma}_1^2\mathbf{U}^T, \end{aligned} \tag{2.127}$$

so the right inverse of $\mathbf{A}$ is $\mathbf{A_R}$ and is given by

$$\begin{aligned} \mathbf{A_R} &= \mathbf{A}^T(\mathbf{A}\mathbf{A}^T)^{-1} \\ &= \begin{bmatrix}\mathbf{V}_1 & \mathbf{V}_2\end{bmatrix}\begin{bmatrix}\boldsymbol{\Sigma}_1 \\ \mathbf{0}\end{bmatrix}\mathbf{U}^T\mathbf{U}\boldsymbol{\Sigma}_1^{-2}\mathbf{U}^T \\ &= \begin{bmatrix}\mathbf{V}_1 & \mathbf{V}_2\end{bmatrix}\begin{bmatrix}\boldsymbol{\Sigma}_1^{-1}\mathbf{U}^T \\ \mathbf{0}\end{bmatrix} \\ &= \mathbf{V}_1\boldsymbol{\Sigma}_1^{-1}\mathbf{U}^T, \end{aligned} \tag{2.128}$$

and the *minimum-norm solution* is given by

$$\begin{aligned} \hat{\mathbf{x}} &= \mathbf{A_R}\mathbf{z} \\ &= \mathbf{V}_1\boldsymbol{\Sigma}_1^{-1}\mathbf{U}^T\mathbf{z}. \end{aligned} \tag{2.129}$$

The estimate for $\mathbf{z}$ given $\hat{\mathbf{x}}$ is

$$\begin{aligned} \hat{\mathbf{z}} &= \mathbf{A}\hat{\mathbf{x}} \\ &= \mathbf{A}\mathbf{A_R}\mathbf{z} \\ &= \mathbf{U}\begin{bmatrix}\boldsymbol{\Sigma}_1 & \mathbf{0}\end{bmatrix}\begin{bmatrix}\mathbf{V}_1^T \\ \mathbf{V}_2^T\end{bmatrix}\mathbf{V}_1\boldsymbol{\Sigma}_1^{-1}\mathbf{U}^T\mathbf{z} \\ &= \mathbf{z}, \end{aligned} \tag{2.130}$$

i.e., the exact value of $\mathbf{z}$ since $\mathbf{z} \in R(\mathbf{A}) = \mathbb{R}^m$. The estimation error is equal to zero.

The full solution is thus

$$\hat{\mathbf{x}} + \mathbf{x}_{N(\mathbf{X})} = \mathbf{V}_1\boldsymbol{\Sigma}_1^{-1}\mathbf{U}^T\mathbf{z} + \mathbf{V}_2\mathbf{V}_2^T\mathbf{w} \tag{2.131}$$

for any $\mathbf{w} \in \mathbb{R}^n$. The second term is in the null space of $\mathbf{A}$, $N(\mathbf{A}) = \mathbb{R}(\mathbf{V}_2\mathbf{V}_2^T)$.

With these, we conclude some important preliminary mathematical machineries with *real* vectors and matrices. In the following section, we shall introduce another type of matrix—*boolean* or *binary* matrix, which is very useful for diagnosis and prognosis of *Rule-Based Systems* (RBSs) and *discrete-event systems*.

2.5 BOOLEAN MATRICES

In mathematics, particularly matrix theory, a *boolean* matrix or a $(0,1)$ matrix is one whose entries are either *zero* or *one*.

Example 2.4: The following 2×2 real matrix

$$\begin{bmatrix} 1 & 1 \\ 1 & 0 \end{bmatrix} \tag{2.132}$$

is a 2×2 boolean matrix as its entries are only zero or one.

Binary matrices are extremely useful in modeling binary relations, logical IF-THEN-ELSE rule bases, graphs, as well as control of event-triggered systems and discrete-event systems, etc. In this section, we shall introduce some applications that exist in current literature.

2.5.1 BINARY RELATION

Consider two elements of a system s_i and s_j. Suppose it is possible to that say s_i and s_j are related in a certain way, or they are not. Using the notation and example in [56], the *relation* **R** between s_i and s_j can be modeled as either

$$s_i\, \mathbf{R}\, s_j, \tag{2.133}$$

or

$$s_i\, \overline{\mathbf{R}}\, s_j, \tag{2.134}$$

where the bar over **R** is the negation.

As such, one can model this system of binary relations in a set of k elements $\mathbf{S} = \begin{bmatrix} s_1, & s_2, & \cdots, & s_k \end{bmatrix}$ as a binary square matrix. The binary matrix will have an entry of one in position (i, j) if (2.133) is true, and zero if (2.134) is true.

Example 2.5: Given a binary relation matrix **M** as

$$\mathbf{M} = \begin{matrix} \\ s_1 \\ s_2 \\ s_3 \\ s_4 \end{matrix} \begin{matrix} \begin{matrix} s_1 & s_2 & s_3 & s_4 \end{matrix} \\ \begin{bmatrix} 1 & 0 & 1 & 1 \\ 0 & 1 & 1 & 0 \\ 1 & 0 & 0 & 1 \\ 0 & 0 & 1 & 0 \end{bmatrix} \end{matrix}. \tag{2.135}$$

The corresponding graph representing the binary relation matrix **M** is shown in Figure 2.2 [56].

The graph in Figure 2.2 is also known as an *undirected* graph, since there is no direction in the connections of the nodes or vertices. More examples and definitions of graphs will be covered in the next section.

Due to the nature of binary matrices, little matrix operations exist. For the binary relation matrix **M** shown in (2.135), one possible operation is the *permutation*

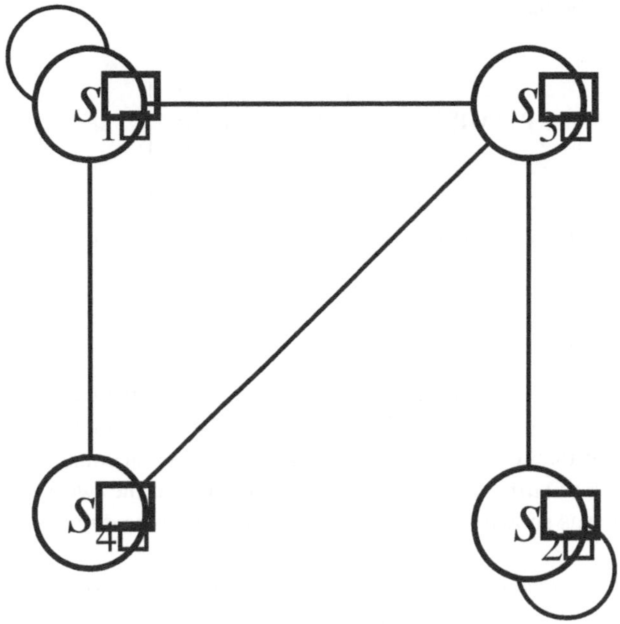

FIGURE 2.2 Graph of binary relation matrix **M**.

operation **P** given by [56]

$$\mathbf{P} = \begin{bmatrix} 1 & 0 & 0 & 0 \\ 0 & 0 & 0 & 1 \\ 0 & 1 & 0 & 0 \\ 0 & 0 & 1 & 0 \end{bmatrix}. \tag{2.136}$$

Carrying out the operation **MP** using Boolean multiplication (logical AND) and addition (logical OR), one obtains the result

$$\mathbf{MP} = \begin{array}{c} \\ s_1 \\ s_2 \\ s_3 \\ s_4 \end{array} \begin{array}{c} \begin{array}{cccc} s_1 & s_3 & s_4 & s_2 \end{array} \\ \left[\begin{array}{cccc} 1 & 1 & 1 & 0 \\ 0 & 1 & 0 & 1 \\ 1 & 0 & 1 & 0 \\ 0 & 1 & 0 & 0 \end{array} \right] \end{array}. \tag{2.137}$$

The matrix **P** has an inverse $\mathbf{P}^{-1}$ since it is non-singular. If this matrix is multiplied into **MP**, one obtains

$$\mathbf{P}^{-1}\mathbf{MP} = \begin{array}{c} \\ s_1 \\ s_3 \\ s_4 \\ s_2 \end{array} \begin{array}{c} \begin{array}{cccc} s_1 & s_3 & s_4 & s_2 \end{array} \\ \left[\begin{array}{cccc} 1 & 1 & 1 & 0 \\ 1 & 0 & 1 & 0 \\ 0 & 1 & 0 & 0 \\ 0 & 1 & 0 & 1 \end{array}\right] \end{array}, \tag{2.138}$$

and the order of the nodes in the graphs are permutated without any change in logical connectivities.

2.5.2 GRAPHS

Graph Theory is the study of graphs (or mathematical structures) which are used to model relations between a set of objects of interest. A *graph* in this context refers to a collection of vertices or nodes and a collection of edges that connect pairs of vertices. A graph may be *undirected*, i.e., no distinction between the two vertices associated with each edge as shown in Figure 2.2, or its edges may be directed from one vertex to another.

As such, a *directed graph* or *digraph* for short, is a set of points $P_1, P_2, \cdots, P_n$ and *ordered* pairs of points P_iP_j. A digraph G is represented as $G = (V, E)$ where [57]

- V is a set, whose elements are called *vertices* or *nodes*,
- E is a set of ordered pairs of vertices, called *edges*.

An example of a digraph is shown in Figure 2.3.

Sometimes a digraph is called a *simple* digraph to distinguish it from a directed multigraph, where the arcs constitute a multiset rather than a set of ordered pairs of vertices. In a simple digraph, *loops* are disallowed. A loop is an arc that pairs a vertex to itself. On the other hand, some contexts allow loops, multiple arcs, or both, in a digraph.

A boolean matrix which is widely used to represent and model a digraph is the *adjacency matrix*. Let G be a digraph with n vertices or nodes. The *adjacency matrix* of A is $A(G)$, and is the $n \times n$ matrix with $A(G)_{ij}$ equal to one if P_iP_j is an edge and zero otherwise. As such, the adjacency matrix of the digraph shown earlier in Figure 2.3 is

$$A(G) = \left[\begin{array}{ccccc} 0 & 1 & 0 & 0 & 0 \\ 0 & 0 & 1 & 1 & 1 \\ 0 & 0 & 0 & 0 & 0 \\ 0 & 1 & 0 & 0 & 0 \\ 0 & 0 & 0 & 1 & 0 \end{array}\right]. \tag{2.139}$$

2.5.3 DISCRETE-EVENT SYSTEMS

In discrete-event systems, tasks and jobs are *event-triggered*, as opposed to conventional continuous or discrete linear systems, which are *time-triggered*. As such,

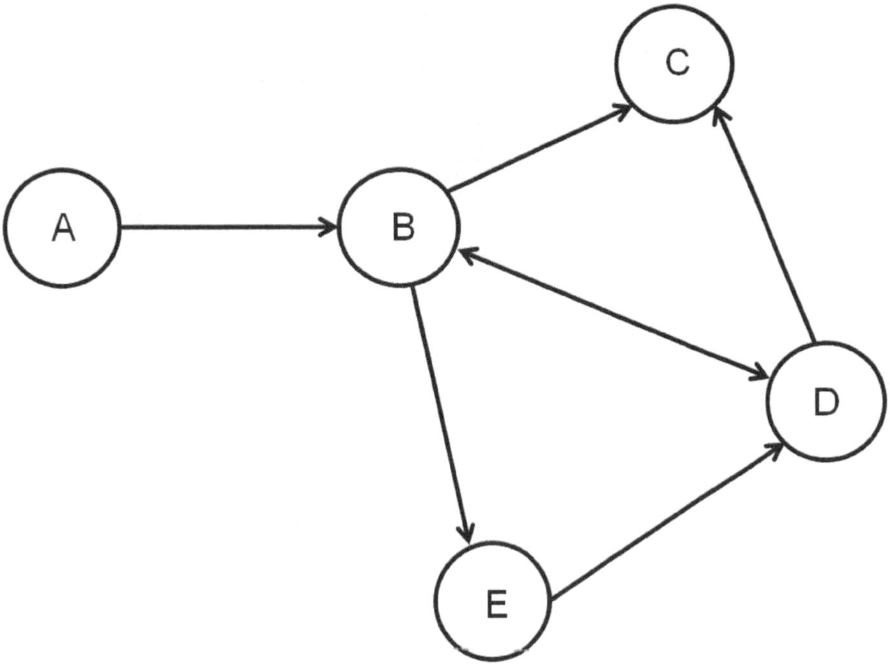

FIGURE 2.3 Directed graph.

matrix-based *Discrete Event Control* (DEC) has been widely used for modeling and analysis of complex interconnected discrete event systems with shared resources, routing decisions, and dynamic resource management, etc., [58]. This formulation gives a very direct and efficient technique for both computer simulation and actual online supervisory control of discrete-event systems.

The matrix approach provides a rigorous, yet intuitive mathematical framework to represent the dynamic evolution of discrete-event systems according to linguistic *IF-THEN* rules. Consider an ith rule in a rule-based system of the form

Rule i: IF <conditions hold> THEN <consequences>.

A possible rule in a realistic engineering application could be

Rule i: IF <sensor 1 has completed task 1> AND <robot 2 is available> AND <a chemical alert is detected>
THEN <robot 2 starts task 5> AND <sensor 2 takes measure>

in a Wireless Sensor Network (WSN). The overall architecture of a DEC using the example of WSN is shown in Figure 2.4 [59].

Let **r** be the vector of resources used in the system—e.g., mobile robots and Unattended Ground Sensors (UGSs), **v** be the vector of tasks that the resources can

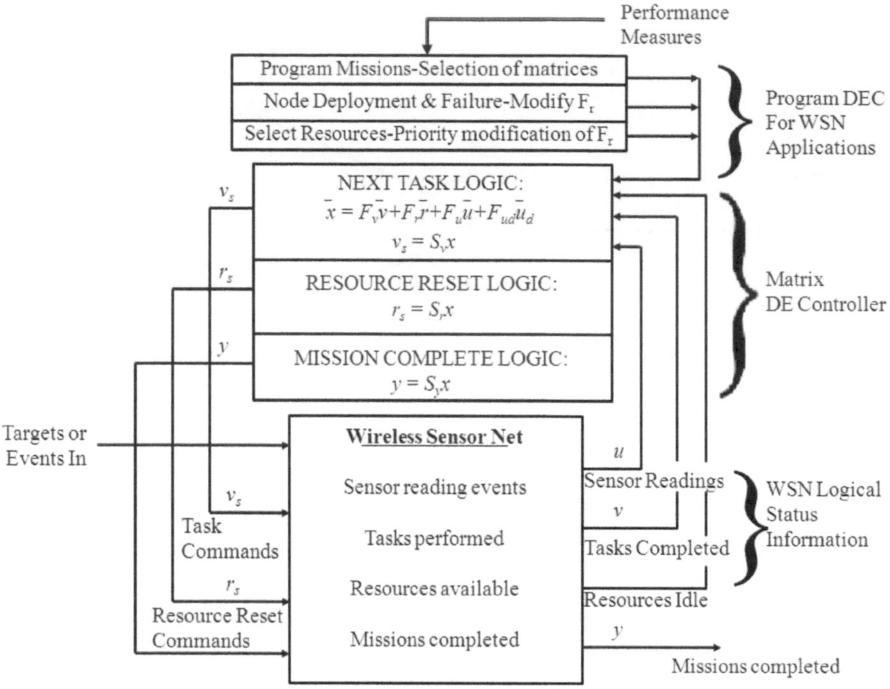

FIGURE 2.4 Complete architecture of DEC in a WSN.

perform—e.g., go to a prescribed location, take a measurement, retrieve and deploy UGSs, etc., and $\mathbf{u}$ the vector of input events, occurrence of sensor detection events, and node failures, etc. We define a *mission* as a prescribed sequence of tasks programmed into the DEC for completion based on sensor readings.

Also, let $\mathbf{y}$ be the vector of completed missions or outputs. Finally, let $\mathbf{x}$ be the state logical vector of the rules of the DEC, whose entry of "1" in position i denotes that rule i of the supervisory-control policy is currently activated.

We then define two different sets of logical equations, one for checking the conditions for the activation of rule i or *matrix controller-state equation*, and one for defining the consequences of the activation of rule i or *matrix controller output equations*. Note that, all matrix operations are defined to be in a specialized *or/and* algebra, where "+" denotes logical "OR" and "×" denotes logical "AND."

The *matrix controller-state equation* is then computed as following [58]

$$\bar{\mathbf{x}} = \mathbf{F_v}\bar{\mathbf{v}} + \mathbf{F_r}\bar{\mathbf{r}} + \mathbf{F_u}\bar{\mathbf{u}} + \mathbf{F_{ud}}\bar{\mathbf{u}}_\mathbf{d}, \tag{2.140}$$

where $\mathbf{F_v}$ is the *Task-Sequencing Matrix* (TSM), $\mathbf{F_r}$ is the *Resource Assignment Matrix* (RAM), and $\mathbf{F_u}$ is the *input matrix*. $\mathbf{F_{ud}}$ is the conflict-resolution matrix and $\mathbf{u_d}$ is the conflict-resolution vector, used to avoid simultaneous activation of conflicting rules. The current status of the discrete event system includes a task vector $\mathbf{v}$, whose entries of "1" represent "completed tasks," resource vector $\mathbf{r}$ whose entries of "1" represent "resources currently available," and the input vector $\mathbf{u}$ whose entries of "1" represent occurrences of certain predefined events (fire alarm, intrusion, etc.). The overbar in (2.140) denotes a logical negation, so that the tasks completed and resources released are represented by "0" entries in $\bar{\mathbf{v}}$ and $\bar{\mathbf{r}}$, respectively.

The TSM $\mathbf{F_v}$ has element (i, j) set to "1" if the completion of task v_j is an immediate prerequisite for the activation of logic state x_i. The RAM $\mathbf{F_r}$ has element (i, j) set to "1" if the availability of resource j is an immediate prerequisite for the activation of logic state x_i.

After calculating the matrix controller-state equation and obtaining the state vector x_i, the DEC has to sequence according to the following *matrix controller output equations* in the next iteration

$$\mathbf{v_s} = \mathbf{S_v x}, \tag{2.141}$$

$$\mathbf{r_s} = \mathbf{S_r x}, \tag{2.142}$$

$$\mathbf{y} = \mathbf{S_y x}, \tag{2.143}$$

where

- $\mathbf{S_v}$ is the *task-start matrix* and has element (i, j) set to "1" if logic state x_j determines the activation of task i;
- $\mathbf{S_r}$ is the *resource-release matrix* and has element (i, j) set to "1" if the activation of logic state x_j determines the release or resetting of resource i;
- $\mathbf{S_y}$ is the *output matrix* and has element (i, j) set to "1" if the activation of logic state x_j determines the completion of mission i;
- $\mathbf{v_s}$ is the *task vector* whose "1" entries denote which tasks are to be started;
- $\mathbf{r_s}$ is the *resource vector* whose "1" entries denote which resources are to be released, represent the commands sent to the discrete event system by the controller; and
- $\mathbf{y}$ contain entries of "1" entries which missions have been successfully completed.

These equations form the rule base of supervisory control of discrete-event systems. All the coefficient matrices are composed of boolean elements, and realtime computations are easily carried out even for large-scale interconnected networked systems.

The DEC framework also has a direct state-to-state correspondence with *Petri Nets* (PNs) [60]. The complete dynamical description of the system according to the new approach is described using the marking transitions of a PN by the PN *transition equation*

$$\mathbf{m}(t+1) = \mathbf{m}(t) + (\mathbf{S}' - \mathbf{F})'\mathbf{x}(t), \tag{2.144}$$

where $\mathbf{S}$ and $\mathbf{F}$ are the PN *output* and *input incidence matrices*, respectively. This equation gives a useful insight on the dynamic of discrete-event systems, but does not provide a complete dynamical description. However, if we relate $\mathbf{x}$ as the vector associated with the PN transitions and $\mathbf{u}$, $\mathbf{v}$, $\mathbf{r}$, and $\mathbf{u_d}$ as associated with the places in a PN, it follows from [58] that

$$\begin{aligned}
\mathbf{m}(t) &= [\ \mathbf{u}(t)', \quad \mathbf{v}(t)', \quad \mathbf{r}(t)', \quad \mathbf{u_d}(t)'\ ]', \\
\mathbf{S} &= [\ \mathbf{S}'_{\mathbf{u}}, \quad \mathbf{S}'_{\mathbf{v}}, \quad \mathbf{S}'_{\mathbf{r}}, \quad \mathbf{S}'_{\mathbf{u_d}}, \quad \mathbf{S}'_{\mathbf{y}}\ ]', \\
\mathbf{F} &= [\ \mathbf{F}'_{\mathbf{u}}, \quad \mathbf{F}'_{\mathbf{v}}, \quad \mathbf{F}'_{\mathbf{r}}, \quad \mathbf{F}'_{\mathbf{u_d}}, \quad \mathbf{F}'_{\mathbf{y}}\ ]',
\end{aligned} \tag{2.145}$$

and we can use (2.143) to generate the allowable firing vector to trigger transitions in (2.140). The combination of the DEC (2.140) and the PN marking transition equation provides a *complete* dynamical description of the system.

In order to take into account the time durations of the tasks and the time required for resource releases, we can split $\mathbf{m}(t)$ into two vectors, namely, one representing available resources and current finished tasks with $\mathbf{m_a}(t)$ and the other representing the tasks in progress and idle resources with $\mathbf{m_p}(t)$. As such, we can rewrite $\mathbf{m}(t)$ as

$$\mathbf{m}(t) = \mathbf{m_a}(t) + \mathbf{m_p}(t). \tag{2.146}$$

This is equivalent to introducing timed-places in a PN and dividing each place into two parts, one relative to the pending states (task in progress, resource idle) and the other relative to the steady states (task completed and resource available). As a consequence, we can also split (2.146) as

$$\mathbf{m_a}(t+1) = \mathbf{m_a}(t) - \mathbf{F}'\mathbf{x}(t), \tag{2.147}$$

$$\mathbf{m_p}(t+1) = \mathbf{m_p}(t) + \mathbf{S}\mathbf{x}(t). \tag{2.148}$$

When a transition fires, a *token* is moved from $\mathbf{m_p}(t)$ to $\mathbf{m_a}(t)$ where it may be used to fire subsequent transitions. Therefore, the above-mentioned DEC framework represents a complete description of the dynamical behavior of the *entire* discrete-event system, and can be implemented for the purposes of computer simulations using any programming language, e.g., MATLAB or C [58]. In the case of a mobile WSN where experiments on wide and hostile areas can be very complex and challenging, it allows one to perform extensive simulations of the control strategies and before experimental implementations. More details on treatment of DEC will be covered in future chapters.

2.6 CONCLUSION

In this chapter, we review the fundamental concepts of vectors and matrices and their applications to linear systems. Two important matrix decompositions, i.e., eigenvalue decomposition and Singular Value Decomposition (SVD), are also introduced to compute the required eigenvalues and singular values, respectively. Besides real matrices, Boolean or binary matrices are also discussed, leading to their applications in Discrete Event Control (DEC) of event-triggered discrete-event systems which are automated by a logical IF-THEN-ELSE rule base. These concepts are also illustrated with examples, along with applications to realistic engineering applications.

Measured data from sensors (direct sensing) or features (indirect or inferential) sensing are commonly packed into vectors or concatenated into matrices for signal processing, control, and treatment in many industrial applications. In the following chapter, we propose a Modal Parametric Identification (MPI) procedure to identify critical parameters of mechanical actuators in mechatronic systems from measured frequency response data vector. The identified parameters allow us to quickly visualize notorious resonant modes, and these diagnostic information are stored in an online repository for rapid data sharing and prognostic rework of mechanical actuators in enhanced mechatronic Research and Development (R&D) from the integrated systems design framework.

3 Modal Parametric Identification (MPI)

To compress Research and Development (R&D) cycle times of high-tech mechatronic products with conformance performance metrics, improved management of R&D projects, thereby allowing engineers from electrical, mechanical, and mechatronic disciplines to receive real-time design feedback and assessment is essential. In this chapter, we propose a systems design procedure to integrate mechanical design, structural prototyping, and servo evaluation through careful comprehension of the servo-mechanical-prototype production cycle commonly employed in the R&D processes of mechatronic industries. Our approach focuses on the Modal Parametric Identification (MPI) of key feedback parameters for fast exchange of design specifications and information. This enables the efficient conduct of product design evaluations, and supports schedule compression of the R&D project life cycle in the highly competitive consumer electronics industry.

Using the commercial Hard Disk Drive (HDD) as a classic example of mechatronic systems, we demonstrate how our approach allows interdisciplinary specifications to be communicated amongst engineers from different backgrounds to speed up the R&D process for the next generation of intelligent manufacturing. In particular, the developed MPI procedure uses the *measured frequency response data vector* of mechanical actuators and allows for fast estimation of residues, damping ratios, and natural frequencies of the critical resonant modes for Linear-Time-Invariant (LTI) flexible mechanical systems. The essential *diagnostic* information in terms of coefficients and residues of the transfer function model are optimized via two subsequent Least Squares (LS) error optimization criteria in the modal summation form. The proposed methodology is verified with experimental frequency responses measured from a commercial dual-stage HDD using the Laser Doppler Vibrometer (LDV). Our results show the effectiveness and robustness of the proposed system identification algorithm for estimating the modal parameters of flexible mechanical structures in mechatronic systems. This provides project management teams with powerful *prognostic* decision-making tools for project strategy reformulation, and improvements in project outcomes are potentially massive because of the low costs of change.

3.1 INTRODUCTION

Mechatronic systems—the integration of mechanical, electrical, computational, and control systems—have pervaded products ranging from larger scale anti-braking systems in cars to small scale microsystems in portable mobile phones. To sustain market shares in the highly competitive consumer electronics industries, continual improvements in servo evaluation and performance of products are essential, in particular for portable electronic devices requiring ultra-high data storage capacities and ultra-strong disturbance rejection capabilities. Moreover, pressures to minimize time-to-market of new products imply a necessarily high schedule compression of the end-to-end servo-mechanical design and evaluation cycle, i.e., from mechanical structural designing, prototyping, to servo control systems design in order to meet targeted specifications.

Currently, in a typical production life cycle for mechatronic systems, the Business Unit (BU) in the manufacturing industries determines the next generation specifications, targeted deliverables, and customers' needs, etc., which are all driven by technological and market forces. These requirements are directly translated into R&D efforts for road-map driven technologies, which will further be decomposed into targeted specifications and requirements such as dimensions, natural frequencies, and mode shapes, etc. A highly integrative and cross-disciplinary approach for managing such R&D is hence essential, since traditional silo approaches focusing on individual components is clearly insufficient in today's large-scale, interwoven, and networked environment. In particular, mechatronic systems R&D encompass a wide range of technical disciplines and expertise, and are hence necessarily highly integrative and collaborative in nature.

At the implementation level, integrated design of mechatronic systems can be facilitated through the perusal of mechanical and electrical Computer-Aided Design (CAD) systems [61]. One approach to achieve such integration is through the propagation of specifications and targeted deliverables. From a theoretical systems point of view, Iwasaki proposed a systematic integrated system design method to achieve certain desired separation properties [62]. Hara discussed dynamical system designs with integrated control, dividing it into analysis and synthesis [63]. An ideal limit of spillover effects guaranteeing highest joint controllability and observability of all modes in bandwidth of interests is discussed in [64]. An integrated approach to design mechatronic systems with cross-disciplinary constraint modeling is proposed in [65]. A centralized R&D [66] research team design and management is proposed in [67] to focus on big impact and long-term innovation. A systems approach to manage R&D projects is also proposed in [68].

However, in practice such collaborative development is often unsatisfactory, inefficient, and error-prone. This is primarily due to the fact that many existing R&D environments do not support *sufficient* and *efficient* flow of design and *diagnostic* manufacturing information across *different* engineering domains. Clearly, sufficiency of information does not necessarily correlate with volume of data-sharing. Indeed, high-volume data transfer across functionalities may yield little relevant information and knowledge-sharing. Practitioners often query about the efficiency of such information flows, i.e., the marginal benefits of intensive data-mining efforts versus the gain in information and learning.

Hence, while a highly-integrative or collaborative environment for managing complex systems R&D is essential and much sought-after, a key enabler to its success is the ability to identify critical *diagnostic* feedback parameters (or metrics) flowing to and from different sub-systems. In particular, highly-integrated mechatronic systems need to be developed in an environment where both electrical and mechanical engineering domain experts receive relevant real-time *prognostic* feedback for propagation of design modifications throughout the entire R&D process. This feedback information transparency is instrumental for manufacturing optimization and high productivity in development of mechatronic products, for instance in a Totally Integrated Automation (TIA) intelligent factory [69].

Furthering our earlier research works in improving yield and identification of diagnostic critical resonant modes for prognosis with mechanical actuator redesign *a priori* [32][34], we enhance and detail the proposed Modal Parametric Identification (MPI) algorithm for integrative systems design approach in managing mechatronics R&D projects in this chapter [35]. This approach is motivated by life cycle analysis [70], and the strength of our approach is our emphasis on the identification of the key information specific to different teams required for feedback. In view of the above discussion, we endeavour to achieve a *lean* information feedback and interface, providing maximum knowledge transfer (sufficiency) and minimum mining resources (efficiency) across R&D teams. The end objective of the designed management system is to enable the realization of rapid product development targets at low cost and high quality. This also allows real-time sharing, fast visualization, and evaluation of the system status for all project stake-holders. Furthermore, the rigor and simplicity of our proposed approach implies its easy adaptation to manage any mechatronic systems design project. In the long run this can provide support for R&D managers and engineers in developing new systems engineering processes, solutions, and industrial standards, etc., and to manage the end-to-end system design process. Novel methodologies and engineering policies can also be generated to increase CAPital EXpenditure (CAPEX) savings, while reducing ongoing maintenance as well as improving process and control system performance to increase OPerations EXpense (OPEX) savings.

With these objectives in mind, the proposed MPI is essential for rapid visualization of diagnostic modal parameters of prototyped mechanical structures, to be deposited into an online information database or *repository* which are accessible to mechanical and servo designers for corresponding prognostic design re-evaluations and refinements. This improves the reliability and yield of the mass-produced mechatronic systems. These modal parameters are usually not easily accessible as they are coupled into transfer function representations during system identification. The proposed methodology aims to shorten the total time taken for mechanical-servo evaluation and failure analysis, where prototypes or current products can be quickly analyzed in terms of critical notorious resonant modes which cause control performance degradation due to sampled-data control in mechatronic systems. The developed algorithm is tested on the measured frequency responses of the actuators in a commercial dual-stage HDD, and our results show accurate modal identification of the parameters of the resonant modes with robustness considerations.

The rest of the chapter is organized as follows. Section 3.2 introduces the essential background of R&D functions in a typical production cycle of mechatronic products in industries. In Section 3.3, we present an integrated systems design approach for managing the complexities of a mechatronic R&D project. In particular, the approach enables a rigorous process of R&D workflow design synthesis and evaluation; the outcome of which is a proposed system designed to achieve the end objectives. Based on this, we then focus in Section 3.4 the development of the proposed MPI algorithm for feedback information identification, which is key to the success of the proposed integrative servo-mechanical systems design. Section 3.5 presents a case example of the application of our proposed MPI to the implementation of dual-stage actuation in current HDD industries with robustness considerations. Section 3.6 presents the results of our proposed methodology and discusses some problems at large. Our conclusion and future works are summarized in Section 3.7.

3.2 SERVO-MECHANICAL-PROTOTYPE PRODUCTION CYCLE

The goal of any profiteering manufacturing is to make its manufacturing processes cost-efficient such that minimal resources, e.g., human layer intervention, is required. With decisions made on the next generation of technology to be adopted from the BUs, the targeted specifications received are translated by operational task forces to decide on various issues such as equipment selection, technology chosen, and control optimization, etc. In current mechatronic industries, a typical servo-mechanical-prototype cycle depicting various R&D teams for servo (control) engineers, mechanical engineers, and prototype (facility) engineers, is shown in Figure 3.1.

The main engineering functions executed by the various teams can be classified as follows:

- Servo system analysis
    1. Control system design and simulation;
    2. Proposal of mechanical structures' specifications;
    3. Stability analysis of manufactured prototypes;
    4. Design of servo technologies for next generation of mechatronic systems; and
    5. Forecast of target achievables based on received specifications.
- Mechanical structure design
    1. Modeling of mechanical systems using Computed Aided Design (CAD) systems;
    2. Finite Element Modeling (FEM) and FE Analysis (FEA);
    3. Modal analysis of manufactured prototypes;
    4. Provision of solutions for problems in current prototypes; and
    5. Design of mechanical actuators and structures for next generation of mechatronic systems.
- Prototype manufacture and evaluation
    1. Realize product-driven design based on research and business proposals;
    2. Experimental modal analysis of prototypes;
    3. Anticipation of problems during mass production;

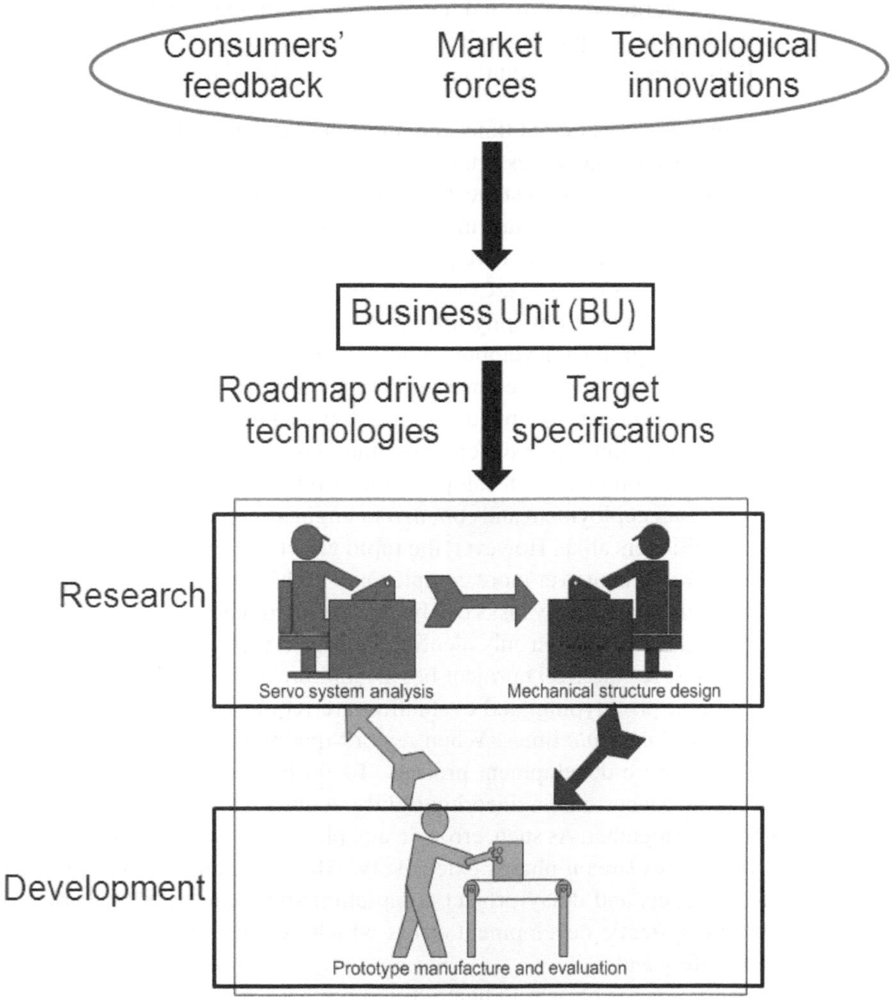

FIGURE 3.1 R&D teams in a typical servo-mechanical-prototype production cycle in mechatronic industries.

4. Define targeted servo and mechanical specifications based on roadmap development model; and
5. Provide engineering solutions to implementation issues.

Due to aggressive market conditions and short product technology life cycles in competitive mechatronic industries, manufacturing firms are forced to develop increasingly high-quality and low cost mechatronic products under high schedule pressures. In the traditional project management framework, the R&D process for mechatronic systems is a sequence of iterative product development stages with scheduled initiation and completion toll gates. Consequently, information generated from one function transfers to the next one only after its completion as shown in Figure 3.1. However, many R&D product development projects fail to deliver their planned objectives in the face of the heightened market competition, substantially overrunning both cost and schedule targets, mainly due to the slow transfer of effective diagnostic and prognostic information across different engineering functions.

To achieve highly-compressed development lead-time targets, paradigms such as *cross-functional* team deployment and *concurrent* engineering are increasingly being espoused by practitioners alike. However, the rapid growth of technical intensity also makes product development ever more complex and highly uncertain. Consequently, at each development phase, many tasks can be incorrectly done in the completion and rework processes. These are often only identified through review and testing activities of the end products. As the R&D project begins and progresses, many iterations of laboratory research, prototyping, and evaluation are required as some of the tasks need to be reworked *multiple* times. When rework quality is low, the rework cycle *can* dominate the entire development process. To further exacerbate the problem, paradigms such as concurrent engineering tightly couple the schedules of different development stages together. As such, errors in any phase can corrupt progress, which propagates in all downstream phases extensively. All these undesirable effects can inflate project budgets and delay project completion significantly. More seriously, they can obscure systemic development errors which might result in consequences related to user-safety and risk.

The above discussions clearly indicate the importance of accurate and efficient feedback information flow in a highly-integrative design environment. In particular, errors discovered and diagnosed in any phase or process must be rapidly identified and eradicated by all concurrent and non-concurrent phases alike. This implies not only the necessity of real-time information sharing, but also the establishment of common data representations that can be quickly comprehended by all parties. We address this issue in Section 3.4. Indeed, due to the differences in technical background and expertise of the members in each of the teams (i.e., servo, mechanical, and prototyping), there is often no common technical jargon or "language" which the teams share in common to improve technical communication and hence throughput. We illustrate a classic example of this difference (yet same subject matter) below.

Consider a general LTI mechanical structure $P(s)$ in a typical mechatronic system consisting of the rigid body mode $\bar{P}(s)$ and flexible body modes $\tilde{P}(s)$ (which include high frequency resonant poles and anti-resonant zeros). As such, the LTI mechanical system $P(s)$ can be represented by

$$\begin{aligned} P(s) &= e^{-T_d s}\left[\bar{P}(s)+\tilde{P}(s)\right] \\ &= e^{-T_d s}\hat{P}(s), \end{aligned} \tag{3.1}$$

where T_d is the input time delay of $P(s)$. This representation allows decoupling of the rigid and flexible body modes into a two time-scale framework, and is advantageous for independent controller design and implementation according to the singular perturbation theory. An example of such a successful application is shown in [71], where the rigid body and flexible body modes of a HDD with mounted PZT active suspension are identified using a two time-scale framework for singular perturbation control.

Currently, system identification for control engineering applications has been well developed theoretically for many different types of systems in both time and frequency domains in [72][73], and coded into a MATLAB system identification toolbox in [74]. Using experimental frequency response measurement data and corresponding frequency vectors, reduced order models can be identified with the frequency domain subspace algorithm in [75] for multivariable systems with input delay. For mechatronic applications, a system identification of the Voice Coil Motor (VCM) in HDDs is shown in [76]. System identification of high frequency critical resonant modes using Nyquist plots to improve servo control performance in HDDs is also studied in [77][78].

For our formulation of the proposed modal parametric identification procedure, the time delay T_d of $P(s)$ is assumed to be known *a priori*, and removed from $P(s)$ to isolate the DC gain and mechanical modes in $\hat{P}(s)$. Motivated by [79][80] with an extension of the LS error optimization framework, a Modal Parametric Identification (MPI) procedure is proposed for direct estimation of residues, damping ratios, and natural frequencies of critical high frequency resonant modes in flexible mechanical systems via solving two subsequent LS optimizations. In this section, the commonly used representations in system identification of transfer function of $\hat{P}(s)$ will be reviewed. The merits and disadvantages of the different forms are also discussed.

3.2.1 MODAL SUMMATION

In mechanical structural designs employed by mechanical engineers, the representation in terms of a linear combination of resonant modes is preferred in order to understand the contribution of the main in-plane, out-of-plane, in-phase, and out-of-phase vibratory resonant modes [81]. As such, $\hat{P}(s)$ is commonly written in modal summation form as

$$\hat{P}(s) = \sum_{i=1}^{N} \frac{R_i}{s^2+2\zeta_i(2\pi f_i)s+(2\pi f_i)^2}, \tag{3.2}$$

where R_i, ζ_i, and f_i are the residues, damping ratios, and natural frequencies of the total number of resonant poles N to be considered, respectively. (3.2) allows fast visualizations of the contributions of each mechanical resonant mode (including rigid body mode and flexible body modes) and their corresponding mode shapes that are captured from simulation and verified with experimental frequency response functions, and redesign of mechanical parameters to meet design specifications and budget. In this chapter, we are interested in identifying transfer function representations of flexible mechanical structures in this form as depicted by (3.2).

During conventional system identification of R_i, ζ_i, and f_i using (3.2), the natural frequencies f_i of each resonant pole are first identified from the frequency responses, and parametric fitting is then done by tuning each residue R_i and damping ratio ζ_i. A setback of this methodology is that residual changes at one frequency usually affect the magnitude and phase at many other frequencies, especially for the dominant modes with large residues or small damping ratios. Also, the phase behavior of the anti-resonant zeros cannot be directly modeled which prolongs the modeling time in obtaining a satisfactory model, as well as effective servo controller design and evaluation. While the presence of anti-resonant zeros is of interest to servo control researchers in controller designs, the in-phase or out-of-phase behaviors of the remaining resonant modes with respect to the first resonant mode are usually of interest to mechanical researchers during structural design and optimization.

3.2.2 POLE-ZERO PRODUCT

Alternatively, $\hat{P}(s)$ can be rewritten in pole-zero product form as

$$\hat{P}(s) = P_{\text{DC}} \prod_{i=1,j=1}^{N,M} \frac{4\pi^2 f_i^2}{4\pi^2 f_j^2} \frac{s^2 + 2\zeta_j(2\pi f_j)s + (2\pi f_j)^2}{s^2 + 2\zeta_i(2\pi f_i)s + (2\pi f_i)^2}, \tag{3.3}$$

where P_{DC} is the DC gain and $M \leq N$ is the number of anti-resonant zeros under consideration. The representation in (3.3) models and considers the magnitude (and phase) contributions of the anti-resonant zeros directly, where the information is absent in (3.2). While the resonant poles can be arbitrarily placed if the LTI mechanical system $P(s)$ is controllable, the anti-resonant zeros are unshifted by feedback control and hence cause signal blocking at the frequencies of the zeros, which degrades tracking performances [82]. The damping ratios and natural frequencies of both the resonant poles and anti-resonant zeros are important for servo controller designs, and an application of a control strategy to compensate for anti-resonant zeros with low sensitivity is shown in [83].

During conventional system identification using (3.3), the zeros can be inserted in the identification process directly using the same concept as with resonant poles, i.e., identifying the damping ratios and natural frequencies of the zeros as well. System identification process now involves tuning the damping ratios (ζ_i and ζ_j) and natural frequencies (f_i and f_j) of the resonant poles and anti-resonant zeros directly. The parametric adjustments of one pole-zero pair is independent of the previously set pairs, and the proposed pole-zero product form is effective in modeling the resonant modes with very small damping ratios or even compound and split modes.

3.2.3 LUMPED POLYNOMIAL

Finally, the commonly used lumped polynomial representation of $\hat{P}(s)$ for system identification algorithms is given by

$$\hat{P}(s) = \frac{\sum_{i=0}^{L-1} m_i s^i}{s^L + \sum_{i=0}^{L-1} n_i s^i}, \tag{3.4}$$

where $L = 2N$ and the coefficient of the largest power of s has been normalized to unity. This representation is preferred as the coefficients of $\hat{P}(s)$ can be easily written in Linear-In-The-Parameter (LITP) form for easy problem formulation and solution optimization [84]. The system identification process is now equivalent to parametric estimation of the numerator coefficients m_i and denominator coefficients n_i, and has been well studied. Interested readers are kindly referred to [73] for more in-depth discussions.

However, this approach blinds any inherent information and physical properties of the mechanical system $P(s)$. The contributions and concepts of physical parameters such as damping ratios ζ_i, residues R_i, and natural frequencies f_i, are lumped into the coefficients in the polynomial form using (3.4). In a typical system identification procedure used by engineers in prototype manufacture and evaluation team where numerator coefficients m_i and denominator coefficients n_i are being identified in (3.4), this approach blinds any inherent information and physical properties of the mechanical system $P(s)$ though this type of system identification has been well established and studied, e.g., in [73]. Obviously while it is easy to expand (3.2) or (3.3) to form (3.4), the reverse is generally not possible to obtain the required residues, damping ratios, and natural frequencies information of resonant poles (for mechanical engineers) and anti-resonant zeros (for servo researchers), in mechanical structure and servo controller design, respectively. This renders the information m_i and n_i not helpful to engineers in other teams.

In all, it is straightforward to rewrite (3.3) and (3.2) into (3.4) but the reverse is generally not possible nor direct. The concepts of resonant poles, anti-resonant zeros, and residues are hence only "local" information to the respective team members, and the information cannot be easily exchanged for tighter integration and faster evaluations. Currently, the control of key attributes and parameters, e.g., dimensions and natural frequencies, etc., are obtained via iterations of this servo-mechanical-production cycle from repeated sampling, offline analysis, and reporting results, etc., which result in significant delays arising from waits and holds. Moreover, much manual intervention is also required, thereby increasing CAPEX. It is hence of paramount importance to identify which *diagnostic* parameters in each of the teams are crucial for feedback to other teams to "close" the data loop for corresponding *prognostic* measures and actions.

3.3 SYSTEMS DESIGN APPROACH

In today's product development, functional participation takes place through the formation of teams consisting of representatives from the functions involved. In this section, we incorporate life cycle analysis [70] to study mechatronic R&D projects and propose a systems design approach to manage the complexities of such projects.

Generally, a systems design approach can be defined as the process of defining the architecture, components, modules, interfaces, and data, etc., for a system to satisfy specified user requirements. Consequently it can be viewed as the application of systems theory to product or service development. We feel that a systems design approach is particularly appropriate for the design of management of mechatronic projects, for the following reasons. Firstly, a systems design approach requires at the outset that all design architecture must be directed to the end goals of the user-requirements. Secondly, it embraces the system complexity and does not attempt to over-simplify issues as in traditional project management approaches. Last but not least, there is strong focus on designing the interface and data between sub-systems or modules.

In the academic literature and industrial practice, there are many variations of systems design (engineering) approaches [85][86][87][88][89][90][91], most of which are similar in concept and application. One such simplified four-phase systems design roadmap is illustrated in Figure 3.2. A systems design approach typically begins with an *Identification of Needs* phase. Here end-user needs ("voice of customer") and stakeholder objectives are solicited, in possibly qualitative formats. This forms the effort focus and purpose that guides the entire system development. For a mechatronic product, this could be for instance, "high data storage capacity," "ease of portability," "high reliability," "competitive pricing," "shorter data access time," and "higher bandwidth," etc. Stakeholder objectives may include agenda such as "market innovation," "manufacturability," "low rework," and in particular relevant for competitive markets, "short time-to-market," and "short product development cycle," etc.

In the second phase (*Establish Technical Performance Metrics*), the qualitative customer needs are formulated into structured operational requirements and quantitative Technical Performance Metrics (TPMs). This concretizes the performance benchmarking and evaluation standards of design alternatives. For mechatronic systems, typical TPMs may be "mean-time-to-failure," "weight and dimension ratios," "total manufacturing cost," "development lead time," "natural frequencies," and "damping ratios," etc. In the face of many conflicting objectives, a popular tool used for ranking performance criteria is the Quality Function Deployment [92][93].

The third phase *Functional and Requirements Analysis* is an essential element of early conceptual and preliminary design. The purpose is to develop a functional description of the system directed at meeting the identified system requirements. The primary facilitating tool for this phase is the Functional Flow Block Diagram (FFBD) [85][86][90]. The functional descriptions can be developed from the system level, to the subsystem level and as far down the hierarchical structure as necessary to identify design criteria and constraints for the various system elements. Figure 3.3 is a simplified first level view of the end-to-end FFBD for the concept, design, and manufacture of mechatronic products. One may view functions 1.0 and 2.0 as corresponding to the role of the BU organization as shown in Figure 3.1. Similarly, an example of a realization of function 3.0 is the servo-mechanical-prototype cycle

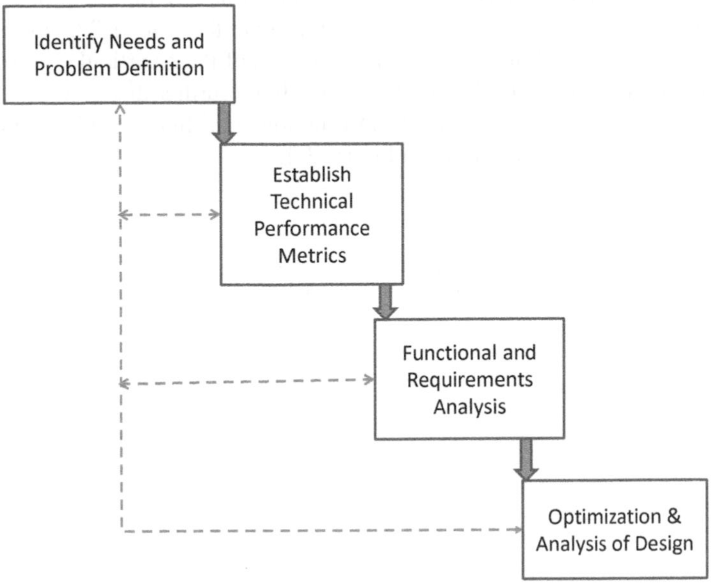

FIGURE 3.2 A four-phase systems design approach.

in Figure 3.1. However, we emphasize here that the functional analysis is focused on identifying the "Whats" of the system, i.e., what are the functional requirements of the system to meet the objectives, rather than the "Hows." Consequently at this juncture, one should not make presuppositions of their particular realizations. Each of the indexed function blocks can obviously be further disaggregated if necessary. In this work, we focus on the R&D function (i.e., function block 3.0), where the corresponding FFBD is shown in Figure 3.4.

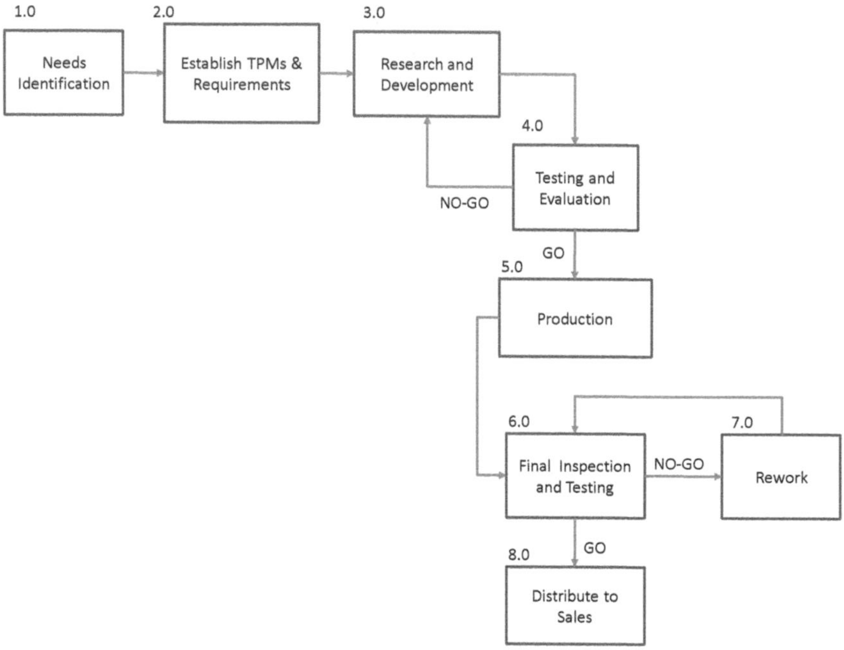

FIGURE 3.3 Top-level functional flow block diagram for mechatronic product development.

Once a sufficiently detailed functional description is developed, requirements analysis and allocation are performed. The purpose of requirements analysis is to identify a partitioning of system components into subsystems and "packages" to reduce external system complexity. In the R&D function of Figure 3.4, the identified subsystems consist of a servo-engineering team (functions 3.1, 3.3, 3.5, and 3.7), a mechanical design engineering team (functions 3.2, 3.4, and 3.9) and a prototype fabrication team (functions 3.6 and 3.8). Furthermore, requirements allocation is often conducted to apportion the identified target system performance requirements to individual subsystems. For instance, portability requirements will be largely apportioned to the mechanical design and prototype development subsystems. Budget requirements would be accordingly apportioned to each of the cost-centers.

In the final phase of *Optimization and Analysis of Design*, different configurations

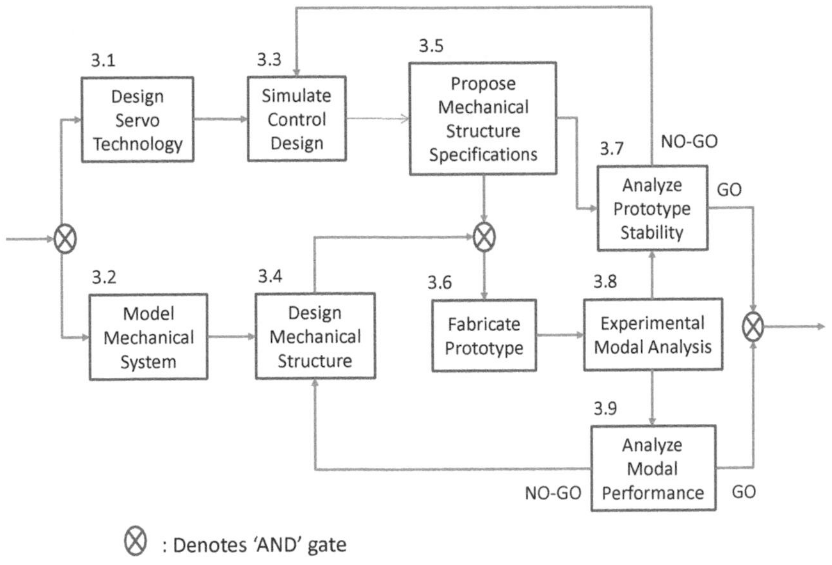

FIGURE 3.4 Functional flow block diagram for mechatronic R&D.

of the subsystems are synthesized and evaluated. Hence, this is a design stage where the "Hows" of the system are defined. Figure 3.5 shows a configuration akin to a more conventional work-flow cycle of the three subs-systems (similar to Figure 3.1). In this configuration the servo-team relays specifications to the mechanical team, who then in turn develops the structural design plans. These are then transmitted to the Fabrication team who then develops the prototypes and conducts the essential experimental modal analysis. The servo team retrieves this information and updates their control systems design. The work-flow is *iterated* until the prototype conforms to the targeted servo and mechanical specifications set forth by the BU *a priori*.

A quick evaluation of the design in Figure 3.5 in relation to the system requirements of short development cycle time and low rework rate, however, reveals its deficiency. Firstly, information generated from one function transfers to the next one only after its completion *sequentially* in a conventional project management implementation of the design. This however, is highly inefficient, and can cause the schedule to overrun the lead-time targets substantially. In a concurrent engineering implementation, development changes in one stage are only sensed in other stages after a substantial lag, and this can cause considerable rework due to the error propagation and late error discovery, leading to long development times. Furthermore, the ineffective communication due to differences in technical expertise between the engineering teams

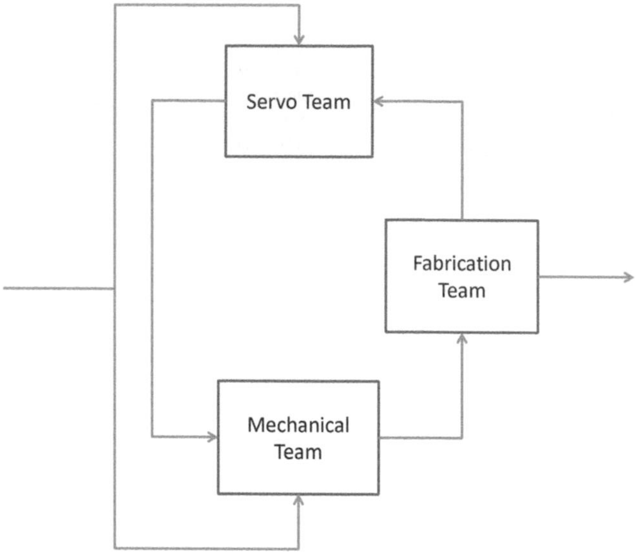

FIGURE 3.5 Initial synthesized design for mechatronic R&D workflow.

as highlighted in the previous section can increase chances of development errors, potentially corrupting all subsequent developments and again inducing an increase in scrapping and rework. This is unacceptable in current competitive manufacturing industries.

In contrast to Figure 3.5, the proposed design in Figure 3.6 mitigates some of the above undesirable effects. In the design, a new system component "Information Repository" is identified. This component provides the common cross-talk interface among all three sub-systems, so that relevant data and diagnostic information from each team can be shared *simultaneously*. On the other hand, the conventional channels of information relays are also retained for system redundancy. The new workflow design is highly amenable to a concurrent engineering implementation, since any development revisions and updates are identified rapidly across all stages. Note that this new design necessitates the functional and requirements analysis of the new "Information Repository" component. Consequently, the design process can iterate back and forth between functional analysis and design optimization.

In the next section, we identify in the design the set of critical information parameters most relevant for the "Information Repository" component. However, at this juncture, we highlight the following insights. With the many process improvement or project management tools available in the market, one can be easily "sold" into the

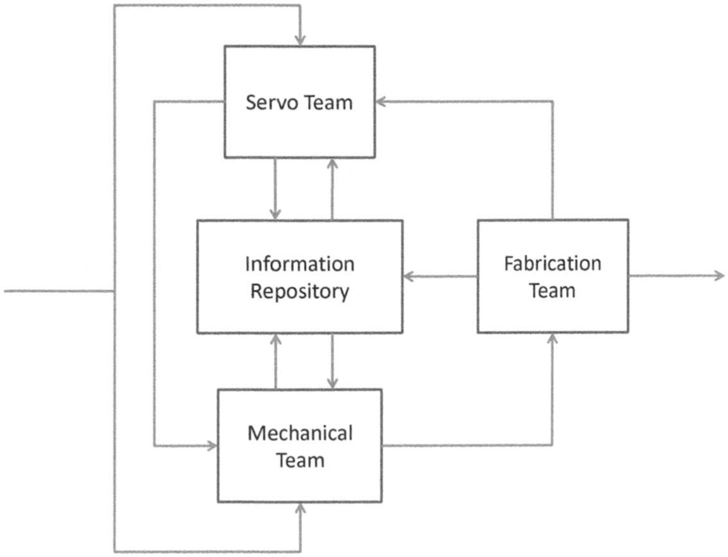

FIGURE 3.6 New proposed design for mechatronic R&D workflow.

belief that such improvements can be realized conveniently and effortlessly by simply adapting existing workflow scheduling systems or changing management styles. This is a typical pitfall of many observed practices in the industry. Figure 3.5 is a simplified but nevertheless lucid example of how attempting to improve the performance (product development cycle time) of a bad design by merely changing work scheduling (e.g., task concurrency) can yield marginal results at best, if not detrimental ones. A systems design approach, on the other hand, offers two fundamentally important strengths. Firstly, it recognizes at the outset the importance and long-term impact of system structure, and emphasizes the overriding concern of achieving its targets. User requirements and stakeholder objectives are explicitly *designed* into the system, and every phase of the end-to-end design effort is focused and driven towards meeting these objectives. Secondly, the systems design approach delegates *leadership* and *authority* to the engineering and technical experts in designing and defining the workflow and management structure. This is in contrast to many existing practices where product structure is driven (partially) by R&D and engineering expertise, while management and workflow structure is driven (almost) completely by non-engineering domain personnel. The non-technical personnel might lack the important insights and foresights to evaluate the feasibility of various workflow alternatives in synthesis with the product design itself. This clearly contradicts the fundamental tenets of a *systems* approach, that is to view all aspects of the final delivered mechatronic systems as interrelated entities in a System-of-Systems (SoS) framework [94][95].

As discussed in the previous section, owing to the uncertainties in complex product development processes, the release of preliminary information to downstream or horizontal functions often introduces the need for rework when there is a revision in the released information [68]. Hence, a primary objective of our proposed integrative servo-mechanical systems design framework is to reduce project uncertainties by identifying the potential quality problems and making revisions as early as possible. This directly translates to, at the outset of implementing any integrated or concurrent mechatronic R&D, an identification procedure for the set of critical diagnostic information feedback to be supported across all engineering design teams for prognostic actions.

3.4 MODAL PARAMETRIC IDENTIFICATION (MPI) ALGORITHM

Using the commercial high-performance HDD as a classic example of mechatronic systems, a typical servo-mechanical-prototype production cycle in current HDD industries depicting various R&D teams for servo (control) engineers, mechanical engineers, and prototype (facility) engineers for next generation of dual-stage HDDs is shown in Figure 3.7.

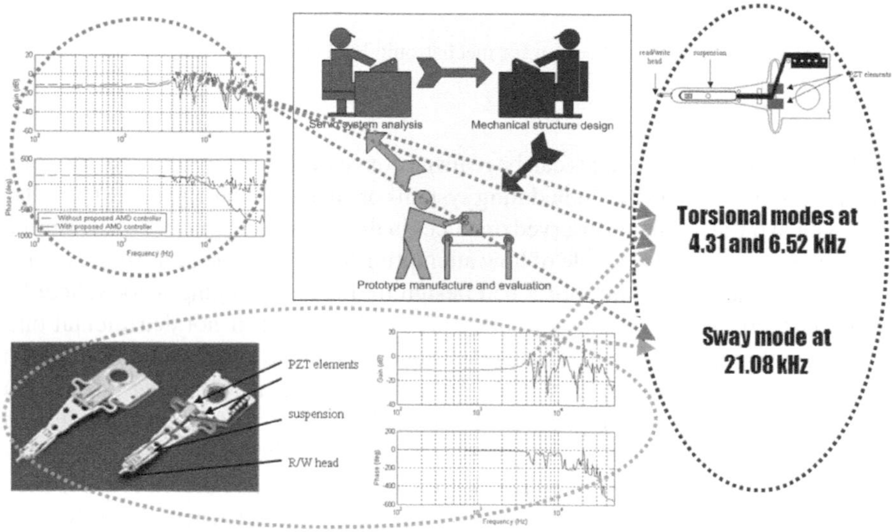

FIGURE 3.7 R&D teams in a typical servo-mechanical-prototype production cycle in HDD industries.

Upon life cycle analysis and applying the systems design approach in Section 3.3, functional and requirements analysis yield the following partitioning of the relevant

HDD R&D functions. The main engineering functions executed by the various teams in a HDD manufacturing industry (with R&D functions) can be classified as follows:

- Servo system analysis
  1. Design of dual-stage control configuration. Carry out extensive simulation studies;
  2. Proposal of VCM's and PZT active suspension's specifications in terms of natural frequencies and damping ratios, etc.;
  3. Stability analysis of proposed digital control system design using mathematical models from manufactured prototypes;
  4. Design of aggressive digital control and advanced servo technologies for next generation of HDDs; and
  5. Forecast of target achievables based on received specifications in terms of bandwidth, areal density, secondary actuator stroke, and sensitivity analysis, etc.
- Mechanical structure design
  1. Design of PZT active suspensions using Computed Aided Design (CAD) systems;
  2. FEM and FEA of designed PZT active suspensions for theoretical;
  3. Modal analysis of manufactured PZT active suspensions in terms of sway, torsional, and bending modes, etc.;
  4. Provision of solutions for problems in current PZT active suspensions via redesign; and
  5. Design of secondary actuators (slider-based or head-based) and structures (MEMS) for next generation of HDDs.
- Prototype manufacture and evaluation
  1. Fabrication of received PZT active suspension drawings from product-driven design based on research and business proposals;
  2. Capturing of experimental mode shapes and frequency response functions of the developed prototypes;
  3. Anticipation of production issues and problems during mass production of PZT active suspensions;
  4. Defining targeted servo and mechanical specifications based on roadmap development model in terms of sampling rate; and
  5. Providing engineering solutions to implementation issues during integration of secondary actuators onto the VCM.

Following the *Optimization and Analysis of Design* phase as shown earlier in Figure 3.6, the proposed MPI is used to identify the critical (yet portable) parameters in the form of (3.2). In this representation, the in-phase and out-of-phase resonant modes can be directly seen from the signs (directions) of the residues R_i. These modal parameters are stored in the *information repository* as shown in Figure 3.6, that all the members of different teams can access readily for carrying out their engineering functions.

To enable an effective and efficient crosstalk interface between the engineering teams, in this section we analyze and identify the set of information feedback to support the synthesized design in Figure 3.6. In view of the issues identified, it is desired to *identify transfer function representations of flexible mechatronic systems $P(s)$ in the form depicted by (3.2)*. In [32], the proposed integrated MPI algorithm exploits the linear-in-the-parameter structure of (3.2) to find the natural frequencies f_i and damping ratios ζ_i of the resonant poles of $\hat{P}(s)$ using a LS formulation, followed by a second LS optimization procedure to identify the corresponding residues R_i. It should be noted that the computation speed of obtaining the essential parametric estimates is important for perusal in mechanical structural designs and servo control evaluation in mechatronic prototyping, as current approaches require much trial-and-error, and will be extremely time consuming and labor intensive when the number of data sets becomes high.

In this section, we detail the MPI procedure which identifies $\hat{P}(s)$ in modal summation form as depicted in (3.2) from the common lumped polynomial representation in (3.4). This integrated system identification algorithm exploits the LITP structure of (3.4) to find the natural frequencies f_i and damping ratios ζ_i of the resonant poles of $\hat{P}(s)$ using a LS formulation, followed by another LS optimization procedure to identify the corresponding residues R_i. It should be noted that the computation speed of obtaining the essential parametric estimates is important for perusal in mechanical structural designs and servo control evaluation in mechatronic prototyping.

3.4.1 NATURAL FREQUENCIES F_I AND DAMPING RATIOS ζ_I

Using the polynomial representation of $\hat{P}(s)$ in (3.4), an LS formulation of the identified transfer function and frequency response measurement at each frequency is formulated. The theoretical fundamentals of the LS formulation in [79][80] are reviewed in a rigorous manner in this subsection.

Prior to identifying the natural frequencies and damping ratios of $\hat{P}(s)$, the desired frequency range and model order L is selected, and L is chosen as an even number for modal summation. Now let the frequency response measurement at the kth frequency point f_k (in Hz) be h_k. As such, the modeling error e_k between the identified model $\hat{P}(s)$ and the experimental measured frequency response h_k at f_k is given by

$$\begin{aligned} e_k &= \hat{P}(j2\pi f_k) - h_k \\ &= \frac{\sum_{i=0}^{L-1} m_i (j2\pi f_k)^i}{(j2\pi f_k)^L + \sum_{i=0}^{L-1} n_i (j2\pi f_k)^i} - h_k, \end{aligned} \tag{3.5}$$

where $s = j2\pi f_k$ is the sinusoidal steady-state of each frequency f_k. Multiplying the denominator to both sides, we define the augmented error $e_{\text{MN},k}$ to be minimized as [79][80]

$$e_{\text{MN},k} = \sum_{i=0}^{L-1} m_i (j2\pi f_k)^i - h_k \left\{ (j2\pi f_k)^L + \sum_{i=0}^{L-1} n_i (j2\pi f_k)^i \right\}. \tag{3.6}$$

Packing all the errors $e_{\mathrm{MN},k}$ over the entire measurement frequency range of K measurement points into a column vector $\mathbf{E_{MN}}$, we can write $\mathbf{E_{MN}}$ as

$$
\begin{aligned}
\mathbf{E_{MN}} &= \begin{bmatrix} e_{\mathrm{MN},1} \\ e_{\mathrm{MN},2} \\ \vdots \\ e_{\mathrm{MN},K} \end{bmatrix} \\
&= \begin{bmatrix} \sum_{i=0}^{L-1} m_i(j2\pi f_1)^i - h_1\left\{(j2\pi f_1)^L + \sum_{i=0}^{L-1} n_i(j2\pi f_1)^i\right\} \\ \sum_{i=0}^{L-1} m_i(j2\pi f_2)^i - h_2\left\{(j2\pi f_2)^L + \sum_{i=0}^{L-1} n_i(j2\pi f_2)^i\right\} \\ \vdots \\ \sum_{i=0}^{L-1} m_i(j2\pi f_K)^i - h_K\left\{(j2\pi f_K)^L + \sum_{i=0}^{L-1} n_i(j2\pi f_K)^i\right\} \end{bmatrix} \\
&= \underbrace{\begin{bmatrix} 1 & (j2\pi f_1) & (j2\pi f_1)^2 & \cdots & (j2\pi f_1)^{L-1} \\ 1 & (j2\pi f_2) & (j2\pi f_2)^2 & \cdots & (j2\pi f_2)^{L-1} \\ \vdots & \vdots & \vdots & \ddots & \vdots \\ 1 & (j2\pi f_K) & (j2\pi f_K)^2 & \cdots & (j2\pi f_K)^{L-1} \end{bmatrix}}_{\mathbf{\Omega}} \underbrace{\begin{bmatrix} m_0 \\ m_1 \\ \vdots \\ m_{L-1} \end{bmatrix}}_{\mathbf{M}} \cdots \\
&\quad - \begin{bmatrix} h_1 & h_1(j2\pi f_1) & h_1(j2\pi f_1)^2 & \cdots & h_1(j2\pi f_1)^{L-1} \\ h_2 & h_2(j2\pi f_2) & h_2(j2\pi f_2)^2 & \cdots & h_2(j2\pi f_2)^{L-1} \\ \vdots & \vdots & \vdots & \ddots & \vdots \\ h_K & h_K(j2\pi f_K) & h_K(j2\pi f_K)^2 & \cdots & h_K(j2\pi f_K)^{L-1} \end{bmatrix} \underbrace{\begin{bmatrix} n_0 \\ n_1 \\ \vdots \\ n_{L-1} \end{bmatrix}}_{\mathbf{N}} \\
&\quad - \underbrace{\begin{bmatrix} h_1(j2\pi f_1)^L \\ h_2(j2\pi f_2)^L \\ \vdots \\ h_K(j2\pi f_K)^L \end{bmatrix}}_{\mathbf{H_\Omega}} \\
&= \mathbf{\Omega M} - \underbrace{\begin{bmatrix} h_1 & 0 & 0 & \cdots & 0 \\ 0 & h_2 & 0 & \cdots & 0 \\ \vdots & \vdots & \vdots & \ddots & \vdots \\ 0 & 0 & 0 & \cdots & h_K \end{bmatrix}}_{\mathbf{H}} \mathbf{\Omega N} - \mathbf{H_\Omega} \\
&= \mathbf{\Omega M} - \mathbf{H\Omega N} - \mathbf{H_\Omega},
\end{aligned}
\tag{3.7}
$$

where $\mathbf{\Omega} \in \mathbb{C}^{K\times L}$ is a matrix of powers of Laplace operators, i.e., differentiators. $\mathbf{M} \in \mathbb{R}^L$ and $\mathbf{N} \in \mathbb{R}^L$ are vectors containing the unknown coefficients m_i of the zero polynomial and n_i pole polynomial of $\hat{P}(s)$ to be determined, respectively. $\mathbf{H_\Omega} \in \mathbb{C}^K$ is a weighted version of the frequency response vector differentiated L times. $\mathbf{H} \in \mathbb{C}^{K\times K}$ is a symmetrical diagonal matrix with the frequency response measurement data as its diagonal entries, i.e., $\mathbf{H} = \mathrm{diag}(h_1, h_2, \cdots, h_k)$. The error vector $\mathbf{E_{MN}}$ is now in the LITP form in terms of unknown vectors **M** and **N**.

With these, we propose the following theorem for LS parametric estimation of **M** and **N**.

Theorem 3.1

The LS estimation of **M** and **N** for $\hat{P}(s)$ is obtained by

$$\begin{bmatrix} \mathbf{M} \\ \mathbf{N} \end{bmatrix} = \begin{bmatrix} \mathbf{\Omega}^H\mathbf{\Omega} & -\Re\{\mathbf{\Omega}^H\mathbf{H}\mathbf{\Omega}\} \\ -\Re\{\mathbf{\Omega}^H\mathbf{H}^*\mathbf{\Omega}\} & \mathbf{\Omega}^H|\mathbf{H}|^2\mathbf{\Omega} \end{bmatrix}^{-1} \begin{bmatrix} \Re\{\mathbf{\Omega}^H\mathbf{H}_{\mathbf{\Omega}}\} \\ -\Re\{\mathbf{\Omega}^H\mathbf{H}^*\mathbf{H}_{\mathbf{\Omega}}\} \end{bmatrix}, \tag{3.8}$$

where $\Re\{.\}$, $(.)^H$, and $(.)^*$ are the real, Hermitian, and conjugate operators, respectively. ■

Proof 3.1

Defining a positive definite scalar error cost function $V_{\mathbf{MN}}$ as

$$
\begin{aligned}
V_{\mathbf{MN}} &= \frac{1}{2}\mathbf{E_{MN}}^H\mathbf{E_{MN}} = \frac{1}{2}(\boldsymbol{\Omega}\mathbf{M} - \mathbf{H}\boldsymbol{\Omega}\mathbf{N} - \mathbf{H}_{\boldsymbol{\Omega}})^H(\boldsymbol{\Omega}\mathbf{M} - \mathbf{H}\boldsymbol{\Omega}\mathbf{N} - \mathbf{H}_{\boldsymbol{\Omega}}) \\
&= \frac{1}{2}(\mathbf{M}^T\boldsymbol{\Omega}^H - \mathbf{N}^T\boldsymbol{\Omega}^H\mathbf{H}^* - \mathbf{H}_{\boldsymbol{\Omega}}{}^H)(\boldsymbol{\Omega}\mathbf{M} - \mathbf{H}\boldsymbol{\Omega}\mathbf{N} - \mathbf{H}_{\boldsymbol{\Omega}}) \\
&= \frac{1}{2}(\mathbf{M}^T\boldsymbol{\Omega}^H\boldsymbol{\Omega}\mathbf{M} - \mathbf{M}^T\boldsymbol{\Omega}^H\mathbf{H}\boldsymbol{\Omega}\mathbf{N} - \mathbf{M}^T\boldsymbol{\Omega}^H\mathbf{H}_{\boldsymbol{\Omega}}\cdots \\
&\quad -\mathbf{N}^T\boldsymbol{\Omega}^H\mathbf{H}^*\boldsymbol{\Omega}\mathbf{M} + \mathbf{N}^T\boldsymbol{\Omega}^H\mathbf{H}^*\mathbf{H}\boldsymbol{\Omega}\mathbf{N}\cdots \\
&\quad +\mathbf{N}^T\boldsymbol{\Omega}^H\mathbf{H}^*\mathbf{H}_{\boldsymbol{\Omega}} - \mathbf{H}_{\boldsymbol{\Omega}}{}^H\boldsymbol{\Omega}\mathbf{M} + \mathbf{H}_{\boldsymbol{\Omega}}{}^H\mathbf{H}\boldsymbol{\Omega}\mathbf{N} + \mathbf{H}_{\boldsymbol{\Omega}}{}^H\mathbf{H}_{\boldsymbol{\Omega}}) \\
&= \frac{1}{2}\mathbf{M}^T\boldsymbol{\Omega}^H\boldsymbol{\Omega}\mathbf{M} + \frac{1}{2}\mathbf{N}^T\boldsymbol{\Omega}^H|\mathbf{H}|^2\boldsymbol{\Omega}\mathbf{N} + \frac{1}{2}\mathbf{H}_{\boldsymbol{\Omega}}{}^H\mathbf{H}_{\boldsymbol{\Omega}}\cdots \\
&\quad -\frac{1}{2}(\mathbf{M}^T\boldsymbol{\Omega}^H\mathbf{H}\boldsymbol{\Omega}\mathbf{N} + \mathbf{N}^T\boldsymbol{\Omega}^H\mathbf{H}^*\boldsymbol{\Omega}\mathbf{M})\cdots \\
&\quad -\frac{1}{2}(\mathbf{M}^T\boldsymbol{\Omega}^H\mathbf{H}_{\boldsymbol{\Omega}} + \mathbf{H}_{\boldsymbol{\Omega}}{}^H\boldsymbol{\Omega}\mathbf{M}) + \frac{1}{2}(\mathbf{N}^T\boldsymbol{\Omega}^H\mathbf{H}^*\mathbf{H}_{\boldsymbol{\Omega}} + \mathbf{H}_{\boldsymbol{\Omega}}{}^H\mathbf{H}\boldsymbol{\Omega}\mathbf{N}) \\
&= \frac{1}{2}\mathbf{M}^T\boldsymbol{\Omega}^H\boldsymbol{\Omega}\mathbf{M} + \frac{1}{2}\mathbf{N}^T\boldsymbol{\Omega}^H|\mathbf{H}|^2\boldsymbol{\Omega}\mathbf{N} + \frac{1}{2}\mathbf{H}_{\boldsymbol{\Omega}}{}^H\mathbf{H}_{\boldsymbol{\Omega}}\cdots \\
&\quad -\frac{1}{2}\left[\mathbf{M}^T\boldsymbol{\Omega}^H\mathbf{H}\boldsymbol{\Omega}\mathbf{N} + (\mathbf{M}^T\boldsymbol{\Omega}^H\mathbf{H}\boldsymbol{\Omega}\mathbf{N})^H\right]\cdots \\
&\quad -\frac{1}{2}\left[\mathbf{M}^T\boldsymbol{\Omega}^H\mathbf{H}_{\boldsymbol{\Omega}} + (\mathbf{M}^T\boldsymbol{\Omega}^H\mathbf{H}_{\boldsymbol{\Omega}})^H\right] \\
&\quad +\frac{1}{2}\left[\mathbf{N}^T\boldsymbol{\Omega}^H\mathbf{H}^*\mathbf{H}_{\boldsymbol{\Omega}} + (\mathbf{N}^T\boldsymbol{\Omega}^H\mathbf{H}^*\mathbf{H}_{\boldsymbol{\Omega}})^H\right] \\
&= \frac{1}{2}\mathbf{M}^T\boldsymbol{\Omega}^H\boldsymbol{\Omega}\mathbf{M} + \frac{1}{2}\mathbf{N}^T\boldsymbol{\Omega}^H|\mathbf{H}|^2\boldsymbol{\Omega}\mathbf{N} + \frac{1}{2}\mathbf{H}_{\boldsymbol{\Omega}}{}^H\mathbf{H}_{\boldsymbol{\Omega}} \\
&\quad -\Re\{\mathbf{M}^T\boldsymbol{\Omega}^H\mathbf{H}\boldsymbol{\Omega}\mathbf{N}\} - \Re\{\mathbf{M}^T\boldsymbol{\Omega}^H\mathbf{H}_{\boldsymbol{\Omega}}\} + \Re\{\mathbf{N}^T\boldsymbol{\Omega}^H\mathbf{H}^*\mathbf{H}_{\boldsymbol{\Omega}}\}, \qquad (3.9)
\end{aligned}
$$

where $(\boldsymbol{.})^T$ is the matrix transpose operator. **M** and **N** are real (i.e., $\mathbf{M}^H = \mathbf{M}^T$ and $\mathbf{N}^H = \mathbf{N}^T$) and **H** is symmetrical (i.e., $\mathbf{H}^H = \mathbf{H}^*$). It should be noted that the identity $\mathbf{Z} + \mathbf{Z}^H = 2\Re\{\mathbf{Z}\}$ for complex symmetric matrices is used.

The partial derivatives of $V_{\mathbf{MN}}$ in (3.9) with respect to the parametric unknowns **M** and **N**, respectively, are given by

$$
\begin{aligned}
\frac{\partial V_{\mathbf{MN}}}{\partial \mathbf{M}} &= \boldsymbol{\Omega}^H\boldsymbol{\Omega}\mathbf{M} - \Re\{\boldsymbol{\Omega}^H\mathbf{H}\boldsymbol{\Omega}\mathbf{N}\} - \Re\{\boldsymbol{\Omega}^H\mathbf{H}_{\boldsymbol{\Omega}}\} \\
&= \boldsymbol{\Omega}^H\boldsymbol{\Omega}\mathbf{M} - \Re\{\boldsymbol{\Omega}^H\mathbf{H}\boldsymbol{\Omega}\}\mathbf{N} - \Re\{\boldsymbol{\Omega}^H\mathbf{H}_{\boldsymbol{\Omega}}\}, \qquad (3.10) \\
\frac{\partial V_{\mathbf{MN}}}{\partial \mathbf{N}} &= \boldsymbol{\Omega}^H|\mathbf{H}|^2\boldsymbol{\Omega}\mathbf{N} - \Re\{\boldsymbol{\Omega}^H\mathbf{H}^*\boldsymbol{\Omega}M\} + \Re\{\boldsymbol{\Omega}^H\mathbf{H}^*\mathbf{H}_{\boldsymbol{\Omega}}\} \\
&= \boldsymbol{\Omega}^H|\mathbf{H}|^2\boldsymbol{\Omega}\mathbf{N} - \Re\{\boldsymbol{\Omega}^H\mathbf{H}^*\boldsymbol{\Omega}\}\mathbf{M} + \Re\{\boldsymbol{\Omega}^H\mathbf{H}^*\mathbf{H}_{\boldsymbol{\Omega}}\}. \qquad (3.11)
\end{aligned}
$$

For $V_{\mathbf{MN}}$ to be a global minimum, the above partial derivative equations are set to $\mathbf{0}_{L\times 1}$ and packed into a matrix for linear algebraic solution

$$\begin{bmatrix} \mathbf{\Omega}^H\mathbf{\Omega} & -\Re\{\mathbf{\Omega}^H\mathbf{H}\mathbf{\Omega}\} \\ -\Re\{\mathbf{\Omega}^H\mathbf{H}^*\mathbf{\Omega}\} & \mathbf{\Omega}^H|\mathbf{H}|^2\mathbf{\Omega} \end{bmatrix} \begin{bmatrix} \mathbf{M} \\ \mathbf{N} \end{bmatrix} = \begin{bmatrix} \Re\{\mathbf{\Omega}^H\mathbf{H}_{\mathbf{\Omega}}\} \\ -\Re\{\mathbf{\Omega}^H\mathbf{H}^*\mathbf{H}_{\mathbf{\Omega}}\} \end{bmatrix}. \tag{3.12}$$

As such, the required coefficients of the identified coefficients of **M** and **N** can be obtained by taking the inverse of the left matrix. This proves Theorem 3.1. ■

$\hat{P}(s)$ is now realized in the lumped polynomial form in (3.4), and the coefficients of **M** and **N** are real, since all the entries in the above linear system are real. Any noisy measurements can be considered as perturbations into the vector $\mathbf{H}_{\mathbf{\Omega}}$ which will be permeated into the right hand side of (3.8). The above formulation and solution can then be readily extended to a stochastic LS framework by taking the expectation operator on (3.7) if the measurement noise is second order stationary and white.

Unstable resonant modes (negative damping ratios) arise sometimes from numerical issues when solving the LS optimization problem when attempting matrix inverses. To avoid ill-conditioned matrix inverses arising from large natural frequencies during parametric identification, complex orthogonal polynomial transformation [96] and Forsythe method [33] in the enhanced MPI framework [34] are used to reduce computation loads in the next subsection.

3.4.2 REFORMULATION USING FORSYTHE'S ORTHOGONAL POLYNOMIALS

Now, the task of identifying the polynomial form representation of the system $\hat{P}(s)$ involves identifying the polynomials in both the numerator and denominator. Using the example of the numerator polynomial, this is a task of formulation of a polynomial $\hat{y}(s)$ of degree $L-1$ to approximate the true function $y(s)$. The denominator polynomial, on the other hand, is a polynomial of degree L. The general form for a curve fitting polynomial to carry out the approximation is

$$\begin{aligned} \hat{y}(s) &= a_0\psi_0(s) + a_1\psi_1(s) + \cdots + a_i\psi_i(s) + \cdots + a_{L-1}\psi_{L-1}(s) \\ &= \sum_{i=0}^{L-1} a_i\psi_i, \end{aligned} \tag{3.13}$$

where $\psi_0, \psi_1, \cdots, \psi_{L-1}$ are L appropriately chosen functions. The original form of the curve fitting polynomial (or direct solution) uses $\psi_0 = 1$, $\psi_1 = s$, $\psi_2 = s^2$, etc., with coefficients $m_i = a_i$ to give

$$\hat{y}(s) = m_0 + m_1 s + \cdots + m_i s^i + \cdots + m_{L-1}s^{L-1}. \tag{3.14}$$

However, the direct solution leads to a very small determinant of coefficients, so the coefficients become very inaccurate when derived. This leads to the ill-conditioned matrices mentioned earlier. A better representation is obtained if the set of polynomials $\psi_0, \psi_1, \cdots, \psi_i, \cdots, \psi_{L-1}$ for $i = 0, 1, \cdots, L-1$, is chosen such that each member of the set is orthogonal to all others in the set over the interval of consideration.

As such, let this set of polynomials be denoted by $\phi_0(s), \phi_1(s), \cdots, \phi_i(s), \cdots, \phi_{L-1}(s)$, so that $\psi_i(s) = \phi_i(s)$ and its coefficients be $c_i = a_i$ to give

$$\hat{y}(s) = c_0\phi_0(s) + c_1\phi_1(s) + \cdots + c_i\phi^i(s) + \cdots + c_{L-1}\phi^{L-1}(s). \tag{3.15}$$

With the Forsythe's method for generating orthogonal polynomials, the weighting function may be used as a set of empirical data, or computed as a function of independent variables. In this application, the weighting function for each data point at the kth frequency is denoted by $(w_k, k = -K, \cdots, K)$ and is unity $(w_k = 1)$. The method generates i^{th} order orthogonal polynomials $Q_{k,i}(s_k)$ at the kth frequency of the form

$$u_i = \frac{1}{D_{i-1}} \sum_{k=-K}^{K} s_k Q_{k,i-1}^2(s_k) w_k^2, \tag{3.16}$$

$$v_{i-1} = \frac{1}{D_{i-2}} \sum_{k=-K}^{K} s_k Q_{i-1}(s_k) Q_{i-2}(s_k) w_k^2, \tag{3.17}$$

$$D_i = \sum_{k=-K}^{K} Q_i^2(s_k) w_k^2. \tag{3.18}$$

Taking advantage of the symmetry properties of these polynomials, the orthogonality conditions can be written in terms of positive frequency values only. The orthogonal polynomials can be represented as a sum of two half-functions, where one half-function $Q_{k,i}^+(s)$ is defined for positive frequencies $k > 0$, and the other half-function $Q_{k,i}^-(s)$ for negative frequencies $k < 0$, i.e.,

$$Q_{k,i}(s_k) = Q_{k,i}^+(s_k) + Q_{k,i}^-(s_k), \tag{3.19}$$

where $k = -K, \cdots, -1, 1, \cdots, K$. The origin is excluded from further developments of the curve fitting algorithm to preserve the symmetric properties of the functions about the origin. The real polynomial half functions can be generated first by applying the following equations

$$\begin{aligned} H_{k,-1}^+(s_k) &= 0, \\ H_{k,0}^+(s_k) &= \frac{1}{\sqrt{2\sum_{k=1}^{K} w_k^2}}, \\ &\vdots \\ H_{k,i}^+(s_k) &= \frac{z_{k,i}^+(s_k)}{D_i}, \\ &\vdots \end{aligned} \tag{3.20}$$

where $i = 0, \cdots, L-1$, and $k = 1, \cdots, K$. Also,

$$
\begin{aligned}
z^{+}_{k,i}(s_k) &= (2\pi f_k)H^{+}_{k,i-1}(s_k) - v_{i-1}H^{+}_{k,i-2}(s_k), &(3.21)\\
v_{i-1} &= 2\sum_{k=1}^{K}(2\pi f_k)H^{+}_{k,i-1}(s_k)H^{+}_{k,i-2}(s_k)w_k^2, &(3.22)\\
D_i &= \sqrt{2\sum_{k=1}^{K}\left[Z^{+}_{k,i}(s_k)\right]^2 w_k^2}. &(3.23)
\end{aligned}
$$

The required complex polynomials can then be represented in terms of real valued polynomials by substituting $s_k = j2\pi f_k$ to give the relationship

$$Q^{+}_{k,i}(s_k) = j^i H^{+}_{k,i}(s_k). \tag{3.24}$$

This process of generating the required complex polynomial applies similarly to both the numerator and denominator polynomials. The only difference in application is that the numerator polynomial is of degree $L-1$ while the denominator polynomial is of degree L. The two different sets of polynomials are then denoted differently, with ϕ_i used for the numerator complex polynomial half functions and θ_i used for the denominator half function.

Now, the reformulation of the polynomial form representation of $\hat{P}(s)$ yields

$$\hat{P}(s) = \frac{\sum_{i=0}^{L-1} c_i\phi_i}{\theta_L + \sum_{i=0}^{L-1} d_i\theta_i}. \tag{3.25}$$

Similarly, the modeling error e_k at the kth frequency point f_k between the identified model $\hat{P}(s)$ and experimental measured frequency response h_k is now given by

$$
\begin{aligned}
e_k &= \hat{P}(j2\pi f_k) - h_k\\
&= \frac{\sum_{i=0}^{L-1} c_i\phi_{k,i}}{\theta_{k,L} + \sum_{i=0}^{L-1} d_i\theta_{k,i}} - h_k. &(3.26)
\end{aligned}
$$

Multiplying both sides with the denominator, we get

$$e_{CD,k} = \sum_{i=0}^{L-1} c_i\phi_{k,i} - h_k\{\theta_{k,L} + \sum_{i=0}^{L-1} d_i\theta_{k,i}\}. \tag{3.27}$$

Like the previous formulation, we assemble all the errors over the entire measurement frequency range of K measurement points into an error vector with a compact vector-matrix form that yields the updated column vector as

$$\begin{aligned}
\mathbf{E_{CD}} &= \begin{bmatrix} e_{CD,1} \\ e_{CD,2} \\ \vdots \\ e_{CD,K} \end{bmatrix} \\
&= \begin{bmatrix} \sum_{i=0}^{L-1} c_i\phi_{1,i} - h_1\{\theta_{1,L} + \sum_{i=0}^{L-1} d_i\theta_{1,i}\} \\ \sum_{i=0}^{L-1} c_i\phi_{2,i} - h_2\{\theta_{2,L} + \sum_{i=0}^{L-1} d_i\theta_{2,i}\} \\ \vdots \\ \sum_{i=0}^{L-1} c_i\phi_{K,i} - h_K\{\theta_{K,L} + \sum_{i=0}^{L-1} d_i\theta_{K,i}\} \end{bmatrix} \\
&= \underbrace{\begin{bmatrix} \phi_{1,0} & \phi_{1,1} & \phi_{1,2} & \cdots & \phi_{1,L-1} \\ \phi_{2,0} & \phi_{2,1} & \phi_{2,2} & \cdots & \phi_{2,L-1} \\ \vdots & \vdots & \vdots & \ddots & \vdots \\ \phi_{K,0} & \phi_{K,1} & \phi_{K,2} & \cdots & \phi_{K,L-1} \end{bmatrix}}_{\mathbf{\Phi}} \underbrace{\begin{bmatrix} c_0 \\ c_1 \\ \vdots \\ c_{L-1} \end{bmatrix}}_{\mathbf{C}} \\
&\quad - \underbrace{\begin{bmatrix} h_1\theta_{1,0} & h_1\theta_{1,1} & h_1\theta_{1,2} & \cdots & h_1\theta_{1,L-1} \\ h_2\theta_{2,0} & h_2\theta_{2,1} & h_2\theta_{2,2} & \cdots & h_2\theta_{2,L-1} \\ \vdots & \vdots & \vdots & \ddots & \vdots \\ h_K\theta_{K,0} & h_K\theta_{K,1} & h_K\theta_{K,2} & \cdots & h_K\theta_{K,L-1} \end{bmatrix}}_{\mathbf{H\Theta}} \underbrace{\begin{bmatrix} d_0 \\ d_1 \\ \vdots \\ d_{L-1} \end{bmatrix}}_{\mathbf{D}} - \underbrace{\begin{bmatrix} h_1\theta_{1,L} \\ h_2\theta_{2,L} \\ \vdots \\ h_K\theta_{K,L} \end{bmatrix}}_{\mathbf{H_\Theta}} \\
&= \mathbf{\Phi C} - \underbrace{\begin{bmatrix} h_1 & 0 & 0 & \cdots & 0 \\ 0 & h_2 & 0 & \cdots & 0 \\ \vdots & \vdots & \vdots & \ddots & \vdots \\ 0 & 0 & 0 & \cdots & h_K \end{bmatrix}}_{\mathbf{H}} \mathbf{\Theta D} - \mathbf{H_\Theta} \\
&= \mathbf{\Phi C} - \mathbf{H_\Theta D} - \mathbf{H_\Theta}. \qquad (3.28)
\end{aligned}$$

A squared error criterion $\mathbf{V_{CD}}$, to be minimized, is formed from the error vector $\mathbf{E_{CD}}$ in (3.28)

$$\begin{aligned}
V_{\mathbf{CD}} &= \frac{1}{2}\mathbf{E}_{\mathbf{CD}}^H\mathbf{E_{CD}} \\
&= \frac{1}{2}\mathbf{C}^T\mathbf{\Phi}^H\mathbf{\Phi}C + \frac{1}{2}\mathbf{D}^T\mathbf{\Theta}^H|\mathbf{H}|^2\mathbf{\Theta D} + \frac{1}{2}\mathbf{H}_\mathbf{\Theta}^T\mathbf{H_\Theta} \\
&\quad -\Re\{\mathbf{C}^T\mathbf{\Phi}^H\mathbf{H\Theta D}\} - \Re\{\mathbf{C}^T\mathbf{\Phi}^H\mathbf{H_\Theta}\} \\
&\quad +\Re\{\mathbf{D}^T\mathbf{\Theta}^H\mathbf{H}^*\mathbf{H_\Theta}\}. \qquad (3.29)
\end{aligned}$$

The partial derivatives of $\mathbf{V_{CD}}$ with respect to the parametric unknowns $\mathbf{C}$ and $\mathbf{D}$ are

$$\begin{aligned}
\frac{\partial V_{\mathbf{CD}}}{\partial \mathbf{C}} &= \boldsymbol{\Phi}^H\boldsymbol{\Phi}\mathbf{C} - \Re\{\boldsymbol{\Phi}^H\mathbf{H}\boldsymbol{\Theta}\mathbf{D}\} - \Re\{\boldsymbol{\Phi}^H\mathbf{H}_{\boldsymbol{\Theta}}\} \\
&= \mathbf{C} - \Re\{\boldsymbol{\Phi}^H\mathbf{H}\boldsymbol{\Theta}\}\mathbf{D} - \Re\{\boldsymbol{\Phi}^H\mathbf{H}_{\boldsymbol{\Theta}}\}, \qquad (3.30)\\
\frac{\partial V_{\mathbf{CD}}}{\partial \mathbf{D}} &= \boldsymbol{\Theta}^H|\mathbf{H}|^2\boldsymbol{\Theta}\mathbf{D} - \Re\{\boldsymbol{\Phi}^H\mathbf{H}^*\boldsymbol{\Theta}\mathbf{C}\} + \Re\{\boldsymbol{\Theta}^H\mathbf{H}^*\mathbf{H}_{\boldsymbol{\Theta}}\} \\
&= \mathbf{D} - \Re\{\boldsymbol{\Phi}^H\mathbf{H}^*\boldsymbol{\Theta}\}\mathbf{C}. \qquad (3.31)
\end{aligned}$$

It is worth noting that $\boldsymbol{\Phi}^H\boldsymbol{\Phi} = \boldsymbol{\Theta}^H|\mathbf{H}|^2\boldsymbol{\Theta} = \mathbf{I}$ and $\Re\{\boldsymbol{\Theta}^H\mathbf{H}^*\mathbf{H}_{\boldsymbol{\Theta}}\} = 0$ due to the orthogonality property of the polynomials in $\boldsymbol{\Phi}$ and $\boldsymbol{\Theta}$. To satisfy the necessary conditions for a minimum point of $\mathbf{V_{CD}}$, the partial derivatives are set to $\mathbf{0}_{L\times 1}$. The unknowns $\mathbf{C}$ and $\mathbf{D}$ are can now be obtained via

$$\begin{bmatrix} \mathbf{I} & -\Re\{\boldsymbol{\Phi}^H\mathbf{H}\boldsymbol{\Theta}\} \\ -\Re\{\boldsymbol{\Phi}^H\mathbf{H}^*\boldsymbol{\Phi}\} & \mathbf{I} \end{bmatrix} \begin{bmatrix} \mathbf{C} \\ \mathbf{D} \end{bmatrix} = \begin{bmatrix} \Re\{\boldsymbol{\Phi}^H\mathbf{H}_{\boldsymbol{\Theta}}\} \\ \mathbf{0} \end{bmatrix}. \qquad (3.32)$$

With the identified coefficients in $\mathbf{C}$ and $\mathbf{D}$, we can express the fitted polynomial for the numerator in terms of orthogonal polynomials by the Forsythe's method ϕ_i with coefficients c_i as

$$\bar{y}(s) = \sum_{i=0}^{L-1} c_i\phi_i(s), \qquad (3.33)$$

where

$$c_i = \frac{1}{D_i}\sum_{k=-K}^{K} y_k\phi_i(s_k)w_k^2, \qquad (3.34)$$

and the fitted polynomial in terms of the normal direct solution polynomial is

$$\bar{y}(s) = \sum_{j=0}^{L-1} m_j^{L-1}s^j. \qquad (3.35)$$

To convert coefficients of the orthogonal polynomial c_i to those of normal direct solution polynomial m_j, the following relationship is used

$$m_j^L = \sum_{i=j}^{L-1} c_i b_j^i, \qquad (3.36)$$

where

$$b_j^i = \begin{cases} 0, & j<0 \\ 0, & j>i \\ 1, & j=i \\ b_{j-1}^{i-1} - u_i b_j^{i-1} - v_{i-1}b_j^{i-2}, & 0\le j<i \end{cases}. \qquad (3.37)$$

A further understanding of this conversion relationship can be gained by looking at the example of the transformation applied on the numerator polynomial

$$\sum_{j=0}^{L-1} m_j s^j = \sum_{i=0}^{L-1} c_i \phi_{k,i}, \tag{3.38}$$

and the expressions of the individual orthogonal polynomials are of the form, with internal coefficients g_t as

$$\begin{aligned} c_0\phi_{k,0} &= g_0, \\ c_1\phi_{k,1} &= g_1, \\ c_2\phi_{k,2} &= g_2 + g_3 s^2, \\ c_3\phi_{k,3} &= g_4 + g_5 s^3, \\ c_4\phi_{k,4} &= g_6 + g_7 s^2 + g_8 s^2. \\ &\vdots \end{aligned} \tag{3.39}$$

The transformation from the normal polynomial to orthogonal polynomials is now

$$\begin{aligned} &m_0 + m_1 s + m_2 s^2 + \cdots + m_{L-1} s^{L-1} \\ &= c_0\phi_{k,0} + c_1\phi_{k,1} + c_2\phi_{k,2} + \cdots + c_{L-1}\phi_{k,L-1}, \end{aligned} \tag{3.40}$$

and the coefficients of the normal polynomial m_j are made up of the components g_t which are redistributed to make up the internal coefficients of the orthogonal polynomials

$$\begin{aligned} &\Big(\overbrace{g_0 + g_2 + g_6 + \cdots}^{m_0}\Big) + \Big(\overbrace{g_1 + g_4 + \cdots}^{m_1}\Big) s \\ &+ \Big(\overbrace{g_3 + g_7 + \cdots}^{m_2}\Big) s^2 + \Big(\overbrace{g_5 + \cdots}^{m_3}\Big) s^3 + \Big(\overbrace{g_8 + \cdots}^{m_4}\Big) s^4 + c\ldots \\ = \; &\overbrace{g_0}^{c_0\phi_{k,0}} + \overbrace{g_1}^{c_1\phi_{k,1}} + \overbrace{g_2 + g_3 s^2}^{c_2\phi_{k,2}} + \overbrace{g_4 s + g_5 s^3}^{c_3\phi_{k,3}} + \overbrace{g_6 + g_7 s^2 + g_8 s^2}^{c_4\phi_{k,4}} + \cdots \\ = \; &c_0\phi_{k,0} + c_1\phi_{k,1} + c_2\phi_{k,2} + c_3\phi_{k,3} + c_4\phi_{k,4} + \cdots. \end{aligned} \tag{3.41}$$

The required natural frequencies f_i and damping ratios ζ_i of the resonant poles can now be obtained from the denominator polynomials by partial fraction expansion of (3.4) using the identified coefficients in **N**.

3.4.3 RESIDUES R_I

With the identified natural frequencies f_i and damping ratios ζ_i of the resonant poles, their corresponding residues R_i can now identified using a standard LS projection in this section. Note that although the coefficients of the zero polynomials in **M** are available, they are not used to identify the residues R_i.

Rewriting $\hat{P}(s)$ in (3.2) in partial fractions as

$$\begin{aligned}\hat{P}(s) &= \sum_{i=1}^{N} \frac{R_i}{s^2+2\zeta_i(2\pi f_i)s+(2\pi f_i)^2} \\ &= \sum_{i=1}^{N} R_i \underline{P}_i(s),\end{aligned} \tag{3.42}$$

where $\underline{P}_i(s) = \frac{1}{s^2+2\zeta_i(2\pi f_i)s+(2\pi f_i)^2}$, the damping ratios ζ_i, and natural frequencies f_i for individual $\underline{P}_i(s)$ using the identified vectors **N** were obtained by partial fraction expansion previously.

Define another modeling error $e_{\mathrm{R},k}$ between the identified model using residues R_i and experimental frequency response h_k at the kth frequency point as

$$\begin{aligned}e_{\mathrm{R},k} &= \sum_{i=1}^{N} R_i \underline{P}_i(j2\pi f_k) - h_k \\ &= \begin{bmatrix} \underline{P}_1(j2\pi f_k) & \underline{P}_2(j2\pi f_k) & \cdots & \underline{P}_N(j2\pi f_k) \end{bmatrix} \mathbf{R} - h_k,\end{aligned} \tag{3.43}$$

where $\mathbf{R} = \begin{bmatrix} R_1 & R_2 & \cdots & R_N \end{bmatrix}^T \in \mathbb{R}^N$ is the vector of residues R_i to be identified. Packing all the errors $e_{\mathrm{R},k}$ over the entire measurement frequency range of K measurement points into a column vector $\mathbf{E_R}$, we can write $\mathbf{E_R}$ as

$$\begin{aligned}\mathbf{E_R} &= \begin{bmatrix} e_{\mathrm{R},1} \\ e_{\mathrm{R},2} \\ \vdots \\ e_{\mathrm{R},K} \end{bmatrix} \\ &= \underbrace{\begin{bmatrix} \underline{P}_1(j2\pi f_1) & \underline{P}_2(j2\pi f_1) & \cdots & \underline{P}_N(j2\pi f_1) \\ \underline{P}_1(j2\pi f_2) & \underline{P}_2(j2\pi f_2) & \cdots & \underline{P}_N(j2\pi f_2) \\ \vdots & \vdots & \ddots & \vdots \\ \underline{P}_1(j2\pi f_K) & \underline{P}_2(j2\pi f_K) & \cdots & \underline{P}_N(j2\pi f_K) \end{bmatrix}}_{\underline{\mathbf{P}}} \begin{bmatrix} R_1 \\ R_2 \\ \vdots \\ R_N \end{bmatrix} - \underbrace{\begin{bmatrix} h_1 \\ h_2 \\ \vdots \\ h_K \end{bmatrix}}_{\mathbf{h}} \\ &= \underline{\mathbf{P}}\mathbf{R} - \mathbf{h},\end{aligned} \tag{3.44}$$

where $\underline{\mathbf{P}} \in \mathbb{C}^{K \times N}$ and $\mathbf{h} \in \mathbb{C}^K$. Similarly, defining a positive definite scalar error cost function $V_{\mathbf{R}}$ as the common sum-of-squared error criterion, we get

$$\begin{aligned}V_{\mathbf{R}} &= \frac{1}{2}\mathbf{E_R}^H \mathbf{E_R} \\ &= \frac{1}{2}\left(\underline{\mathbf{P}}\mathbf{R} - \mathbf{h}\right)^H \left(\underline{\mathbf{P}}\mathbf{R} - \mathbf{h}\right) \\ &= \frac{1}{2}\mathbf{R}^H \underline{\mathbf{P}}^H \underline{\mathbf{P}} \mathbf{R} - \mathbf{R}^H \underline{\mathbf{P}}^H \mathbf{h} + \frac{1}{2}\mathbf{h}^H \mathbf{h}.\end{aligned} \tag{3.45}$$

The residue vector $\mathbf{R}$ to be identified can then be obtained via the standard LS solution by setting $\frac{\partial V_{\mathbf{R}}}{\partial \mathbf{R}}$ to be $\mathbf{0}_{N\times 1}$ as

$$\begin{aligned} \underline{\mathbf{P}}^H\underline{\mathbf{P}}\mathbf{R} - \underline{\mathbf{P}}^H\mathbf{h} &= \mathbf{0} \\ \therefore \mathbf{R} &= \left(\underline{\mathbf{P}}^H\underline{\mathbf{P}}\right)^{-1}\underline{\mathbf{P}}^H\mathbf{h}, \end{aligned} \tag{3.46}$$

and the residues R_i are chosen from $\Re(\mathbf{R})$ since this solution is complex but residues are real. The complex parts are typically of small magnitude in the normal LS plane, and can be omitted. Moreover, this omission is justifiable as the Laplace operator s corresponds directly to differentiation in the time domain, which has little or no effect for resonant poles which are typically at higher frequencies.

3.4.4 ERROR ANALYSIS

With (3.10) and (3.11), the corresponding error induced in the first LS optimization to find $\mathbf{M}$ and $\mathbf{N}$ is given by

$$\begin{aligned} V_{\mathbf{MN}}(\mathbf{M},\mathbf{N}) &= \frac{1}{2}\mathbf{H}_{\Omega}{}^H\mathbf{H}_{\Omega} - \frac{1}{2}\mathbf{M}^T\boldsymbol{\Omega}^H\boldsymbol{\Omega}\mathbf{M} - \frac{1}{2}\mathbf{N}^T\boldsymbol{\Omega}^H|\mathbf{H}|^2\boldsymbol{\Omega}\mathbf{N} + \Re\{\mathbf{M}^T\boldsymbol{\Omega}^H\mathbf{H}\boldsymbol{\Omega}\mathbf{N}\} \\ &= \frac{1}{2}\mathbf{H}_{\Omega}{}^H\mathbf{H}_{\Omega} - \frac{1}{2}\left(\boldsymbol{\Omega}\mathbf{M} - \mathbf{H}\boldsymbol{\Omega}\mathbf{N}\right)^H\left(\boldsymbol{\Omega}\mathbf{M} - \mathbf{H}\boldsymbol{\Omega}\mathbf{N}\right), \end{aligned} \tag{3.47}$$

and the corresponding error induced with the second LS optimization using (3.46) is

$$\begin{aligned} V_{\mathbf{R}}(\mathbf{R}) &= \frac{1}{2}\mathbf{h}^H\underline{\mathbf{P}}(\underline{\mathbf{P}}^H\underline{\mathbf{P}})^{-1}\underline{\mathbf{P}}^H\mathbf{h} - \mathbf{h}^H\underline{\mathbf{P}}(\underline{\mathbf{P}}^H\underline{\mathbf{P}})^{-1}\underline{\mathbf{P}}^H\mathbf{h} + \frac{1}{2}\mathbf{h}^H\mathbf{h} \\ &= \frac{1}{2}\mathbf{h}^H\mathbf{h} - \frac{1}{2}\mathbf{h}^H\underline{\mathbf{P}}(\underline{\mathbf{P}}^H\underline{\mathbf{P}})^{-1}\underline{\mathbf{P}}^H\mathbf{h} \\ &= \frac{1}{2}\mathbf{h}^H(\mathbf{I} - \mathbf{W})\mathbf{h}, \end{aligned} \tag{3.48}$$

where $\mathbf{W} = \underline{\mathbf{P}}(\underline{\mathbf{P}}^H\underline{\mathbf{P}})^{-1}\underline{\mathbf{P}}^H$ is the orthogonal projection matrix of $\mathbf{h}$ onto the LS of parameter plane with $\mathbf{R}$.

As such, the total modeling error E induced by the proposed MPI is given by

$$E = V_{\mathbf{MN}}(\mathbf{M},\mathbf{N}) + V_{\mathbf{R}}(\mathbf{R}) + V_{\Im(\mathbf{R})}, \tag{3.49}$$

where $\Im(\cdot)$ is the imaginary operator. $V_{\mathbf{MN}}(\mathbf{M},\mathbf{N})$ is the error induced by LS estimation using vectors $\mathbf{M}$ and $\mathbf{N}$ using (3.8), $V_{\mathbf{R}}(\mathbf{R})$ is the LS approximation error for R_i using (3.46), and $V_{\Im(\mathbf{R})}$ is the error induced by using only $\Re\{\mathbf{R}\}$ as residues which is coupled in $V_{\mathbf{MN}}(\mathbf{M},\mathbf{N})$ and $V_{\mathbf{R}}(\mathbf{R})$. This completes the proposed enhanced MPI using Forsythe's orthogonal polynomials.

The error E is minimized in $V_{\mathrm{MN}}(\mathbf{M},\mathbf{N})$ and $V_{\mathrm{R}}(\mathbf{R})$ by the two proposed subsequent LS formulations. It is worth noting that a direct linear LS optimization from (3.2) is not possible as the residues R_i are coupled into the identified parameters in $\mathbf{M}$ and $\mathbf{N}$, which cannot be written in LITP form. E is an important index during optimization, if iterative system identification of modal parameters for critical resonant modes is employed to ensure improved accuracies.

With the essential diagnostic modal parameters identified using the proposed MPI and stored in an online information repository as shown earlier in Figure 3.6, efficient communication among BU and engineering teams can be facilitated. It is well known that communication among development teams can reduce project uncertainties, but at the expense of additional time and cost for communication. In general, if information exchange is too frequent, then communication time and cost would increase significantly. As such, the proposed integrative servo-mechanical systems design framework can help management identify appropriate development policies as well.

3.5 INDUSTRIAL APPLICATION: HARD DISK DRIVE SERVO SYSTEMS

In this section, we apply the proposed integrative servo-mechanical systems design approach and enhanced MPI to commercial HDDs, as HDDs are classic examples of high-performance mechatronic systems. Previously integrated servo-mechanical design for head-positioning control [97][98] and bandwidth estimation [99] are also proposed. In essence, our proposed design approach calls for the development of a "closed" system solution by using life cycle analysis to understand and identify the essential diagnostic feedback information, for integration into transformation or life cycle work for prognostic product life cycle management.

The effectiveness of the proposed MPI is illustrated with frequency response measurement data of a commercial dual-stage HDD consisting of the VCM—primary actuator, and PZT active suspension—secondary actuator. Currently, HDDs employ the so-called single-stage actuation, where the sole actuator VCM rotates about a pivot to carry the magnetic heads to their desired positions on the disks to perform read and write operations. To shorten the data access time and increase data storage capacities, secondary actuation has been proposed by appending another actuator onto the VCM. A picture of the VCM with appended PZT active suspension in a dual-stage actuator control configuration is shown in Figure 3.8. The PZT active suspension from NHK Spring [100] has identified torsion modes at 4.31 and 6.52 kHz, and a sway mode at 21.08 kHz [101]. Raw frequency response measurement data is pre-treated offline to identify and remove the time delay T_D. It is worth noting that the proposed procedure is applicable to any LTI flexible mechanical system in general.

By exciting the input of the PZT active suspension at u_M with swept sine from 400 Hz to 50 kHz, the displacement of the secondary actuator can be obtained at the tip of the suspension at y_M with a LDV non-intrusively. y_M and u_M are channelled into a Dynamic Signal Analyzer (DSA), and the experimental frequency response can be obtained and is shown in Figure 3.9.

Inspecting the experimental frequency response in Figure 3.9, we can see that six distinct resonant poles are observed. As such, the order number of $N = 6$, i.e., $L = 12$, is chosen for our modal identification. Using (3.8) and (3.46), the required modal parameters of residues R_i, damping ratios ζ_i, and natural frequencies f_i are obtained

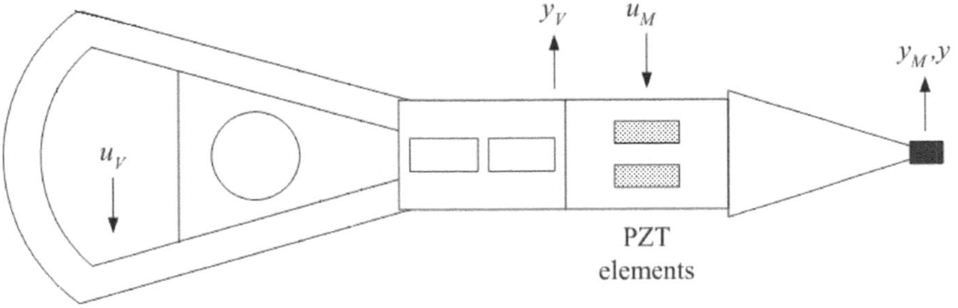

FIGURE 3.8 Picture of VCM with appended PZT active suspension in a dual-stage configuration for HDDs. The downward arrows represent the input signals, and the upward arrows represent the corresponding outputs.

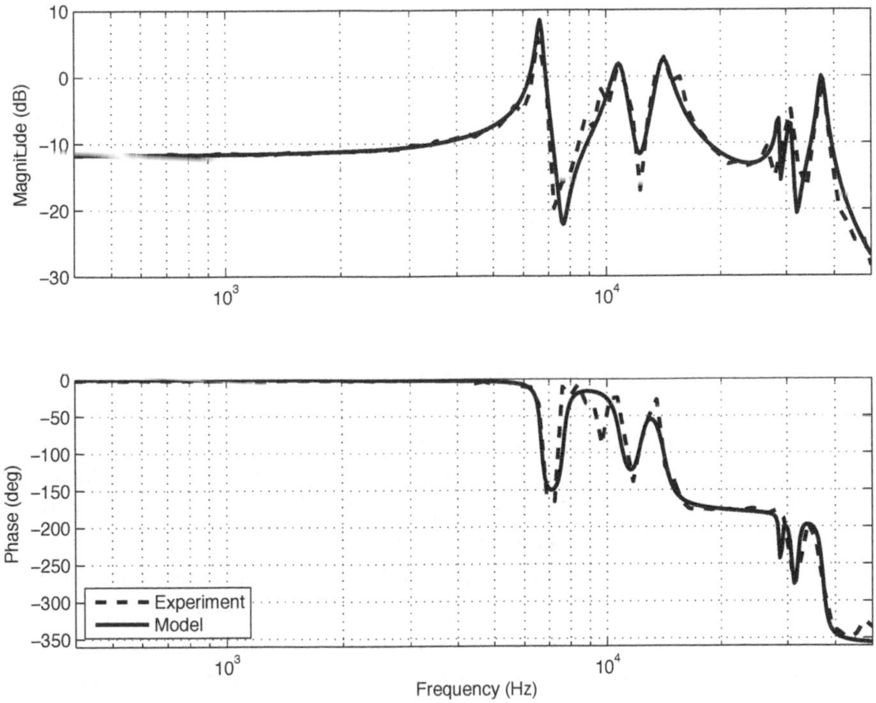

FIGURE 3.9 Frequency responses. Dashed: LDV measurement data of PZT active suspension. Solid: From identified model using proposed modal identification algorithm.

and displayed in Table 3.1. The frequency response of the identified transfer function of the PZT active suspension $\hat{P}_{PZT}(s) = \frac{y_M(s)}{u_M(s)}$ is also shown in Figure 3.9.

TABLE 3.1
Identified Modal Parameters of PZT Active Suspension

i	$R_i(1\times10^9)$	$f_i(1\times10^4)$/Hz	ζ_i
1	0.1698	0.6669	0.0185
2	0.4779	1.0862	0.0438
3	0.9394	1.4136	0.0460
4	−0.1854	2.8676	0.0071
5	−0.4808	3.0827	0.0160
6	−2.2756	3.7117	0.0208

In this representation, the in-phase and out-of-phase resonant modes can be directly seen from the signs (directions) of the residues R_i identified in Table 3.1. The contribution of each resonant mode can be quickly identified for actuator redesign and servo evaluation, and compared with mode shape measurements as well as simulations when available.

Similarly, by exciting the input of the VCM at u_V with swept sine from 100 Hz to 10 kHz, the displacement of the primary actuator can be obtained at the tip at y_V with a LDV non-intrusively. y_V and u_V are also channelled into a DSA, and the experimental frequency response can be obtained and is shown in Figure 3.10.

Inspecting the experimental frequency response in Figure 3.10, we can observe the rigid body mode (double integrator at low frequencies) and along with that four distinct resonant poles. However, the high frequency resonant modes from spillover or mode-splitting are not obvious from the magnitude response due to small Signal-to-Noise Ratio (SNR) measurements at high frequencies from the high gain roll-off at more than -40 dB/dec. As such, the order number of $N=9$, i.e., $L=18$, is chosen for our modal identification. Using (3.8) and (3.46), the required modal parameters of residues R_i, damping ratios ζ_i, and natural frequencies f_i can be obtained and displayed in Table 3.2. The frequency response of the identified transfer function of the VCM $\hat{P}_{VCM}(s)=\frac{y_V(s)}{u_V(s)}$ is shown in Figure 3.10.

The in-phase and out-of-phase resonant modes with respect to the rigid body mode of the VCM can be seen directly from the signs of the residues R_i identified in Table 3.2. The contribution of each resonant mode can be directly identified for actuator redesign, and compared with mode shape measurements as well as simulations when available.

It is worth noting that the modeling errors can be further reduced by performing the proposed MPI in segments of smaller frequency ranges or selection of a higher order $\hat{P}(s)$, i.e., a larger N and hence L. The identified parameters can now be used for dual-stage controller design by servo researchers, which can be quickly reverted to mechanical engineers on the targets achieved and specifications to be met. More importantly, the phase relationships of the flexible resonant modes with respect to the first rigid body mode allow low order phase-stabilized controllers to be designed for improved disturbance rejection in dual-stage HDDs [102].

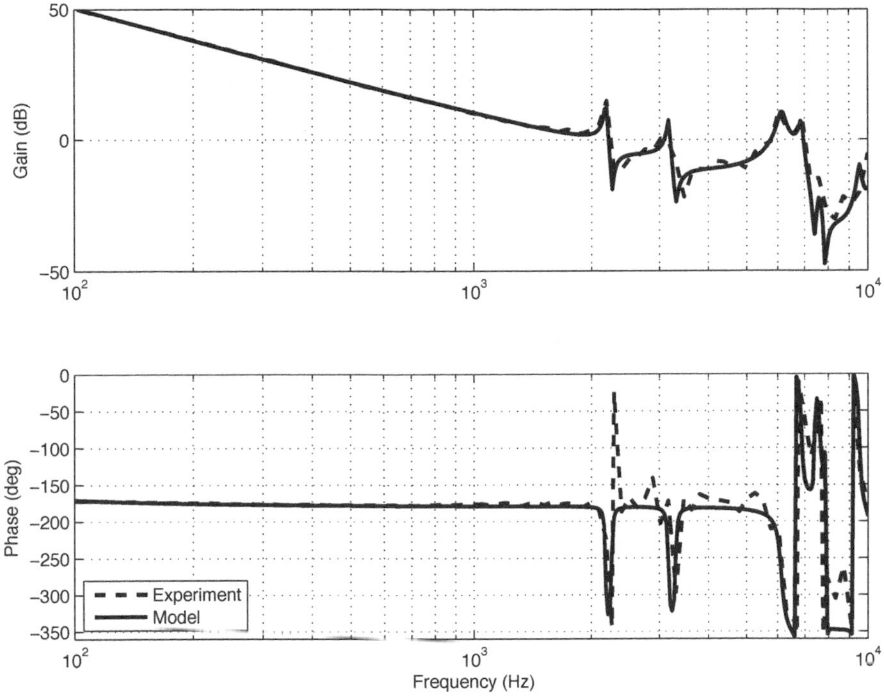

FIGURE 3.10 Frequency responses. Dashed: LDV measurement data of VCM. Solid: From identified model using proposed modal identification algorithm.

TABLE 3.2
Identified Modal Parameters of VCM

i	$R_i(1 \times 10^8)$	$f_i(1 \times 10^4)$/Hz	ζ_i
1	1.2242	0.0030	0.2569
2	−0.1090	0.2180	0.0052
3	−0.1266	0.3165	0.0070
4	−1.9202	0.6125	0.0186
5	0.7997	0.6872	0.0092
6	0.0412	0.7632	0.0107
7	−0.0531	0.9318	0.0443
8	0.2133	0.9533	0.0085
9	−0.1574	1.0221	0.0089

To illustrate the robustness of the proposed modal parametric identification algorithm, the above experimental frequency responses of the VCM and PZT active suspension are perturbed by $\pm 5\%$ from their nominal frequencies of the resonant modes and re-identified in this section. By varying the nominal frequencies of the resonant modes in the PZT active suspension by $\pm 5\%$, the perturbed experimental frequency responses are shown in Figure 3.11.

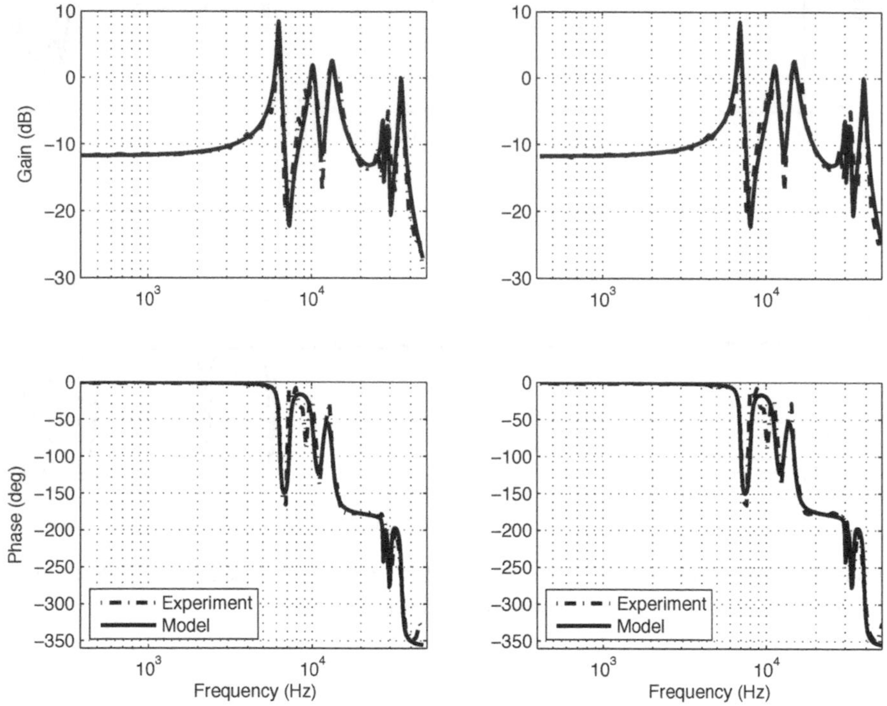

FIGURE 3.11 Frequency responses. Left: -5% shift in resonant frequencies. Right: $+5\%$ shift in resonant frequencies. Dashed: LDV measurement data of PZT active suspension. Solid: From identified model using proposed modal identification algorithm.

With the same order number of $N = 6$, i.e., $L = 12$ chosen for modal identification, the required modal parameters of residues R_i, damping ratios ζ_i, and natural frequencies f_i are obtained using (3.8) and (3.46), and displayed in Table 3.3. The frequency responses of the identified transfer functions of the perturbed PZT active suspension are also shown in Figure 3.11.

It can be seen from Figure 3.11 and Table 3.3 that the perturbed natural frequencies f_i and associated modal parameters are accurately captured.

Similarly, by varying the nominal frequencies of the resonant modes in the primary actuator VCM by $\pm 5\%$, the perturbed experimental frequency responses are shown in Figure 3.12.

With the same order number of $N = 9$, i.e., $L = 18$ chosen for modal identification, the required modal parameters of residues R_i, damping ratios ζ_i, and natural frequen-

TABLE 3.3
Identified Modal Parameters of Perturbed PZT Active Suspension

	−5% shift in f_i			+5% shift in f_i		
i	$R_i(1\times10^9)$	$f_i(1\times10^4)$/Hz	ζ_i	$R_i(1\times10^9)$	$f_i(1\times10^4)$/Hz	ζ_i
1	0.1532	0.6335	0.0184	0.1872	0.7002	0.0183
2	0.4313	1.0319	0.0432	0.5269	1.1405	0.0436
3	0.8478	1.3429	0.0457	1.0357	1.4843	0.0459
4	−0.1673	2.7243	0.0070	−0.2044	3.0110	0.0072
5	−0.4339	2.9286	0.0158	−0.5301	3.2369	0.0168
6	−2.0537	3.5261	0.0203	−2.5089	3.8972	0.0210

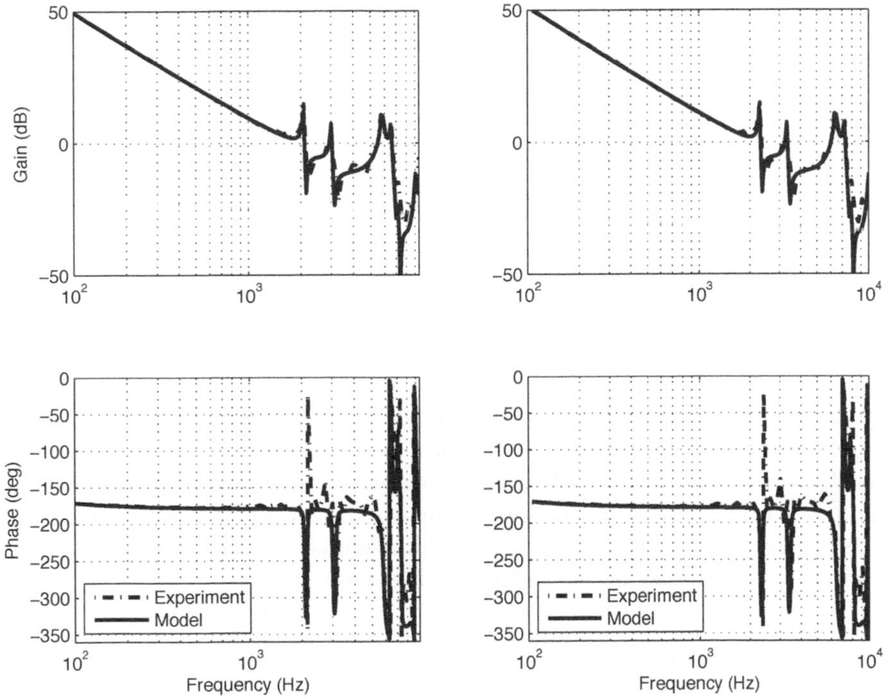

FIGURE 3.12 Frequency responses. Left: −5% shift in resonant frequencies. Right: +5% shift in resonant frequencies. Dashed: LDV measurement data of VCM. Solid: From identified model using proposed modal identification algorithm.

cies f_i are obtained using (3.8) and (3.46), and displayed in Table 3.4. The frequency responses of the identified transfer functions of the perturbed VCM are also shown in Figure 3.12.

TABLE 3.4
Identified Modal Parameters of Perturbed VCM

	−5% shift in f_i			+5% shift in f_i		
i	$R_i(1\times10^8)$	$f_i(1\times10^3)$/Hz	ζ_i	$R_i(1\times10^8)$	$f_i(1\times10^4)$/Hz	ζ_i
1	1.1048	0.0282	0.2569	1.3496	0.0031	0.2612
2	−0.0983	2.0708	0.0052	−0.1201	0.2289	0.0048
3	−0.1141	3.0067	0.0070	−0.1394	0.3323	0.0063
4	−1.7321	5.8188	0.0182	−2.1159	0.6431	0.0190
5	0.7180	6.5281	0.0098	0.8771	0.7215	0.0099
6	0.0289	7.1081	0.0117	0.0353	0.7856	0.0123
7	−0.0445	8.6784	0.0425	−0.0544	0.9592	0.0431
8	0.1546	9.0564	0.0081	0.1888	1.0010	0.0076
9	−1.8262	9.7096	0.0094	−2.2309	1.0732	0.0090

Similarly, it can be seen from Figure 3.12 and Table 3.4 that the perturbed natural frequencies f_i and corresponding modal parameters are accurately identified. The order numbers N for the same batch of mass-produced actuators will be identical, and variations in residues R_i and damping ratios ζ_i will also be robustly identified when subjected to parametric perturbations.

The contribution of each resonant mode can now be quickly identified for actuator redesign and servo evaluation, and compared with mode shape measurements as well as simulations when available. As such, we can improve the cohesiveness and technological outputs of the various skilled teams through this integrative approach from the common parameters which are available quickly in real time. This will greatly reduce the risk and other production overheads before actual implementation and product development, as well as reduce the time taken for mechanical redesign while allowing the testing of different dual-stage control algorithms. Furthermore, the proposed MPI technology is highly industrializable and implementable, and can be easily transferred to commercial mechatronic manufacturing industries with minimal intrusion into on-going design and production processes.

3.6 RESULTS AND DISCUSSIONS

The product development literature often provides open loop and single-link linear relationships which are usually lists of over-simplified rules for managing projects. This can usually be determined through analysis of the project budgets and schedules. However, it is also necessary to capture the management policies and constraints that will drive the project behavior. Examples of these include the management team's relative priorities between cost and schedule, and resourcing strategies (e.g., overtime versus hiring). From a technical point of view, it is also crucial to determine what diagnostic data is instrumental to be "feedback" to other teams *in situ* to speed up the R&D process via prognostic rework.

Successful new product developments through R&D are critical for the survival of all companies. As motivated in the previous sections, high pressures on time-to-market of new products have fostered a tighter integration and concurrency of various R&D stages in order to meet delivery schedule targets. Furthermore, in today's information explosion era, advances in engineering technologies through R&D efforts are no longer solely bounded by individual domains in traditional science, physics, and mathematics, but rather a conglomerate of different inter-disciplinary fields with medical, social, and even psychological discipline considerations. All of these are a clear indication of an increasing need and emphasis on a systems-of-systems orientation by studying not individual components, but synergetic integration of composite systems to be considered simultaneously [94]. This emerging System-of-Systems (SoS) concept describes the large-scale integration of many independent, self-contained systems in order to satisfy a global need [95].

Currently, most works on Condition-Based Monitoring (CBM), Fault Detection and Isolation (FDI), or even Structural Health Monitoring (SHM), consider solely the integrity of independent modules, even when the complex integrated industrial process consist of several mutually interacting components interwoven together. However, an intelligent R&D architecture should consider crosstalk to facilitate actions and decisions amongst the synergetic integration of composite systems simultaneously.

Our proposed systems design approach for mechatronic systems emphasizes the synergetic integration of different operating processes through identification of which "local" signals are to be feedback in such a real-time distributed control system through life cycle analysis. From a systems point of view, the effective capability can be realized through the integration of existing and future manufacturing and industrial processes and systems. This "big-picture" approach will also limit intrinsic uncertainties and variabilities within the components, and suppress any extrinsic uncertainties from human intervention. The future of engineering systems will require their incorporation into increasingly integrated complex SoS.

In this chapter, we demonstrated the potential of our proposed systems design approach using MPI to manage complexity and uncertainty arising in a mechatronics R&D project. This is especially important during the startup phase of the project, thus potentially reducing the rework efforts considerably. Our proposed approach hence provides indispensable support for management looking to implement concurrent product development. The proposed methodology is also applicable to other processes and operation units on a larger scale in different domains. For our future works, we target to identify more information and available and implementable industrial technologies with most economical benefits, while best for our environment.

The developed tools also allow for higher level decision making and command in synergetic integration between several industrial processes and stages, for shorter time in failure and fault analysis in the entire industrial production life cycle. In this way, we apply life cycle analysis to manufacturing strategies to determine transform current batch-based manufacturing to an integrated collocated facility-based manufacturing. With this, concepts of green energy, energy-efficient production, and even green data storage, etc., producing less carbon footprints may also be simultaneously incorporated to allow management make important decisions and policies, as well as educate fellow researchers and the public about the advantages of various sustainable technologies.

3.7 CONCLUSION

In this chapter, we propose an integrative servo-mechanical systems design approach for mechatronics R&D using life cycle analysis. Using the commercial Hard Disk Drives (HDDs) as an example, we analyzed the various teams in an integrative approach with crosstalk, and identified the key parameters to be isolated and feedback online to other teams for next generation of dual-stage HDDs using the proposed Modal Parametric Identification (MPI).

The proposed MPI identifies the essential diagnostic information in terms of residues, damping ratios, and natural frequencies of critical resonant modes in a modal summation format using Forsythe's orthogonal polynomials for quick feedback into the depository online, thereby allowing fast prognostic rework. The information is coupled in common transfer function representations, and can be obtained via two subsequent minimization of Least Squares (LS) error criterions. The modal summation form combines the required information for mechanical structural and servo controller designs essential for shorter servo-mechanical integration life cycle, and the proposed algorithm is tested on measured frequency responses of actuators in a commercial dual-stage HDD. Our results produce frequency responses of identified mathematical models which match closely with the experimental data obtained from experimental frequency responses—both nominal and perturbed at $\pm 5\%$ of the natural frequencies of the resonant modes, of the VCM as well as the PZT active suspension.

The proposed intelligent systems design approach in an integrated servo-mechanical framework allows for higher level decision making and command in synergetic integration between several industrial processes and stages, and shorter time in failure and fault analysis in the entire industrial production life cycle. With this methodology, direct bandwidth and secondary actuator stroke prediction, as well as identification of critical resonant modes at desired sampling frequency, etc., can be incorporated. This empowers us to decide on the next stage of intelligent factory control, and the manufacturing design of the next generation of HDDs with green considerations [103].

In the following chapter, we propose a Dominant Feature Identification (DFI) algorithm which reduces the dimension of a real matrix containing data logged from sensors (direct) or features (inferential or indirect) which are monitoring an industrial system or process of interest. The proposed DFI selects the dominant sensors or features, and is proven rigorously to be of minimum Least Squares (LS) errors in *both* estimation and clustering.

4 Dominant Feature Identification (DFI)

Identification and online prediction of the lifetime of industrial machines using minimal and cheap sensors are crucial to reduce production costs and down-time in engineering systems. In this chapter, we provide a formal decision software tool in the so-called Dominant Feature Identification (DFI) framework, using matrix theory to extract the dominant features thereby enabling failure prediction and fault isolation. Selection of dominant features and sensors are important, as retaining only essential features and sensors allows reduced signal processing and cuts costs.

The proposed DFI decision-making tool is based on a formal mathematical approach that selects dominant features using the Singular Value Decomposition (SVD) of real-time measurements. Using Computer Numerical Control (CNC) milling machines as an example, we apply the proposed DFI to estimate the wear of an industrial cutting tool from force measurements using dynamometers and Acoustic Emissions (AE) using acoustic sensors, since AE sensors are cheap, non-intrusive, and coupled with fast dynamic responses. It is shown that the proposed method of dominant feature selection is optimal in the sense that it minimizes both Least-Squares (LS) estimation and clustering errors. A reduced feature subset is then selected to reduce signal processing and number of sensors required. Tool wear is predicted using the Recursive Least Squares (RLS) algorithm to identify parameters in forecasting the time series of cutting tool wear using a static Multiple Regression Model (MRM) and a dynamic Auto-Regressive Moving Average with eXogenous inputs (ARMAX) model based on the reduced features. Experimental results on an industrial high speed milling machine show the effectiveness in predicting the tool wear using only the dominant features, and a reduction at least 16.83% in Mean Relative Error (MRE) is observed when compared to other methods proposed in the literature. When DFI is combined with Neural Networks (NNs) in a two-stage framework, the reduced features set can be used for industrial Fault Detection and Isolation (FDI) applications. Our experimental results also show a high FDI accuracy with a great reduction in number of sensors and features when this software tool is applied to an industrial machine fault simulator.

4.1 INTRODUCTION

In an era of intensive competition, the new challenges faced by industrial manufacturing processes include maximizing productivity, ensuring high product quality, and reducing the production time while minimizing the production cost simultaneously. As such, it is crucial that asset usage and plant operating efficiency be maximized, as unexpected downtime due to machinery failure has become more costly than before. Predictive maintenance is thus actively pursued in the manufacturing industry in recent years, where equipment outages are predicted and maintenance is carried out only when necessary. Also, one of the causes of delay in networked manufacturing processes is machine down time or failure of the machining tools when faults arise. Much manpower is also required to identify, locate, and rectify the faults for large-scale interconnected processes, thereby increasing OPerating EXpenditure (OPEX).

To ensure successful condition-based maintenance, it is necessary to detect, identify, and classify different kinds of failure modes in the manufacturing process. One of the causes of delay in manufacturing processes is machine down-time or failure of the machining tools. Failure of machine tools can also affect the production rate and quality of products, and detection of the tool state becomes a pivotal role in manufacturing [36].

When predicting tool wear in industrial cutting machines, advanced sensing and signal processing technologies are necessary in implementing the corresponding fault detection systems. There are currently two major approaches used in the sensing technologies; namely the direct method (or *direct sensing*)—which measures and evaluates the physical quantity of the fault from a single direct sensor, and the indirect method (*indirect* or *inferential sensing*)—that adopts different kinds of sensors at different positions to sense signals *correlated* to the different failure modes. In many applications, the direct method is not possible for online realization, since tool wear cannot be measured without having the milling or cutting operations stopped and the tool be extracted for visual inspection. This results in machine downtime and human user intervention costs. On the other hand, failure to remove worn tools can lead to their failure, with concomitant damage to expensive parts.

Many methods exist for Tool Condition Monitoring (TCM), and indirect methods that rely on the relationship between tool conditions and measurable signals (such as force [104][105][106][107][108], acoustic emission [109][110], vibration [111][112], electrical current [113], etc.), for detecting tool conditions have been extensively studied. Research works that have been devoted to tool wear monitoring in the machining process include tool wear sensing and metal-removing rate control [109], using of acoustic emission signal to detect abnormal event in the machining process [114], and using current-sensor-based feed cutting force estimates [113]. The indirect method may work as an on-line systematic technique as it measures the operating parameters during the manufacturing process without shutting down the machine, but the amount of sensors could be large and determining failure could mean taking minutes or hours to analyze data from all the sensors. In spite of these efforts, realization of acceptable prediction of tool wear still needs improvement.

Among these sensing signals, the Acoustic Emission (AE) signal is very effective for indirect methods in TCM because of its non-intrusiveness, ease of operation, and fast dynamic response [110]. The AE frequency range is also higher than that of machine vibration and environmental noise [115]. Another advantage of using the AE signal is that an AE sensor is small and can be installed easily, and the sampling process does not interrupt the machining operations [116]. The costs of AE sensors and force sensors also differ greatly, and using AE signals is less expensive than using force signals. Typical advantages of AE are high sensitivity, early and rapid detection of defects for leaks and cracks, etc. Also, the AE signal is more convenient to use in real time TCM.

Fundamentally, AE signals can be divided into continuous and burst types. The continuous-type AE signal waveform is similar to Gaussian random noise but its amplitude varies with AE activity. The burst type is a short duration pulse due to the discrete release of high amplitude strain energy [117]. As the tool becomes worn, the AE signal produces multiple bursts [118]. AE can then be used to detect the chip formation mechanism, and friction between the chip and tool rake face in both turning and milling operations [119].

The characteristics of an AE signal to be used in TCM are studied by researchers in various aspects. In [120], the authors used spectral, statistical, and time series analyses to analyze the tool wear. Liao et al. observed that there was a general increasing trend of AE root-mean-square (rms) with cutting speed and depth of cut, and the use of oil based lubricant reduced the mean AE rms [115]. They also observed that the measurement of the peak count ratio displayed a constant rate of decrease with gradual wear. Ravindra et al. found that there was a significant change in skew and kurtosis of AE signals with tool failure [121]. Kannatey and Dornfield proposed a relationship between skew, kurtosis, and tool wear [122]. In these efforts, various features have been proposed for monitoring the AE signals in milling processes. Zhu et al. [123] selected cutting force features using Fisher's linear discriminate analysis for tool wear monitoring. Binsaeid et al. [124] used a correlation-based feature selection technique to evaluate the significance of the features from multiple sensor signals. In spite of all these efforts, perusal of AE signals for feature selection affecting tool wear is still a problem addressed for further studies.

After data collection of the essential sensor signals, *features* are often extracted and selected to analyze the signals from all these embedded sensors to assess the condition of the system in the indirect method, as they can be realized without interrupting the production process for signal collection, signal processing, and classification, etc., *in situ*. As such, features are sometimes also known as *soft* sensors; being computed out of Digital Signal Processing (DSP) techniques as compared to real physical sensors. The processed signal features are then related to the observed values of wear by a suitable wear model [125]. However, a large number of features is usually computed in many industrial applications and this leads to an increase in time and computational space complexity of the recognition process in order for the system to make a reasonably accurate deduction. The choice of features also affects several aspects of the recognition process such as accuracy, learning time, and essential sample size, etc., and it has been observed that beyond a certain threshold, inclusion of additional features actually leads to a worse performance.

As such, it is obvious that the main goal of feature subset selection is to reduce the number of features used in classification without compromising on accuracy. Data compression schemes include Principal Component Analysis (PCA) [126][127][128] and Principal Feature Analysis (PFA) [129][130] which had been applied to pattern recognition to identify key features of original data. Both PCA and PFA require computationally intensive calculations of eigenvectors (and their corresponding eigenvalues) based on the correlation or covariance matrix to derive a linear transformation matrix for feature selection to compress the raw data into lower dimensions. Our proposed DFI on the other hand, uses an inner product matrix of a lower dimension, and reduces the Least Squares Error (LSE) induced when selecting the principal components and clustering to identify the dominant features. Compared with PCA and PFA, DFI is numerically efficient, and reduces the complexity of feature selection greatly while reducing the LSE during dominant feature selection and clustering automatically.

Industries also always have an urgent need for prediction of fault progression and remaining use life. Prediction leads to improved management and hence effectiveness of equipment, and multifaceted guarantees are increasingly being given for industrial machines, products, and services [131]. Time series prediction has been extensively studied using various methods, e.g., (a) linear methods including classical estimation theory using time- or frequency-domain methodologies, post-modern $\mathcal{H}_2$ and $\mathcal{H}_\infty$ theories, etc., (b) non-linear estimation and function approximations as well as identification and fault diagnosis using Neural Networks (NNs) [132][133], or (c) artificial intelligent methods using wavelet transforms [134], support vector machines [135], diagnosis model based on fault trees [136], etc. These methods normally segregate raw data into a "training" set and "validation" set—the former used to train the underlying prediction mechanism, and the latter used to test if the identified model is of satisfactory accuracy [73]. These methods are typically offline approaches and require extensive computations.

In this chapter, we present a formal decision software tool for reasoning with data from installed sensors to predict tool wear in industrial cutting machines continuously using an indirect approach. A rigorous mathematical framework for Dominant Feature Identification (DFI) is developed that provides an autonomous rule-base for data and sensor reduction. We use Singular Value Decomposition (SVD) to decompose the inner product matrix of collected data from the sensors monitoring the wear of an industrial cutting tool. The principal components affecting the machine wear are optimized in a least squares sense in a certain reduced space, and the dominant features are extracted using the K-means clustering algorithm [137][138]. This DFI framework uses a formal mathematical analysis to select dominant features. A numerically efficient scheme for implementation is presented that is based on the inner product matrix of the collected data, not the correlation (outer product) matrix. It is proven that the proposed method of dominant feature selection is optimal in the sense that it minimizes the total least-squares estimation and clustering errors.

Moving on, the DFI framework is enhanced into a new DFI methodology along with measurement of AE signals to predict tool wear in a ball nose cutter and dynamic Autoregressive Moving-Average with eXogenous inputs (ARMAX) modeling [73] is utilized to predict the behavior of the performance degradation of machines or cutters. ARMA modeling for performance degradation assessment and prediction has been previously studied [139][140]. We present a software decision tool for selecting the dominant features that are most essential in predicting time series of tool wear in industrial milling machines using an online, real-time, and indirect approach, with data from installed AE sensors.

The proposed DFI methodologies are tested on the ball nose cutter of an industrial high speed milling machine. Force measurements are taken over a time period using a three axis dynamometer, and AE measurements were taken over a time period using an AE sensor. During the measuring period, the tool was periodically extracted from the chuck and tool wear was measured using an LECIA MZ12.5 microscope. This yielded a baseline time plot of actual tool wear versus time. Sixteen features, commonly used for machinery monitoring in industries, were computed from the measured force and AE data. The DFI method was then used to select various sets of dominant features, which were used a regression model to predict the baseline observed tool wear using the ARMAX model. The DFI performance was evaluated based on the accuracy of prediction of the actual tool wear. Comparisons are made with another technique for feature selection in the literature [129][130], and using force sensors with a dynamometer with recursive least squares in [37].

Rotary machines-—consisting of numerous components such as electric motors, shafts, bearings, gears, and belt drives, etc., are commonly used in many industries in various scales, ranging from cooling fans to vehicle engines and even power plant generators in ascending order of scale. A fault might occur at each of these components, which include stator winding faults [141], rotor [142], misaligned shafts [143], loose belts [144], cracked bearings [145][146], and gears, etc. These faults are by no means exhaustive, which increase exponentially with the complexity of rotary machines. As such, vast amounts of time and effort have been invested to develop rotary machines with higher efficiencies and improved robustness to prevent down time and loss of productivity. However, mechanical faults arise due to wear and tear from prolonged operations over time, causing the machines to operate at lower efficiencies with undesirable effects such as excessive vibrations and noises. In more severe instances, the machines might even experience critical failure and breakdown, even posing possible work hazards to their operators.

Numerous machine parameters have been used to form the basis of machine health assessment, including motor speeds, input currents, transmitted torques, and vibration parameters such as accelerations, velocities, and displacements, etc. Each of these parameters might produce characteristic features that can be used for fault detection at the data processing phase. Also, many methods like statistical moment computation and spectral analysis of various machine parameters [147], Hilbert-Huang Transform (HHT) [148], minimum variance cepstrum, amplitude modulation and envelope analysis [149], Fast Fourier Transform (FFT), statistical computations, wavelets [150][151], etc., have been used for feature calculation and extraction. In current literature, the common types of faults include imbalance fault, loose belt fault, bearing faults, and variations in resonant frequencies. Much research has been conducted to improve the detection and identification of rotary machine faults to ensure diagnosis and prognosis can be taken in time before a critical machine failure. The components that are causal to the faults are also isolated, eliminating the need of unnecessary replacements of other functional components and human intervention.

With these motivation in mind, we will apply the proposed DFI to these realistic industrial systems for intelligent fault diagnosis and prognosis as well as FDI. The rest of the chapter is organized as follows. Section 4.2 reviews the theoretical preliminaries of PCA using SVD. Section 4.3 presents the proposed Dominant Feature Identification (DFI) algorithm, using SVD to decompose the inner product matrix and then clustering to identify the dominant features. It is proven that the proposed method of DFI is optimal in the sense that it minimizes the total least-squares estimation error. Section 4.4 applies the proposed DFI methodology on an industrial cutting tool for tool wear prediction. The dominant features selected from the force signals are used for tool wear times series prediction using the Recursive Least Squares (RLS) algorithm and static models. In Section 4.5, we present the usage of a dynamic ARMAX model with Extended Least Squares (ELS) for tool wear prediction in the new DFI framework. Detailed discussions on the improved experimental setup with the AE sensor and corresponding calculation of dominant features are also provided, together with experimental results and comparisons with various previous works. DFI and Neural Networks (NNs) are used in a two-stage framework for industrial Fault Detection and Isolation (FDI) applications in Section 4.6. Our conclusion and future work directions are summarized in Section 4.7.

4.2 PRINCIPAL COMPONENT ANALYSIS (PCA)

In this section, the theoretical fundamentals of PCA using SVD are reviewed in a rigorous manner, which is essential for selecting dominant features in Section 4.3. For simplicity but without loss of generality, all vectors are denoted in lower case and all matrices are denoted in upper case.

The SVD of a linear transformation X where $X \in \mathbb{R}^{m \times n}$ of rank $n < m$ is

$$X = U \Sigma V^T, \tag{4.1}$$

with $U \in \mathbb{R}^{m \times n}$ and $V \in \mathbb{R}^{n \times n}$, such that $U^T U = V^T V = I_n$ with $I_n \in \mathbb{R}^{n \times n}$ being an identity matrix of dimension n. $\Sigma \in \mathbb{R}^{n \times n}$ is a diagonal matrix whose elements are corresponding singular values (principal gains) arranged in descending order, i.e., with $\Sigma = \text{diag}(\sigma_1, \sigma_2, \cdots, \sigma_n)$ and $\sigma_1 \geq \sigma_2 \geq \cdots \geq \sigma_n > 0$.

4.2.1 APPROXIMATION OF LINEAR TRANSFORMATION X

X can be regarded as a transformation from feature space $\mathbb{R}^n$ into data space $\mathbb{R}^m$. Note that $X = \sum_{i=1}^{n} \sigma_i u_i v_i^T$, where u_i are the column vectors of U and v_i^T are the row vectors of V^T, respectively. Partition the SVD of X according to

$$\begin{aligned} X &= \begin{bmatrix} U_1 & U_2 \end{bmatrix} \begin{bmatrix} \Sigma_1 & 0 \\ 0 & \Sigma_2 \end{bmatrix} \begin{bmatrix} V_1^T \\ V_2^T \end{bmatrix} \\ &= U_1 \Sigma_1 V_1^T + U_2 \Sigma_2 V_2^T, \end{aligned} \tag{4.2}$$

with $q < m$ as the desired number of singular values to be retained in Σ_1 for data space of dimension m. As such, Σ_2 contains the $n - q$ discarded singular values.

Obviously, $U_1 \in \mathbb{R}^{m\times q}$, $U_2 \in \mathbb{R}^{m\times(n-q)}$, $\Sigma_1 \in \mathbb{R}^{q\times q}$, $\Sigma_2 \in \mathbb{R}^{(n-q)\times(n-q)}$, $V_1^T \in \mathbb{R}^{q\times n}$, and $V_2^T \in \mathbb{R}^{(n-q)\times n}$.

Now the approximation $\hat{X}$ to X is

$$\hat{X} = U_1\Sigma_1 V_1^T. \tag{4.3}$$

Then $\hat{X} = \sum_{i=1}^{q} \sigma_i u_i v_i^T$ contains the columns u_i of U_1 and the rows v_i^T of V_1^T. The dominant singular values, i.e., the q retained singular values, and their associated columns of U are called principal components in PCA [126]. PCA is also commonly known as the Karhunen-Loève Transform (KLT) in signal processing and communication applications.

The error induced $\tilde{X}$ by the approximation $\hat{X}$ of the linear transformation X is given by

$$\begin{aligned} \tilde{X} &= X - \hat{X} \\ &= U_2\Sigma_2 V_2^T \\ &= \sum_{i=q+1}^{n} \sigma_i u_i v_i^T. \end{aligned} \tag{4.4}$$

The covariance matrix of the approximation error is

$$\begin{aligned} P_{\tilde{X}} &= (X - \hat{X})(X - \hat{X})^T \\ &= U_2\Sigma_2 V_2^T V_2 \Sigma_2 U_2^T \\ &= U_2\Sigma_2^2 U_2^T, \end{aligned} \tag{4.5}$$

and the 2-norm of the approximation error is given by

$$\begin{aligned} \mathrm{tr}\{P_{\tilde{X}}\} &= \mathrm{tr}\{U_2\Sigma_2^2 U_2^T\} \\ &= \mathrm{tr}\{\Sigma_2^2 U_2^T U_2\} \\ &= \mathrm{tr}\{\Sigma_2^2\} \\ &= \sum_{i=q+1}^{n} \sigma_i^2, \end{aligned} \tag{4.6}$$

where $\mathrm{tr}\{.\}$ denotes the trace operation. This is the sum of squares of the neglected singular values. It can be shown that this SVD approximation gives the Least-Square Error (LSE) of any approximation to X of rank q.

4.2.2 APPROXIMATION IN RANGE SPACE BY PRINCIPAL COMPONENTS

Now, we identify $\mathbb{R}^n$ as the feature space and $\mathbb{R}^m$ as the range space of X which we term the data space. Consider an arbitrary vector $x \in \mathbb{R}^n$ being mapped onto a vector $z \in \mathbb{R}^m$ by X according to $z = Xx$. As such, z in the singular value space (range space) of $\mathbb{R}^m$ can also be represented according to the partitioned singular value matrix in (4.2) as

$$\begin{aligned} z &= Xx \\ &= U_1\Sigma_1 V_1^T x + U_2\Sigma_2 V_2^T x, \end{aligned} \tag{4.7}$$

with Σ_1 containing the retained q singular values of X.

An approximation to z is $\hat{z}$ given in terms of the q retained singular values as

$$\begin{aligned} \hat{z} &= U_1 \Sigma_1 V_1^T x \\ &= \hat{X} x, \end{aligned} \tag{4.8}$$

with $\hat{X}$ being the approximation of X. Note that $\hat{z} = \sum_{i=1}^{q} u_i(\sigma_i v_i^T x)$ which expresses $\hat{z}$ as a linear combination of principal components u_i with coefficients $(\sigma_i v_i^T x)$.

The approximation error is given by

$$\begin{aligned} \tilde{z} &= z - \hat{z} \\ &= (X - \hat{X})x \\ &= U_2 \Sigma_2 V_2^T x, \end{aligned} \tag{4.9}$$

and the approximation error 2-norm is given by

$$\begin{aligned} \tilde{z}^T \tilde{z} &= x^T V_2 \Sigma_2^2 V_2^T x \\ \therefore ||\tilde{z}||^2 &= \mathrm{tr}\{\Sigma_2^2 V_2^T x x^T V_2\} \\ &\leq \sigma_{q+1}^2 x^T V_2 V_2^T x \\ &= \sigma_{q+1}^2 ||x||^2 \\ &\leq \mathrm{tr}\{\Sigma_2^2\} ||x||^2. \end{aligned} \tag{4.10}$$

4.3 DOMINANT FEATURE IDENTIFICATION (DFI)

In this section, the proposed Dominant Feature Identification (DFI) methodology of using SVD to identify the dominant features is detailed. It is shown that the proposed method of DFI is optimal in the sense that it minimizes the total least-squares estimation error. Note that traditional PCA is performed with respect to data space $\mathbb{R}^m$, but the features, however, reside in $\mathbb{R}^n$. It is important to select only the most relevant features for data compression, since the correct selection will result in reduced signal processing requirements, and even in elimination of some sensors that do not yield relevant information.

4.3.1 DATA COMPRESSION

Select

$$Y = U_1^T X \in \mathbb{R}^{q \times n}, \tag{4.11}$$

so that $x \in \mathbb{R}^n$ is mapped to $y = Yx = U_1^T X x \in \mathbb{R}^q$. Then vectors $z = Xx \in \mathbb{R}^m$ may be approximated in terms of vectors $y \in \mathbb{R}^q$ according to

$$\hat{z} = U_1 y. \tag{4.12}$$

Combining with (4.2), we get an approximation of $z \in \mathbb{R}^m$ in terms of vectors in the reduced space $\mathbb{R}^q$ depicted by (4.11). *viz.*

$$\begin{aligned}\hat{z} &= U_1 y \\ &= U_1 U_1^T X x \\ &= (U_1 U_1^T)(U_1 \Sigma_1 V_1^T + U_2 \Sigma_2 V_2^T)x \\ &= U_1 \Sigma_1 V_1^T x \\ &= \hat{X} x, \end{aligned} \tag{4.13}$$

with $\hat{X}$ depicted in (4.3).

It is well known in the literature that (4.11) and (4.13) provide the best approximation of data vectors in $\mathbb{R}^m$ in terms of reduced vectors $y \in \mathbb{R}^q$. The approximation error $\tilde{z}$ is given by

$$\begin{aligned}\tilde{z} &= (X - \hat{X})x \\ &= \tilde{X} x \\ &= U_2 \Sigma_2 V_2^T x, \end{aligned} \tag{4.14}$$

which is exactly as (4.9). In fact, note that

$$\begin{aligned}\hat{z}^T \tilde{z} &= (U_1 \Sigma_1 V_1^T x)^T U_2 \Sigma_2 V_2^T x \\ &= x^T V_1 \Sigma_1 U_1^T U_2 \Sigma_2 V_2^T x \\ &= 0, \end{aligned} \tag{4.15}$$

i.e., $\hat{z}$ is orthogonal to $\tilde{z}$ which implies that (4.11) is the optimal choice of $\mathbb{R}^q$ for Least Squared Error (LSE) approximation of vectors in $\mathbb{R}^m$ by (4.12).

Note that in the approximation of $z \in \mathbb{R}^m$ using $y \in \mathbb{R}^q$, the reduced space is equivalent to approximation by principal components in (4.8).

4.3.2 SELECTION OF DOMINANT FEATURES

For any $q > 0$, the selection of the first q singular values of X yields a reduced space $\mathbb{R}^q$ generated by the linear transformation $U_1^T : \quad \mathbb{R}^n \to \mathbb{R}^q$ as in (4.11). Moreover, $y = Yx$ and

$$\hat{z} = U_1 y \tag{4.16}$$

best approximates $z = Xx$ in a least squares sense. This allows us to use reduced vectors $y \in \mathbb{R}^q$ to compute approximations to data vectors $z \in \mathbb{R}^m$, instead of using the full feature vector $x \in \mathbb{R}^n$.

Now we wish to approximate further by selecting the dominant features in $\mathbb{R}^n$. That is, it is desired to select the most important basis vectors from $\mathbb{R}^n$ to approximate the data vectors z. To do so, it is instrumental to note the little-realized fact that

$$\begin{aligned}Y &= U_1^T X \\ &= U_1^T (U_1 \Sigma_1 V_1^T + U_2 \Sigma_2 V_2^T) \\ &= \Sigma_1 V_1^T. \end{aligned} \tag{4.17}$$

The original basis vectors in feature space $\mathbb{R}^n$ are known as features, with the ith basis vector corresponding to the ith feature. The original basis vectors are denoted in $\mathbb{R}^n$ by $\{e_1, e_2, \cdots, e_n\}$ with e_i being the ith column of I_n, i.e., e_i is an n-vector consisting of zeros except for one in the ith position.

In terms of these notions, the vector generated in $\mathbb{R}^q$ by the ith feature $e_i \in \mathbb{R}^n$ is given by

$$\begin{aligned} Ye_i &= U_1^T X e_i \\ &= \Sigma_1 V_1^T e_i. \end{aligned} \tag{4.18}$$

Recall that the rows of V_1^T are denoted by row vectors v_i^T, i.e., the columns of V_1 are denoted by column vectors v_i. By contrast, denote now the *columns* of $\Sigma_1 V_1^T$ as column vectors $w_i \in \mathbb{R}^q$. Then

$$\begin{aligned} Ye_i &= \Sigma_1 V_1^T e_i \\ &= \begin{bmatrix} w_1 & w_2 & \cdots & w_n \end{bmatrix} e_i \\ &= w_i. \end{aligned} \tag{4.19}$$

Therefore, the ith feature in $\mathbb{R}^n$ maps into the reduced space $\mathbb{R}^q$ as the ith column of matrix $\Sigma_1 V_1^T$. There are therefore n vectors w_i in $\mathbb{R}^q$ corresponding to the n basis axes e_i, i.e., features, in $\mathbb{R}^n$.

The above-mentioned notions of the proposed DFI algorithm are summarized in Figure 4.1.

We now want to select the best features to retain so as to obtain the best approximation to $z \in \mathbb{R}^m$. We call these dominant features. This corresponds to selecting which basis vectors e_i in $\mathbb{R}^n$ to retain, which is equivalent to *selecting the best columns* w_i *of* $\Sigma_1 V_1^T \in \mathbb{R}^q$. This may be accomplished by several methods, including projections; we use clustering methods inspired by [129][130]. Then, $z \in \mathbb{R}^m$ will be approximated using the selected p dominant features within $\mathbb{R}^n$.

Note that we will cluster the n columns w_i of $\Sigma_1 V_1^T$, as dictated by (4.19). This is in contrast to [129][130] who clustered the columns of U_1^T in (4.11).

Clustering is the classification of n objects in a data set into p different subsets (clusters), usually by minimizing some norms or pre-defined performance indices. Data clustering is commonly used in statistical data and random signal analysis, e.g., data mining, pattern recognition, and image processing applications, etc., using Fuzzy c-means [152], quality threshold [153], or NNs [154], etc.

To select the dominant features in $\mathbb{R}^n$, we cluster the n vectors $w_i \in \mathbb{R}^q$ into $n \geq p \geq q$ clusters. For our application, the commonly used K-means algorithm is used [155]. The K-means algorithm minimizes the following positive semi-definite scalar error cost function J iteratively

$$J = \sum_{i=1}^{p} \sum_{w_j \in S_i} (w_j - c_i)^T (w_j - c_i), \tag{4.20}$$

where S_i is the ith cluster set, and c_i is its centroid (or center of "mass") in the cluster space. J is in essence the expectation of the 2-norm (or Euclidian distance) between

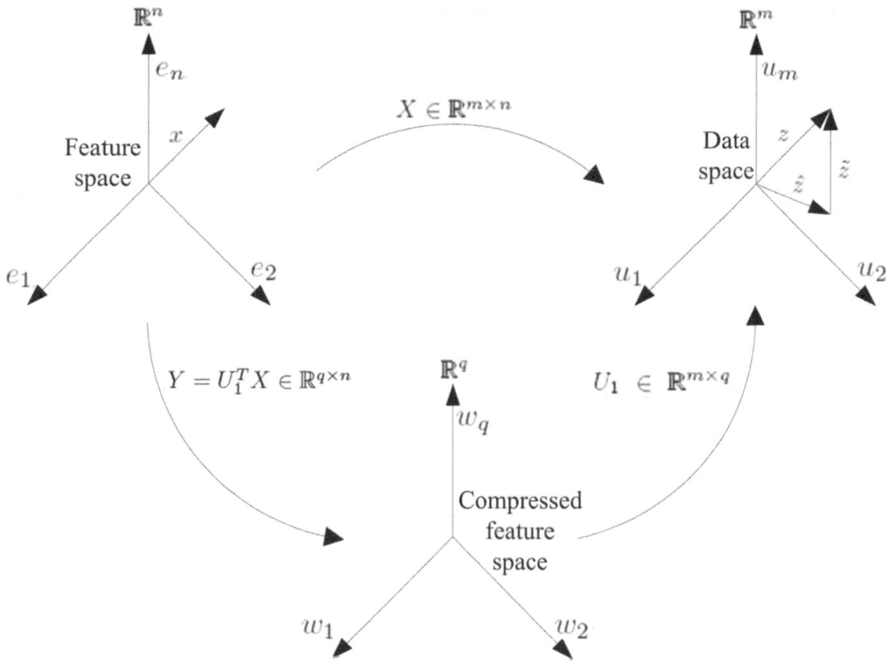

FIGURE 4.1 Proposed DFI algorithm showing feature space $\mathbb{R}^n$, compressed feature space $\mathbb{R}^q$, and data (singular value) space $\mathbb{R}^m$.

the objects in the cluster. Note that the K-means algorithm relies heavily on input parameter p, and a poor choice of p usually results in inferior clustering results, which depends heavily on the variance of the data in each cluster. For good approximations in $\mathbb{R}^m$, one should select $p > q$, the number of retained singular values.

4.3.3 ERROR ANALYSIS

Here we determine the total error induced by retaining only q singular values and by clustering the vectors $w_i \in \mathbb{R}^q$ into p clusters. It is shown that the proposed method of DFI is optimal in the sense that it minimizes the total least-squares estimation error.

For each cluster, we shall select the vector $\bar{w}_i \in \mathbb{R}^q$ closest to the cluster center c_i as representative of each other vector $w_j \in \mathbb{R}^q$ in that cluster. We call this representative vector for cluster i, $\bar{w}_i$, the cluster leader. The p features $\bar{e}_i \in \mathbb{R}^n$ corresponding to the p cluster leaders $\bar{w}_i \in \mathbb{R}^q$ shall be selected as dominant features. This means that the clustering error is given by

$$\hat{J} = \sum_{i=1}^{p} \sum_{w_j \in S_i} (w_j - \bar{w}_i)^T (w_j - \bar{w}_i). \tag{4.21}$$

To summarize these notions, recall that

$$\begin{aligned} Y &= \Sigma_1 V_1^T \\ &= \begin{bmatrix} w_1 & w_2 & \cdots & w_n \end{bmatrix}, \end{aligned} \tag{4.22}$$

and define

$$\hat{Y} = \begin{bmatrix} \hat{w}_1 & \hat{w}_2 & \cdots & \hat{w}_n \end{bmatrix}, \tag{4.23}$$

where $\hat{w}_j = \bar{w}_i$ if $w_j \in S_i$, i.e., each vector $w_j \in S_i$ is replaced by its cluster leader $\bar{w}_i$.

This means that only the corresponding features $\bar{e}_i \in \mathbb{R}^n$ are needed for computation since $\bar{w}_i = \Sigma_1 V_1^T \bar{e}_i$.

Note that by (4.19)

$$\begin{aligned} y &= \Sigma_1 V_1^T x \\ &= Yx \\ &= \sum_{j=1}^{n} w_j x_j \\ &= \sum_{j=1}^{n} x_j \left[\Sigma_1 V_1^T e_j \right], \end{aligned} \tag{4.24}$$

and define

$$\begin{aligned} \hat{y} &= \hat{Y}x \\ &= \sum_{j=1}^{n} \hat{w}_j x_j \\ &= \sum_{i=1}^{p} \left(\sum_{w_j \in S_i} x_j \right) \bar{w}_i. \end{aligned} \tag{4.25}$$

Then an estimate for $z \in \mathbb{R}^m$ taking into account both $q < n$ retained singular values and $p < n$ features is given by

$$\hat{\hat{z}} = U_1 \hat{y}. \tag{4.26}$$

Recall from (4.16) that $\hat{z} = U_1 y$, so the error induced by K-means clustering is

$$\begin{aligned} \hat{z} - \hat{\hat{z}} &= U_1(y - \hat{y}) \\ &= U_1(Y - \hat{Y})x \\ &= U_1 \tilde{Y} x. \end{aligned} \tag{4.27}$$

Therefore, the error norm induced by clustering is

$$\begin{aligned} \|\hat{z} - \hat{\hat{z}}\|^2 &= (\hat{z} - \hat{\hat{z}})^T (\hat{z} - \hat{\hat{z}}) \\ &= x^T \tilde{Y}^T U_1^T U_1 \tilde{Y} x \\ &= \mathrm{tr}\{\tilde{Y}^T U_1^T U_1 \tilde{Y} x x^T\} \\ &\leq \hat{J} \|x\|^2, \end{aligned} \tag{4.28}$$

since $\hat{J} = \text{tr}\{\tilde{Y}^T\tilde{Y}\}$.

The total error induced by neglecting the $n-q$ singular values in Σ_2 and by K-means clustering is then

$$\begin{aligned} z - \hat{\hat{z}} &= (z - \hat{z}) + (\hat{z} - \hat{\hat{z}}) \\ &= U_2\Sigma_2 V_2^T x + U_1\tilde{Y}x. \end{aligned} \tag{4.29}$$

Therefore, the total approximation error norm is

$$||z - \hat{\hat{z}}||^2 \leq (\text{tr}\{\Sigma_2^2\} + \hat{J})||x||^2, \tag{4.30}$$

whose first term depends on the neglected singular values, and the second term is the clustering error, i.e., the neglected features.

We claim that the procedure of first selecting q principle components and then selecting p dominant features yields the minimum overall approximation error in (4.29). For note that

$$\begin{aligned} \tilde{z}^T(\hat{z} - \hat{\hat{z}}) &= (U_2\Sigma_2 V_2^T x)^T U_1\tilde{Y}x \\ &= x^T V_2\Sigma_2 U_2^T U_1\tilde{Y}x \\ &= 0, \end{aligned} \tag{4.31}$$

i.e., the error in neglecting $n-q$ singular values and the clustering error are orthogonal. This means that there is no better way of selecting dominant features than the DFI methodology proposed therein.

4.3.4 SIMPLIFIED COMPUTATIONS

Traditional PCA relies on computations using the correlation matrix $XX^T = U\Sigma^2U^T \in \mathbb{R}^{m\times m}$ [126]. This is computationally expensive since generally $m \gg n$.

Defining the inner product matrix as X^TX, we get

$$\begin{aligned} X^TX &= V\Sigma U^T U\Sigma V^T \\ &= V\Sigma^2 V^T. \end{aligned} \tag{4.32}$$

As $X^TX \in \mathbb{R}^{n\times n}$ with $n \ll m$, the computation of Σ_1 and V_1^T required to find Y in (4.17) is highly simplified using (4.32).

4.4 TIME SERIES FORECASTING USING FORCE SIGNALS AND STATIC MODELS

We now wish to apply DFI to prediction of tool wear in an industrial cutting tool. Tool wear can only be measured by removing the tool and performing visual inspection and measurement, which is tedious and time consuming and results in downtime for the machine. We wish to predict tool wear using signals that are easily monitored in real time.

In our application, sensors are used to measure cutting force online in real time. Then, signal processing is used to compute n features, such as mean force value, maximum force level, standard deviation, third moment (skew), etc. Data is taken over N time steps and stored, and consists of the values of the n features at each time step, along with the tool wear measured (by visual inspection) at each time step.

To put this into the framework just discussed for DFI, define $f_k \in \mathbb{R}^n$ as the feature vector and d_k as the tool wear at time k. It is desired to predict d_k in terms of f_k. One has the standard linear regression form

$$\begin{bmatrix} f_1^T \\ f_2^T \\ \vdots \\ f_k^T \end{bmatrix} \theta = \begin{bmatrix} d_1 \\ d_2 \\ \vdots \\ d_k \end{bmatrix}, \tag{4.33}$$

where $\theta \in \mathbb{R}^n$ is the unknown parameter vector that expresses the tool wear in terms of the collected data as

$$d_k = f_k^T \theta. \tag{4.34}$$

However, we desire to express the tool wear not in terms of all features, but only of the most important features. This is motivated by the fact that selection of only the dominant features allows a reduction in signal processing and possibly also in the number of sensors installed on the cutting tool.

Express (4.33) as

$$Xx = z, \tag{4.35}$$

which is (4.7). This identifies

$$X \equiv \begin{bmatrix} f_1^T \\ f_2^T \\ \vdots \\ f_k^T \end{bmatrix}, \quad z \equiv \begin{bmatrix} d_1 \\ d_2 \\ \vdots \\ d_k \end{bmatrix}, \tag{4.36}$$

and suggests the following means of using dominant features to predict the tool wear.

Partition the collected date set over N time steps into two sets. The data through time $m < N$ is used as a training set to compute the p dominant features and determine the unknown parameter vector θ as a training set. Then the remaining data from time $m+1$ to time N is used to verify the prediction accuracy as a validation set. In our application, $N = 20 \times 10^3$ samples, so a reasonable choice for m is $m = 15 \times 10^3$ samples. The number of features computed is $n = 16$.

4.4.1 DETERMINING THE DOMINANT FEATURES

Define

$$X \equiv \begin{bmatrix} f_1^T \\ f_2^T \\ \vdots \\ f_k^T \end{bmatrix} \in \mathbb{R}^{m \times n} \tag{4.37}$$

in terms of the collected data from the installed sensors through time m. Use the machinery in Sections 4.2 and 4.3 to select the p dominant features.

4.4.2 PREDICTION OF TOOL WEAR

Define $\varphi_k^T \in \mathbb{R}^p$ as the vector containing the measured dominant features at time $k \leq m$. Note that φ_k is a p-subvector of f_k. Then one desires to predict the tool wear d_k in terms of the p dominant features using

$$d_k = \varphi_k^T \theta. \tag{4.38}$$

To do so, one can estimate the parameter vector θ using the measured data as

$$\begin{bmatrix} \varphi_1^T \\ \varphi_2^T \\ \vdots \\ \varphi_k^T \end{bmatrix} \theta = \begin{bmatrix} d_1 \\ d_2 \\ \vdots \\ d_k \end{bmatrix} \tag{4.39}$$

using LS techniques. Note that only the p dominant features are used, i.e., $\theta \in \mathbb{R}^p$.

Define

$$\Phi_k \theta \equiv \begin{bmatrix} \varphi_1^T \\ \varphi_2^T \\ \vdots \\ \varphi_k^T \end{bmatrix} \theta = \begin{bmatrix} d_1 \\ d_2 \\ \vdots \\ d_k \end{bmatrix} \equiv D_k \tag{4.40}$$

in terms of the data collected. The estimation error through time k is

$$E_k = D_k - \Phi_k \theta, \tag{4.41}$$

where $\theta \in \mathbb{R}^p$ is the vector unknowns to be regressed for time series forecast of tool wear.

An LS estimation of θ which minimizes the error norm $E_k^T E_k$ is given by the standard unique batch solution

$$\theta = (\Phi_k^T \Phi_k)^{-1} \Phi_k^T D_k, \tag{4.42}$$

if there is sufficient persistent excitation, i.e., $\Phi_k^T \Phi_k$ is invertible.

To compute θ using efficient on-line recursive means, one may use Recursive Least Squares (RLS) instead of (4.42). The RLS algorithm is governed by the following equations

$$\begin{aligned} \theta(k) &= \theta(k-1) + K(k)[d_k - \varphi_k^T \theta(k-1)], \\ K(k) &= P(k-1)\varphi_k[I + \varphi_k^T P(k-1)\varphi_k]^{-1}, \\ P(k) &= [I - K(k)\varphi_k^T]P(k-1), \end{aligned} \tag{4.43}$$

where $P(k) = [\Phi^T(k)\Phi(k)]^{-1}$ and $\theta(k)$ is the estimate of θ at time k.

4.4.3 EXPERIMENTAL SETUP

In our experiment, we used a röders TEC vertical milling machine as our test bed as shown in Figure 4.2. A ball nose cutter as shown in Figure 4.3 was selected for our testing. The cutting process was performed by predefined procedures. After each cutting process, tool snapshots were taken to measure the amount of tool wear. An LEICA MZ12.5 high performance stereomicroscope was used to measure the tool wear of the cutting tool.

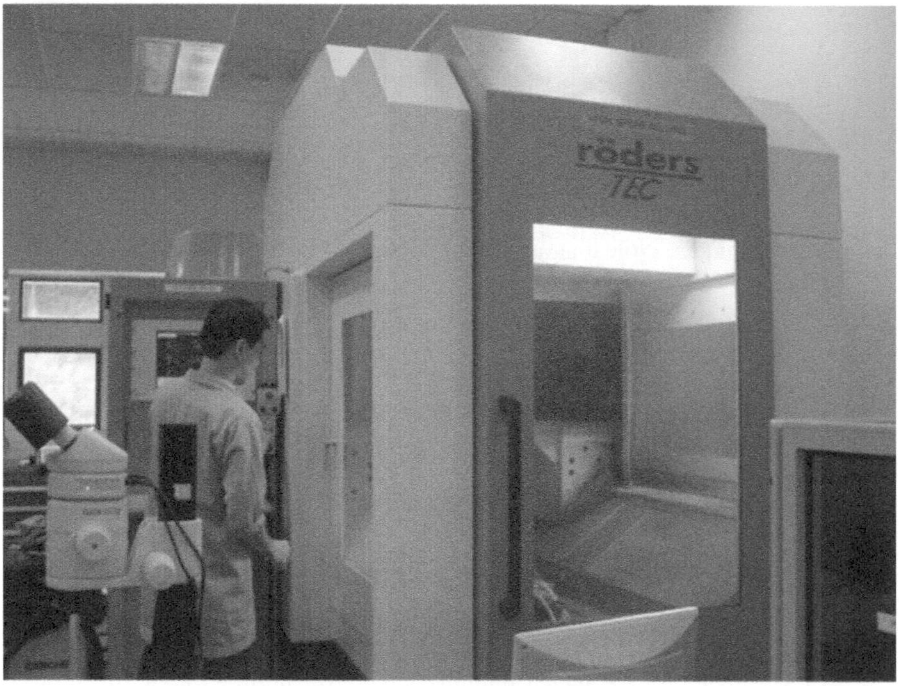

FIGURE 4.2 High speed milling machine.

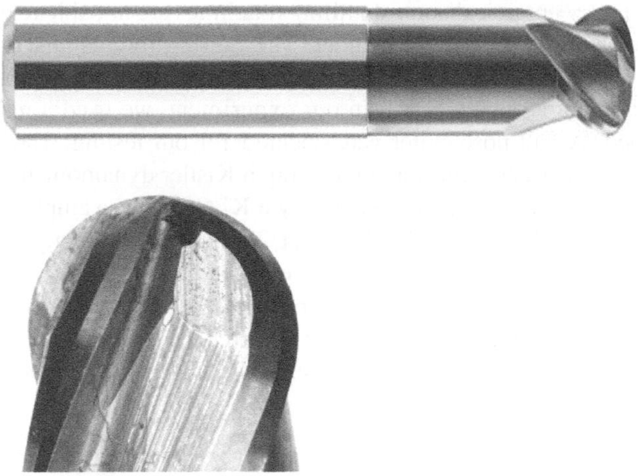

FIGURE 4.3 Ball nose cutter.

A case study is carried out to verify the usability of the method. Tool condition is an important factor in the high speed machining process. Tool wear and tool failure may result in a loss in surface finish and dimensional accuracy of the finished parts, and even possible damage to the work piece and machine [104]. An image of tool wear is shown in Figure 4.4.

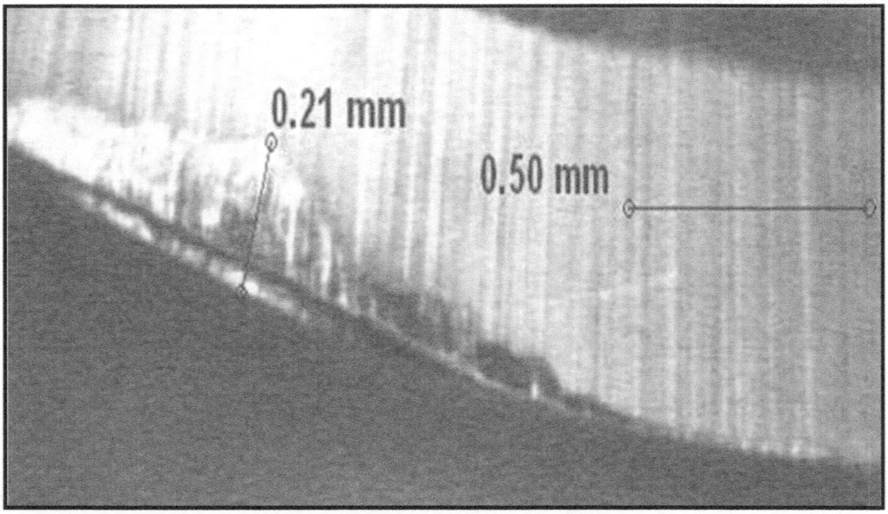

FIGURE 4.4 Flank wear at cutting edge.

Most computer numerical control milling machines are not able to detect machining tool's wear-out in an on-line manner. The cutting force signal is instead used to establish usable models due to its high sensitivity to tool wear, low noise, and good measurement accuracy [156]. In our experiment, we used a milling machine as the test bed. A ball nose cutter was selected for our testing. The cutting forces along the X, Y, and Z axes were captured using a Kistler dynamometer in the form of charges, which were converted to voltages by a Kistler charge amplifier. The voltage signal was sampled by a PCI 1200 board at 2000 Hz and directly streamed to the hard disk of a computer. The flank wear of each individual tooth of the cutting tool was measured with an Olympus microscope. Details of the experimental setup and feature extraction methodologies have been reported in [105], and are omitted here for brevity but without loss of generality. The experimental setup is shown in Figure 4.5 and its components are listed in Table 4.1.

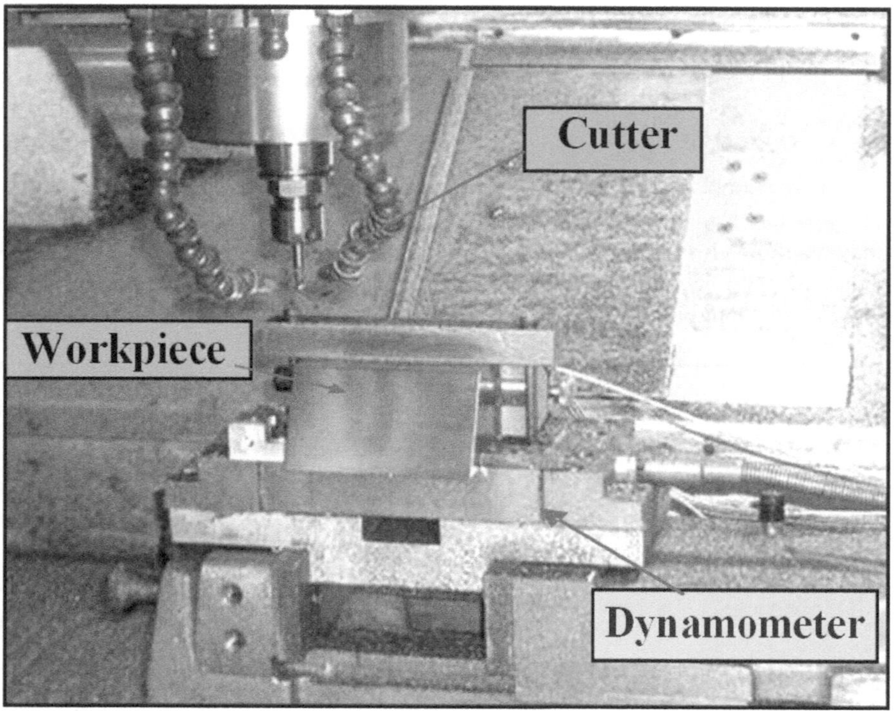

FIGURE 4.5 Experimental setup.

An example of the cutting force signal in three axes (F_x, F_y, F_z) is shown in Figure 4.6.

Various features have been proposed for monitoring the force signals in milling processes. Altintas [157] found that when a breakage occurred, a large cutting-force residual-error was produced between the actual measurement and the predicted value

TABLE 4.1
Experimental Components

Components
Makino CNC milling machine
EGD 4450R cutter with AC325 and A30N inserts
ASSAB718HH workpiece
Kistler 9265B Quartz 3-component dynamometer
Kistler 5019A multichannel charge amplifier
NI-DAQ PCI 1200 board
Olympus microscope
Computer

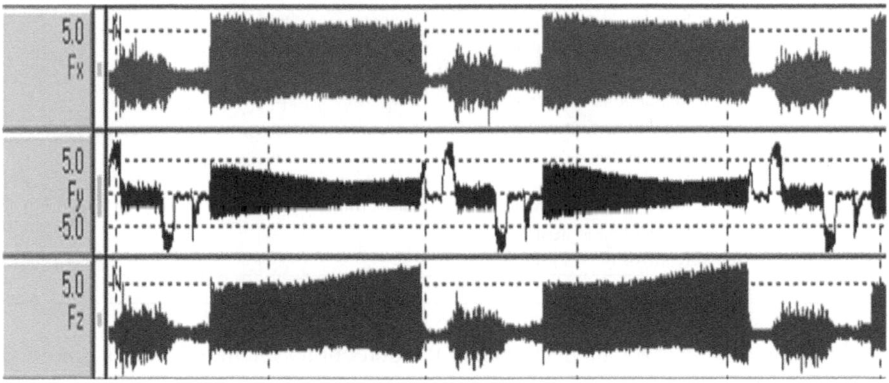

FIGURE 4.6 Three axis cutting force signals (F_x, F_y, F_z).

from the auto regressive model. In another attempt by Altintas et al. [156], time averaged resultant force first and second differencing was used to detect tool failure. In Tarn et al. [158], the four parameters of maximum force level, total amplitude of the cutting force, combined incremental force changes, and the amplitude ratio to monitor the tool conditions were used. Tarng [106] also defined a tool breakage zone as a region located within the frequency range between the DC component and the tooth frequency, and concluded that the force components within this zone correlate with the tool breakage very well. Tansel et al. [107] summed up the squares of the residual errors in each tooth period and correlated it with the tool breakage. Zhang et al. [159] used the peak rate of cutting forces to detect tool breakages. Elbestawi et al. [160] found that the harmonic contents of the cutting forces were sensitive to tool flank wear. Tarng et al. [161] used the average force and the variable force for sensing tool breakage. Leem et al. [108] extracted four statistics from the cutting force, the mean, standard deviation, skew, and kurtosis, to monitor tool wear. These sixteen features from the methodologies mentioned above are summarized in Table 4.2 and form the scope of the feature subset selection.

TABLE 4.2
Features and Nomenclature

No	Feature	Notation	References
1	Residual error	re	[157]
2	First order differencing	fod	[156]
3	Second order differencing	sod	[156]
4	Maximum force level	fm	[158]
5	Total amplitude of cutting force	fa	[158]
6	Combined incremental force changes	df	[158]
7	Amplitude ratio	ra	[158]
8	Standard deviation of force components in tool breakage zone	fstd	[106]
9	Sum of the squares of residual errors	sre	[107]
10	Peak rate of cutting forces	kpr	[159]
11	Total harmonic power	thp	[160]
12	Average force	fca	[161]
13	Variable force	vf	[161]
14	Standard deviation	std	[108]
15	Skew (3rd moment)	skew	[108]
16	Kurtosis (4th moment)	kts	[108]

The signals and raw data displayed in the detection system reflect the conditions and relevance of all the features of the cutting tool. To avoid inefficiency, which results in loss in productivity, the tool wear and part failures are estimated online without ceasing operation of the cutting tool. While it is essential that the information gathered from the input features is sufficient to determine the cutter failure, the usage of redundant input features burdens the training process, as well as lengthens the data pre-processing and recognition times for identifying key features which relate to the deterioration.

As such, there is a need for accurate deduction of the dominant features that contribute to the deterioration of the cutting tool through signals and raw data collected in the detection system. Using the proposed DFI methodology, the dominant features contributing to the deterioration of the cutting tool could be extracted for time series forecast of its wear. This allows the lifespan of the cutting tool to be predicted for precise control of operating conditions and productivity improvements.

Based on the above, the five following steps are involved in the proposed DFI to obtain a feature subset:

1. *Data acquisition.* Collect data on time interval $[0,m]$ from the cutting tools' sensors and compute n features using digital signal processing techniques. Pack n time series, each of length m, as columns into matrix $X \in \mathbb{R}^{m \times n}$.

2. *Initialization*. Detrend the data in X by subtracting the mean (average across each dimension) of each of the dimensions to normalize the data set to zero-mean. This step results in a data set whose mean is zero.
3. *Choose number of principal components and clusters*. Perform SVD on inner product matrix X^TX to obtain Σ_1 and V_1^T. Select the desired number q of principal components in $\sigma_i v_i$ and number of clusters p.
4. *Clustering*. Use the K-means algorithm for clustering to find the centroids c_i of each cluster.
5. *Subset Determination*. Select the vector w_i "nearest" to the centroid of each cluster as its cluster leader $\bar{w}_i$, and corresponding e_i as the dominant feature. Combine the p dominant features to form the reduced feature space.

This procedure is summarized in the flowchart in Figure 4.7 and applied in the experimental results of the next section.

4.4.4 EFFECTS OF DIFFERENT NUMBERS OF RETAINED SINGULAR VALUES Q AND DOMINANT FEATURES P

In the experiment, 52,800 time points of measured force sensor data were captured under the following machine settings: spindle speed 1000 rpm, feed rate 200 mm/min, depth of cut 1 mm, and insert number 2. Periodically during this period of data measurement, the tool was removed from the chuck and the tool wear was measured by hand. Specifically, the flank wear of each individual tooth of the cutting tool was measured with an Olympus microscope. This yields the baseline actual tool wear plot shown in Figures 4.9–4.11.

The measured force data was detrended, i.e., the mean value was subtracted. Based on this data, the sixteen features in Table 4.2 were computed as functions of time. This yields $n = 16$ feature vectors, each of which is a function of time and has $m = 52,800$ data points. This yields a matrix X in (4.37) that has $n = 16$ columns and $m = 52,800$ rows. Next, the SVD of the inner product matrix X^TX in (4.32) is performed.

The resulting singular values are shown in Table 4.3 and plotted in Figure 4.8. It can be seen that the fifth and subsequent singular values are quite small compared to the first four singular values. Therefore, it appears that retaining the first four singular values would be sufficient to capture the relevant trend in tool wear prediction for this experiment. However, it is not clear from this observation which of the sixteen features are the main ones useful for tool wear prediction at this point in time. We therefore use our DFI method to select the most important features.

First, q dominant singular values are selected based on the ratio between the sum of the squares of the first q singular values to the sum of the squares of all the singular values. This yields $\Sigma_1 V_1^T$ in (4.17) and (4.19). The best p columns w_i of $\Sigma_1 V_1^T$ in (4.19) are next chosen for clustering. This will yield p dominant features in $\mathbb{R}^n$, where $n \geq p \geq q$. Clustering of the data set using the K-means algorithm now yields p clusters S_i. The p vectors w_i closest to the centroids of each cluster S_i is chosen as the cluster leader $\bar{w}_i$. The corresponding p features are selected as the dominant features. These dominant features are now used for predicting the time series of failure of the

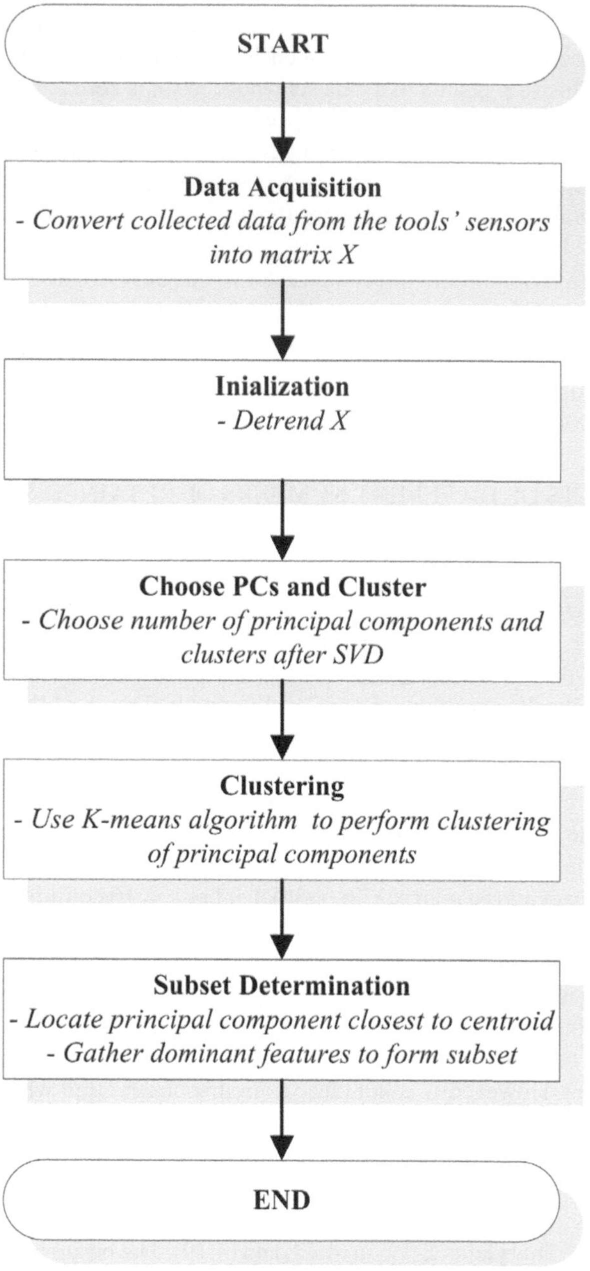

FIGURE 4.7 Flowchart of the proposed DFI for choosing the number of principal components via SVD, and clustering using the K-means algorithm to select the dominant feature subset.

TABLE 4.3
Principal Components and Singular Values

No.	Singular Values
1	31.90702
2	1.082043
3	0.00342
4	0.00026
5	0.00011
6	0.00005
7	0.00005
8	0.00001
9	0.00001
10	3.07857×10^{-6}
11	1.13154×10^{-6}
12	6.45546×10^{-7}
13	4.62940×10^{-7}
14	1.71882×10^{-7}
15	4.60490×10^{-9}
16	2.02272×10^{-9}

cutting tool.

Using the RLS techniques, a Multiple Regression Model (MRM) in (4.38) was identified to predict the baseline measured tool wear using *all sixteen* of the original features. Figure 4.9 shows the actual measured tool wear and the predicted tool wear using this MRM as functions of time. Clearly, the prediction is good. A Mean Relative Error (MRE) of 8.8% is observed for this MRM in Figure 4.9, and represents our best possible prediction of tool wear using this set of sixteen features.

It is desired to use fewer features to predict the tool wear. Figure 4.10 shows the resulting tool wear prediction when four random features in Table 4.2 are randomly selected. The "steps" in Figure 4.10 are due to the usage of these four randomly selected features (and not the four dominant features) for prediction of real measured tool wear. Here, we choose {fm, fod, sod, vf} for illustration purposes but without loss of generality. The MRE is 22.7%, which is unacceptable. Therefore, we are motivated to use our DFI method to select the important features with better justification and results.

Therefore, we select different numbers q of retained singular values and p of dominant features and perform our DFI algorithm in Section 4.3 to select the resulting p dominant features. A smaller Mean Square Error (MSE) and MRE will be obtained when more singular values are retained or more dominant features are used. The best combination of q and p, leading to the minimum number p of dominant features needed in predicting tool wear, depends on the tolerance of MSE and MRE required.

Figure 4.11 shows the result using $q = 3$ retained singular values and $p = 4$ dominant features. The tool wear prediction MRE is 11.12%. This is excellent, and is

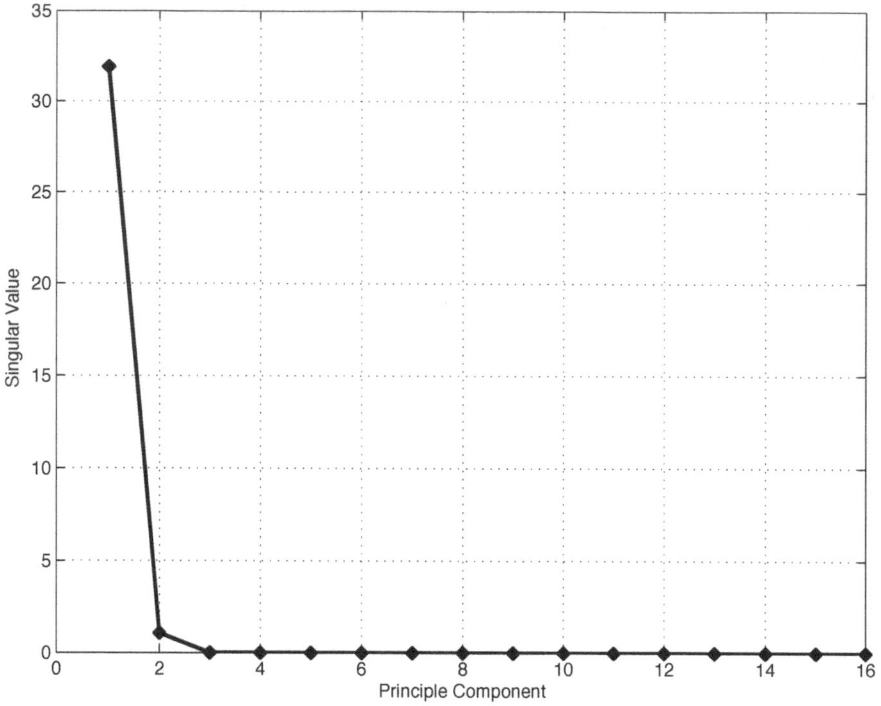

FIGURE 4.8 Plot of principal components vs. singular values.

very close to the MRE of 8.8% obtained retaining all sixteen features in Figure 4.9. An enlarged zoomed-in detailed version is shown in Figure 4.12. The four dominant features turn out to be {fa, fca, fstd, thp}.

4.4.5 COMPARISON OF PROPOSED DOMINANT FEATURE IDENTIFICATION (DFI) AND PRINCIPAL FEATURE ANALYSIS (PFA)

Next, we compared our DFI method to the PFA method of [129][130]. Figure 4.11 shows the actual measured tool wear, and its prediction using the best four features selected by DFI, and the four features selected by PFA. The MRE for DFI is 11.12%, while MRE for PFA is 13.18%. Both PFA and DFI perform better than random selected features (see Figure 4.10), and DFI obtains better performance than using PFA.

Also, comparison studies were carried out using increasing numbers q of retained singular values and p of dominant features. Comparison results of DFI and PFA using three retained singular values are shown in Table 4.4 using DFI methodology and in Table 4.5 using PFA method, respectively.

Results using four retained singular values are shown in Table 4.6 with DFI methodology and in Table 4.7 with PFA method. In these Tables, the first column denotes

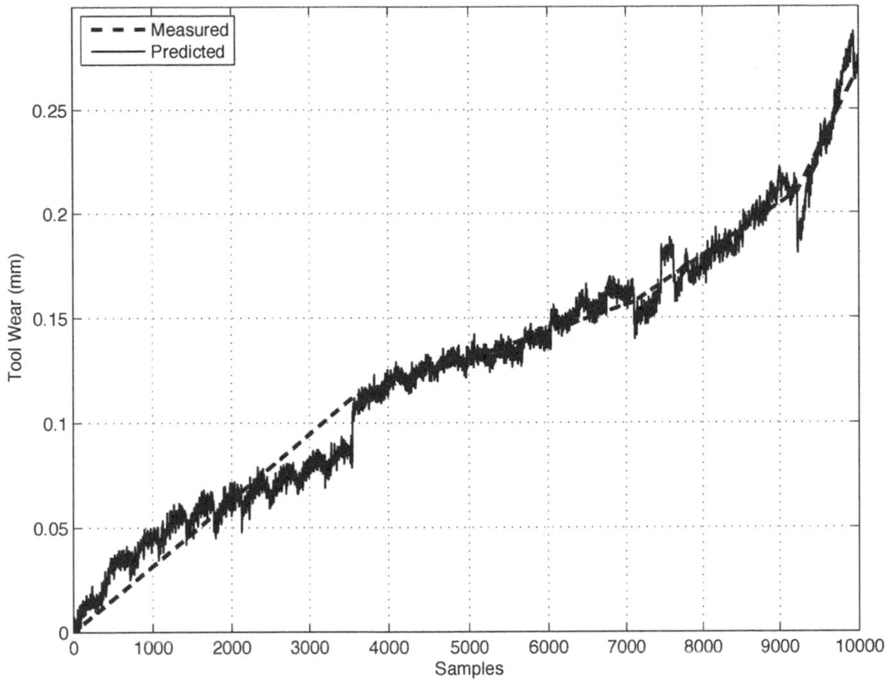

FIGURE 4.9 MRM using sixteen dominant features and the RLS algorithm.

TABLE 4.4
Results of DFI Method Using Three Retained Singular Values

No. of dominant features	Features selected	MSE (mm^2)	MRE (%)
4	fa, fca, fstd, thp	1.262	11.61
5	fa, fca, fm, skew, thp	1.202	11.19
6	fa, fca, fstd, ra, sre, thp	1.111	10.49
7	fa, fca, fstd, ra, skew, sod, thp	1.111	10.40
8	fa, fca, fstd, kts, ra, skew, sod, thp	0.946	8.86
9	fa, fca, fstd, kpr, kts, ra, skew, thp, vf	0.946	8.86

the number of dominant features to be used for tool wear prediction, and the second column displayed the selected dominant features. The MSE and MRE obtained using

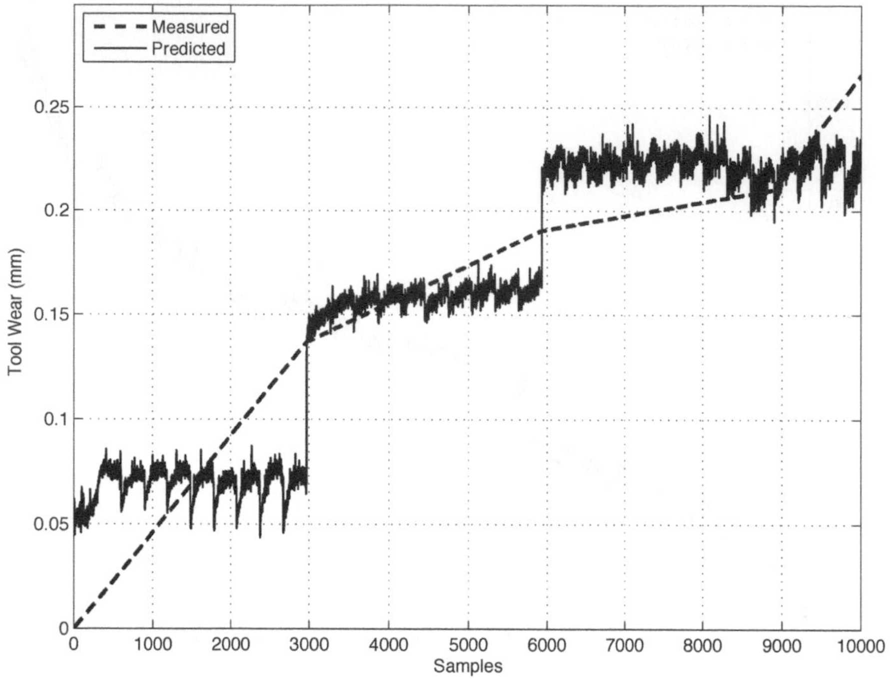

FIGURE 4.10 MRM using random selected four features {fm, fod, sod, vf} and tool wear comparison.

TABLE 4.5
Results of PFA Method Using Three Retained Singular Values

No. of dominant features	Features selected	MSE (mm^2)	MRE (%)
4	fa, fca, ra, hp	1.40	13.18
5	fa, fca, kts, ra, thp	1.365	12.89
6	fa, kts, re, skew, std, thp	1.133	10.53
7	fa, fca, fstd, kts, skew, thp, vf	1.130	10.46
8	fa, fca, fstd, kts, ra, re, skew, thp	0.949	8.88
9	fa, fca, fm, fstd, kts, ra, re, skew, thp	0.948	8.88

these dominant features for tool wear forecast are displayed in columns three and four.

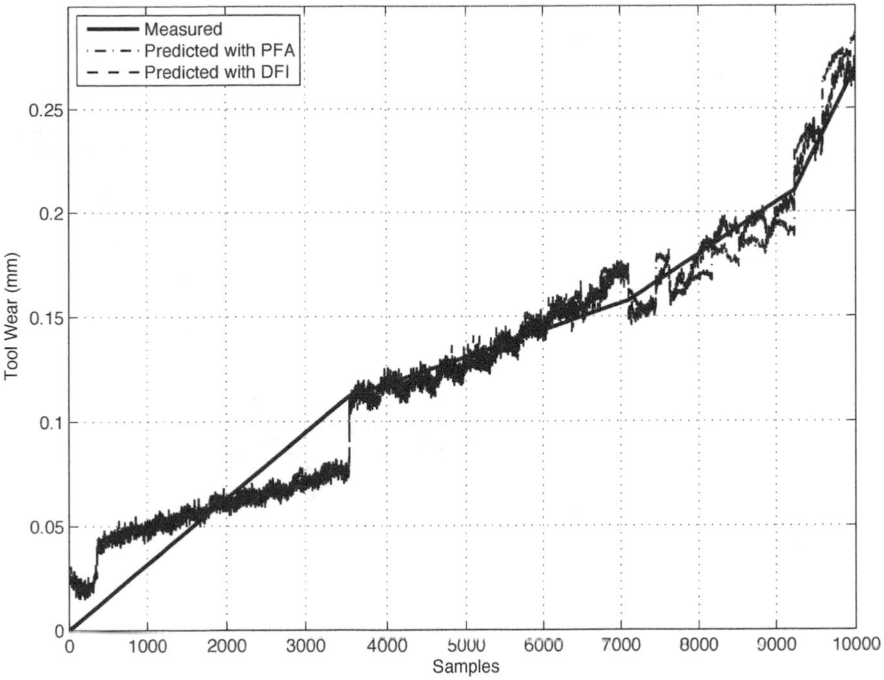

FIGURE 4.11 Examples of MRMs using four dominant features, three principal components, and the RLS algorithm.

TABLE 4.6
Results of DFI Method Using Four Retained Singular Values

No. of dominant features	Features selected	MSE (mm^2)	MRE (%)
5	fa, fca, fm, skew, thp	1.201	11.20
6	fa, fca, fstd, ra, sre, thp	1.111	10.49
7	fa, fca, fstd, ra, skew, sod, thp	1.111	10.40
8	fa, fca, fstd, kts, ra, skew, sod, thp	0.946	8.86
9	fa, fca, fstd, kpr, kts, ra, skew, thp, vf	0.946	8.86

The observations from these tables are as follows:

1. The dominant features are the same using the DFI methodology when dif-

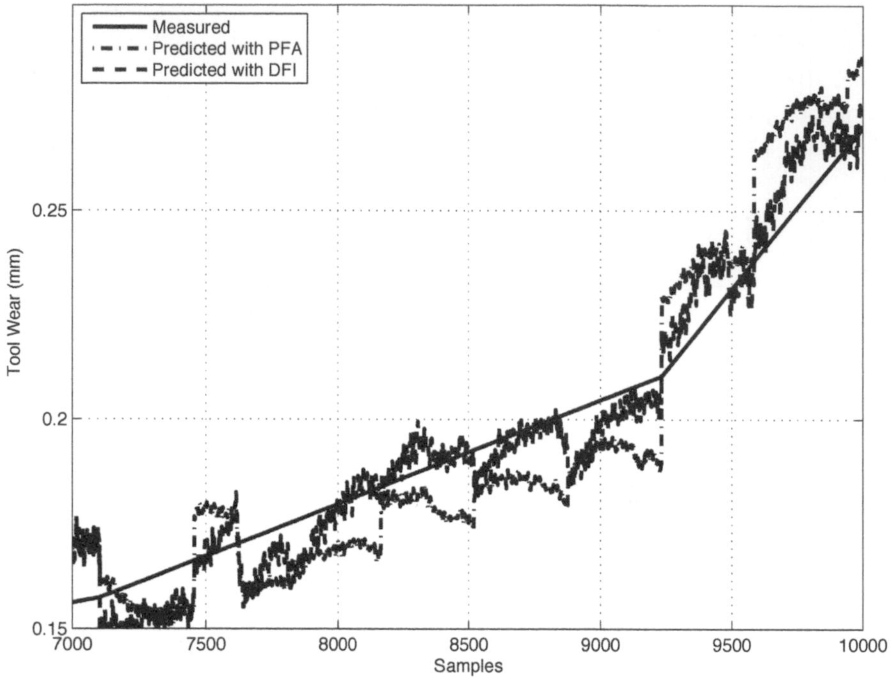

FIGURE 4.12 Examples of MRMs using four dominant features, three principal components, and the RLS algorithm (zoomed).

TABLE 4.7
Results of PFA Method Using Four Retained Singular Values

No. of dominant features	Features selected	MSE (mm^2)	MRE (%)
5	dx, fa, kts, std, thp	1.458	13.64
6	dx, fa, fca, kts, std, thp	1.215	11.19
7	fa, fm, kts, sre, skew, std, thp	1.128	10.45
8	fa, fca, fstd, kts, ra, skew, sod, thp	0.946	8.86
9	fa, fca, fstd, kpr, kts, ra, skew, thp, vf	0.946	8.86

ferent numbers of principal components are used. However, the features are slightly different using the PFA method when different numbers of principal components are used, especially when the numbers of dominant features chosen are four and five;

2. The features chosen are different using the proposed DFI methodology and PFA. DFI gives a smaller MSE and MRE when compared to that using PFA, which provides a better accuracy than PFA in tool wear prediction, especially when the numbers of dominant features chosen are four and five. When using a small number of dominant features, the computational time for features processing is saved. Our experiments also found that the computational time required to build MRM using the original sixteen features takes about five times longer of computational time, when compared to that using four dominant features only; and
3. When the number of dominant features chosen reaches to eight or more, the improvements in MSE and MRE using increasing number of dominant features are insignificant. The MRE of using eight dominant features is 8.86%, which is very close to the MRE value of 8.80% from using all of the sixteen original features. Our experiments also concluded that using the selected eight dominant features with the proposed DFI method to build MRMs saves about 60% of computational time than that is needed to build the models with original sixteen features. As such, using the selected features subset dominant features with the proposed DFI method has the advantage of implementing on-line prediction, as it saves computational time without loss of accuracy.

We have shown that our proposed DFI methodology is superior by far to conventional data compression schemes available in current literature in predicting tool wear for a realistic industrial cutting machine. To further improve the intelligence and effectiveness of its diagnostic and prognostic capabilities, we propose to embed an Acoustic Emission (AE) sensor to the measurement system. A *dynamic* Auto-Regressive Moving Average with eXogenous (ARMAX) model is also used for time series forecasting of tool wear, and the details are presented in the following section.

4.5 TIME SERIES FORECASTING USING ACOUSTIC EMISSION SIGNALS AND DYNAMIC MODELS

In this section, we show how to predict the tool wear using the dominant features (selected using DFI) as in Section 4.3. This can be achieved by using conventional static models or proposed dynamic Auto-Regressive Moving Average with eXogenous (ARMAX) models.

Similarly, we define $\begin{bmatrix} f_1 & f_2 & \cdots & f_n \end{bmatrix}$ as the matrix of features computed from the signal measured by the AE sensor. Each feature f_i is a time signal of length m samples. Select p dominant features using DFI as in the previous section. This selects p columns in X as a reduced matrix $\hat{X} \in \mathbb{R}^{m \times p}$.

It is well known that prediction of tool wear depends heavily on the cutting conditions. In current literature, tool wear is usually predicted with standard static linear prediction models, e.g., in [165][166]. The simplest and most commonly used is the linear Multiple Regression Model (MRM) of the form as shown in the previous section

$$\hat{y}(k) = \phi_k^T \hat{\hat{\theta}}(k), \tag{4.44}$$

where $\phi_k^T \in \mathbb{R}^p$ is the vector containing the measured dominant features at time $k \leq m$ (or a row vector in $\hat{\hat{X}}$ at time k) and $\hat{\hat{\theta}}(k)$ is the vector of unknown coefficients to be regressed for time series forecast of tool wear $\hat{y}(k)$. This is identical to (4.38) with $\hat{y}(k) = d(k)$, $\phi_k^T = \varphi_k^T$, and $\hat{\hat{\theta}}(k) = \theta(k)$, and different notations are intentionally perused in this section to illustrate the effectiveness of the proposed new DFI methodology.

The coefficients in $\hat{\hat{\theta}}(k)$ are estimated by conducting experiments under one cutting condition. The MRM model is not changed or updated in the online monitoring. However, in the real production, cutting conditions (spindle speed, feed rate, and depth of cut, etc.) are varied according to the production requirement. The model that was built using one cutting condition may not represent other cutting conditions. This causes the model to become inaccurate when used at different conditions.

Another common practice is to build different models under different cutting conditions. However as few of the cutting parameters can be changed separately or simultaneously, the combinations of the changes are huge, and it is time consuming to build up many different models that work under different cutting conditions.

To overcome these difficulties, we propose to use dynamic ARMAX model instead of the static MRM modeling in our study for linear regression that depends on current and past values of the features as well as also on past predicted values of tool wear. This gives the prediction model more information about the existing feature and tool wear conditions. Preliminary findings also indicate that this makes tool wear prediction more robust to changes in cutting conditions.

4.5.1 ARMAX MODEL BASED ON DFI

We now desire to predict the real tool wear $y(k)$ at time k in terms of the identified p dominant features using an ARMAX model with Extended Least Squares (ELS) methods. The general ARMAX model has the following structure [73]

$$A(z^{-1})\hat{y}(k) = \mathbf{B}(z^{-1})u(k - n_u) + C(z^{-1})\varepsilon(k), \tag{4.45}$$

where $\hat{y}(k)$ is the predicted tool wear, $u(k) = \begin{bmatrix} u_1 & u_2 & \cdots & u_n \end{bmatrix}^T$ is a column vector of measurements from the p dominant features (or a row vector in $\hat{\hat{X}}$ at time k), and n_u is the input delay; z^{-1} is the unit backward shift operator and $\varepsilon(k)$ is the estimation error. A, $\mathbf{B}$, and C are polynomials of in ascending powers of delays as

$$\begin{aligned}
A(z^{-1}) &= 1 + a_1 z^{-1} + a_2 z^{-2} + \cdots + a_{n_a} z^{-n_a}, \\
\mathbf{B}(z^{-1}) &= \begin{bmatrix} b_{11} + b_{12} z^{-1} + b_{13} z^{-2} + \cdots + b_{1n_b} z^{-n_{b1}+1} \\ b_{21} + b_{22} z^{-1} + b_{23} z^{-2} + \cdots + b_{2n_b} z^{-n_{b1}+1} \\ \vdots \\ b_{p1} + b_{p2} z^{-1} + b_{p3} z^{-2} + \cdots + b_{pn_b} z^{-n_{b1}+1} \end{bmatrix}^T, \\
C(z^{-1}) &= 1 + c_1 z^{-1} + c_2 z^{-2} + \cdots + c_{n_c} z^{-n_c},
\end{aligned} \tag{4.46}$$

where n_a, n_{pb}, and n_c are the orders of A, $\mathbf{B}$, and C, respectively.

Several assumptions are made on the ARMAX model to be valid for our application as follow:

1. $A(z^{-1})$ is assumed to be Hurwitz, i.e., it has zeros strictly inside the unit circle;
2. Multi-input $u(k)$ is assumed to be stationary and persistently exciting;
3. Evolution of coefficients of A, $\mathbf{B}$, and C are slow time-varying when compared to the sampling time of the system under consideration;
4. $\varepsilon(k)$ is uncorrelated with $u(k)$ in a statistical sense; and
5. The orders n_a, n_{pb}, and n_c, are known *a priori*.

If there is no input delay, i.e., $n_u = 0$, we can write the ARMAX model from (4.45) in time series as

$$\hat{y}(k) = \varphi_k^T \theta(k) + \varepsilon(k), \tag{4.47}$$

where

$$\begin{aligned}
\varphi_k^T &= \begin{matrix} [-\hat{y}(k-1) & \cdots & -\hat{y}(k-n_a) & u_1(k) \\ \cdots & u_1(k-n_{b1}+1) & u_2(k) & \cdots \\ u_2(k-n_{b2}+1) & \cdots & u_p(k) & \cdots \\ u_p(k-n_{b2}+1) & \varepsilon(k-1) & \cdots & \varepsilon(k-n_c)], \end{matrix} \\
\theta^T(k) &= \begin{matrix} [a_1 & \cdots & a_{na} & b_{11} \\ \cdots & b_{1n_b} & b_{21} & \cdots \\ b_{2n_b} & \cdots & b_{p1} & \cdots \\ b_{pn_b} & c_1 & \cdots & c_{n_c}]. \end{matrix}
\end{aligned} \tag{4.48}$$

The terms $\varepsilon(.)$ are unknown, but may be approximated by using the prediction errors $\hat{e}(k)$ where

$$\hat{e}(k) = \hat{y}(k) - \hat{\varphi}_k^T \hat{\theta}(k-1), \tag{4.49}$$

and all the terms on the right-hand side of (4.49) consist of past realizable signals at any time instant k. The standard Least Squares (LS) problem can now be reformulated into the Extended LS (ELS) by defining

$$\begin{array}{rcl}\varphi_k^T & = & \begin{array}{llll}[-\hat{y}(k-1) & \cdots & -\hat{y}(k-n_a) & u_1(k) \\ \cdots & u_1(k-n_{b1}+1) & u_2(k) & \cdots \\ u_2(k-n_{b2}+1) & \cdots & u_p(k) & \cdots \\ u_p(k-n_{b2}+1) & \hat{e}(k-1) & \cdots & \hat{e}(k-n_c)],\end{array} \\ \hat{\theta}^T(k) & = & \begin{array}{llll}[\hat{a}_1 & \cdots & \hat{a}_{n_a} & \hat{b}_{11} \\ \cdots & \hat{b}_{1n_b} & \hat{b}_{21} & \cdots \\ \hat{b}_{2n_b} & \cdots & \hat{b}_{p1} & \cdots \\ \hat{b}_{pn_b} & \hat{c}_1 & \cdots & \hat{c}_\varepsilon],\end{array}\end{array} \tag{4.50}$$

where n_ε is the order of $\hat{e}(k)$ used.

As such, collecting past measurable signals as

$$Y_k = \begin{bmatrix} y_1 \\ y_2 \\ \vdots \\ y_k \end{bmatrix}, \tag{4.51}$$

$$\hat{\Phi}_k = \begin{bmatrix} \hat{\varphi}_1^T \\ \hat{\varphi}_2^T \\ \vdots \\ \hat{\varphi}_k^T \end{bmatrix}, \tag{4.52}$$

the unbiased estimate of $\hat{\theta}(k)$ is given by

$$\begin{aligned}\hat{\theta}(k) &= \arg\min_{\hat{\theta}} \left|\left|[Y_k - \hat{\Phi}_k^T \hat{\theta}(k-1)]^T [Y_k - \hat{\Phi}_k^T \hat{\theta}(k-1)]\right|\right| \\ &= (\hat{\Phi}_k^T \hat{\Phi}_k)^{-1} \hat{\Phi}_k^T Y_k\end{aligned} \tag{4.53}$$

via the standard batch solution. To compute $\hat{\theta}(k)$ using efficient on-line recursive means, one may use Recursive Least Squares (RLS) depicted earlier in (4.43). The *combination of the DFI algorithm with the usage of a dynamic ARMAX model using ELS methods* is called the *new DFI* methodology.

The coefficients of the ARMAX model are updated with the real time captured sensor data and estimated using the RLS algorithm depicted earlier in (4.50) and (4.43). The model is constructed dynamically with the latest cutting information to reflect the real cutting conditions more accurately. As such, the problem of building different models under different cutting conditions (or the inaccuracies of using one model for different cutting conditions) can be solved by this approach. For the rest of the section, we will be using the dynamic ARMAX model in the new DFI methodology framework.

4.5.2 EXPERIMENTAL SETUP

Here, both the cutting force and AE signals were used to establish usable models due to their high sensitivity to tool wear, low noise, and good measurement accuracy. AE sensors were also used on a test object's surface to detect dynamic motion resulting from AE events and to convert the detected motion into a voltage-time signal. AE detection is also commonly performed with sensors that use piezoelectric elements for transduction.

For our application, an 8152B211 Piezotron® AE sensor (Kistler) was used for tool condition monitoring in the experiment as shown in Figure 4.13. Due to its small form factor, it can be easily mounted near the source of emission for optimal measurement of the AEs. The AE signal during the machining process is measured by the integrated impedance converter, which measures the AE signal from 100 kHz to 900 kHz.

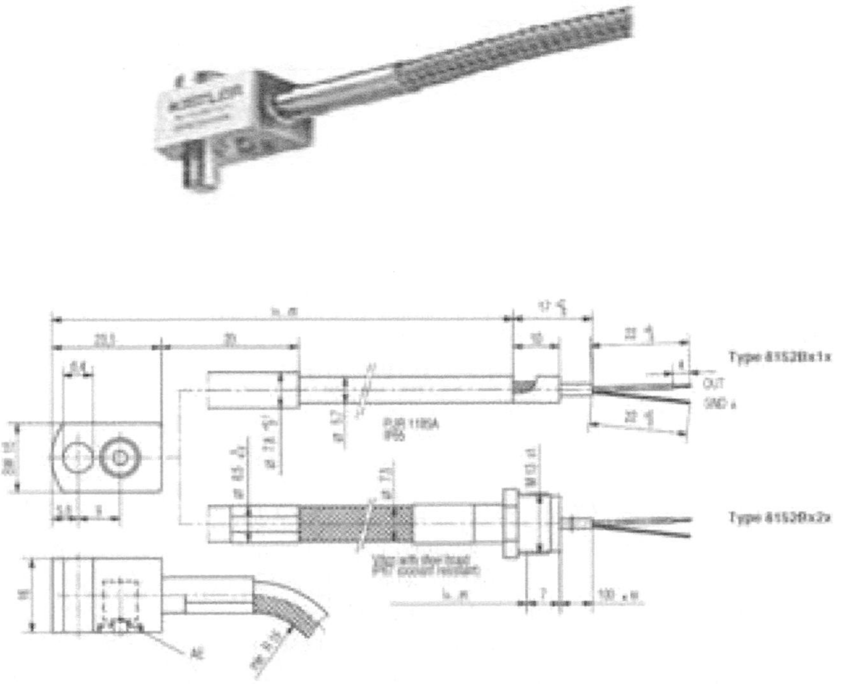

FIGURE 4.13 Piezotron Acoustic Emission Sensor used, 100 to 900 kHz.

The following guidelines suggested by Chen et al. in [162] for sensor fusion in process monitoring are also observed:

1. Measurement point should be as close to the machining point as possible;
2. No reduction in the static and dynamic stiffness of the machine tool;
3. No restriction of working space and cutting parameters;
4. Wear-free and maintenance-free, easy to change, low costs;

5. Resistant to coolant, dirt, chips and mechanical, electromagnetic, and thermal influences;
6. Function independent of tool or work piece;
7. Adequate metrological characteristics; and
8. Reliable signal transmission.

The experimental setup with both the dynamometer (force sensor) and AE sensor deployed in a TEC vertical milling machine is shown in Figure 4.14. A 6-mm ball nose tungsten carbide cutter was chosen to machine a Titanium Ti6Al4V work piece. The experimental components are summarized in Table 4.8. The AE sensor's position was chosen to be near the cutting path in order to efficiently capture the dynamic changes of the cutting process. The mounting position does not change the cutter's static and dynamic stiffness, nor does it add additional constraints on the milling process.

TABLE 4.8
Experimental Components

Components
röders TEC vertical milling machine
6mm ball nose tungsten carbide cutters
Titanium Ti6Al4V workpiece
8152B211 Piezotron® AE sensor (Kistler)
Kistler 5127B11 multichannel charge amplifier
NI-DAQ PCI 6250 M series
LECIA MZ12.5
Computer

An example of the AE signal is shown in Figure 4.15.

Our experiment was carried out with a new cutter until the tool wear became too large, i.e., broken down and damaged. The evolution of measured tool wear during the experimental stage is shown in Figure 4.16.

The feature selection of DFI was performed offline in this study. The collected data were stored in a HDD for signal conditioning and processing. In the online monitoring, only the selected features were extracted, in order to save the computational effort and also for fast data processing. The ARMAX model in (4.45) was then updated dynamically online with the new feature points. AE signals are essentially sound waves generated in solid media, affected by characteristics of the source, path taken from the source to the sensor, sensor's characteristics, and measuring system, etc. Generally, the AE signals are intricate and sensitive, and characterization of the source using AE signals is difficult. As such, information is extracted using simple waveform parameter measurements [118] and artificial intelligence approaches [122]. Advanced signal processing is also necessary for extracting the relevant features from the raw signal. In our study, the signals were firstly truncated to the desired lengths to facilitate signal processing. Next, noise in the raw signals was filtered by the wavelet de-noising

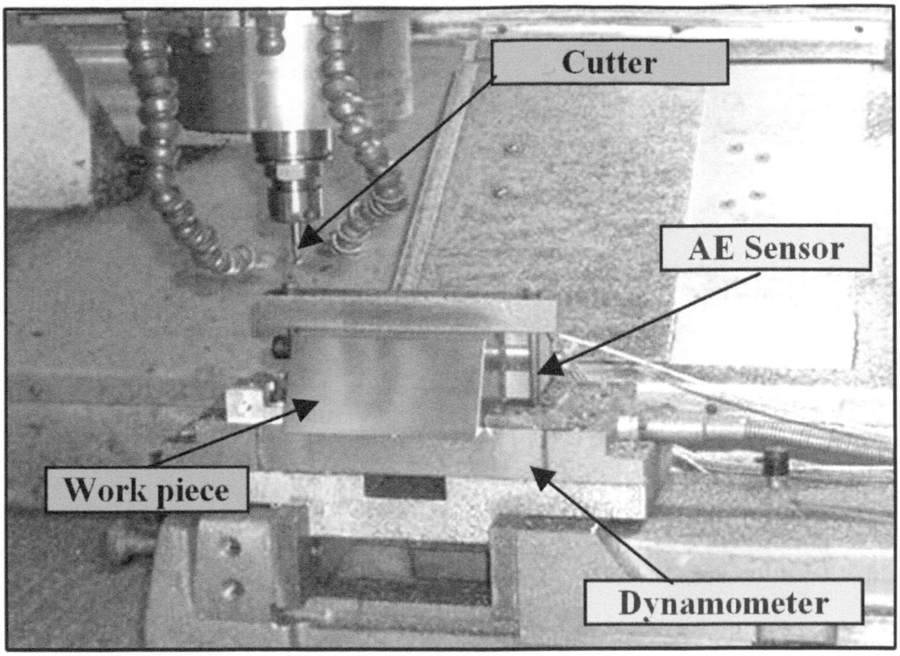

FIGURE 4.14 Experimental setup.

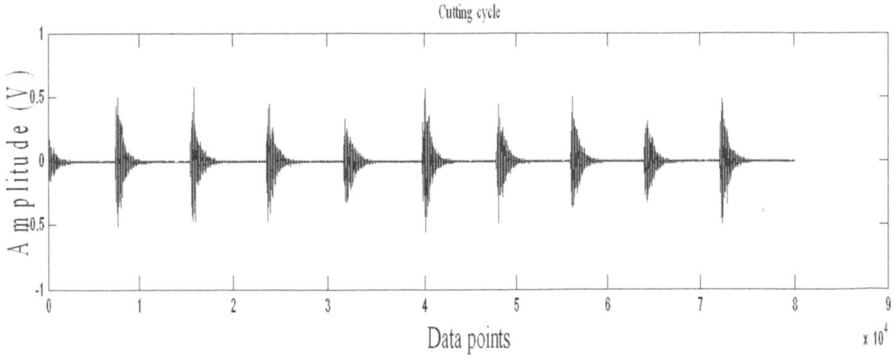

FIGURE 4.15 Unprocessed AE signal during cutting process.

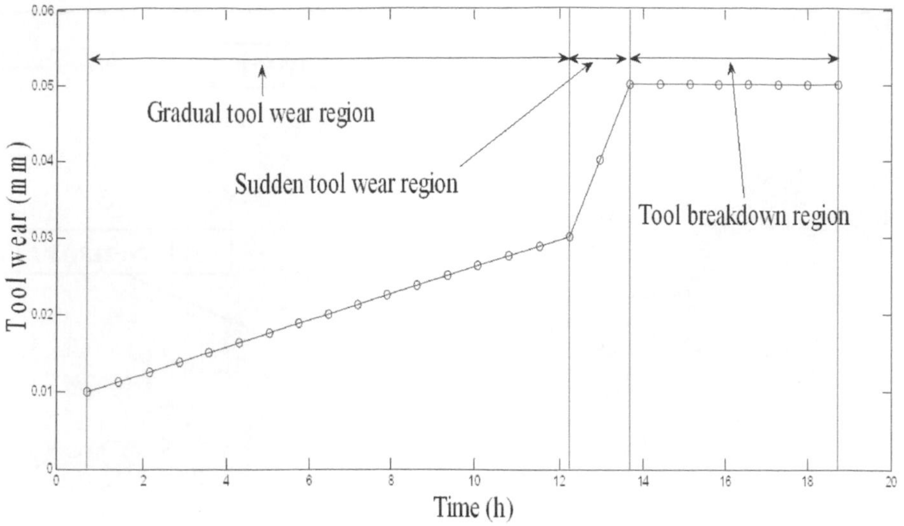

FIGURE 4.16 Stages in evolution of tool wear.

method, which consists of decomposition, identification of threshold detail coefficient, and reconstruction. The processed noise-free signal can then be reconstructed via the inverse wavelet transform of the threshold wavelet coefficients. A third level decomposition was also applied to the signals.

The characteristics of the AE signal that can be used for TCM have been studied by researchers in various aspects, e.g., spectral, statistical, and time series analysis, etc., to analyze tool wear. As such, various computed features using AE signals have been shown to be effective for monitoring milling processes [163][164]. Diniz et al. in [163] extracted eight statistics from the AE, namely skewness, kurtosis, crest-factor, peak, peak to peak, mean of RMS, mean, and standard deviation. In Sun et al. [110], the five parameters of mean of band power, standard deviation of band power, delta (first difference), and the absolute deviation were used to monitor the tool conditions. Ravindra et al. in [121] used the rise time, ring down count to detect tool breakages. Sunilkumar et al. in [164] used area under curve and duration to monitor the tool conditions. As such, these sixteen features from the methodologies mentioned above are summarized in Table 4.9 and form the scope of the feature subset selection.

In the previous sections, force signals from a dynamometer have been shown to produce good prediction of tool wear. The sixteen features derived from force signals earlier in Table 4.2 are extracted for comparisons with the results obtained via AE sensing. The details of the comparison results are discussed in the next section.

TABLE 4.9
AE Features and Nomenclature

No	Feature	Notation	References
1	Skewness	AEskew	[163]
2	Kurtosis	AEkts	[163]
3	Crest-factor	AEcrest	[163]
4	Peak	AEmax	[163]
5	Total amplitude of AE	AEa	[163]
6	Mean of RMS	AErms	[163]
7	Average AE	AEca	[163]
8	Standard deviation	AEstd	[163]
9	Mean of band power	AEmb	[110]
10	Standard deviation of band power	AEstdb	[110]
11	Delta (change in signal)	AEdlt	[110]
12	Absolute deviation	AEad	[110]
13	Ring down count	AEc	[110]
14	Rise Time	AEr	[121]
15	Area under curve	AEca	[164]
16	Duration	AEd	[164]

4.5.3 COMPARISON OF STANDARD NON-DYNAMIC PREDICTION MODELS WITH DYNAMIC ARMAX MODEL

In the experiment, 1250 time points of measured force and AE sensor data were captured under the following machine settings: spindle speed of 7400 rpm, feed rate of 474 mm/min, and depth of cut of 0.1 mm. During the data measurement phase, the tool was removed from the chuck and the tool wear was measured by hand periodically. Specifically, the flank wear of each individual tooth of the cutting tool was measured with an LEICA MZ12.5 high performance stereomicroscope. This yields the baseline actual tool wear plot shown in Figures 4.18–4.21.

The measured force and AE data were detrended, i.e., the mean value was subtracted and normalized. Based on these data, the sixteen features in Tables 4.2 and 4.9 were computed as functions of time for force features and AE features, respectively. This yields two sets of $n = 16$ feature vectors (each of which is a function of time and has $m = 1250$ data points) and X in (4.1) that has $n = 16$ columns and $m = 1250$ rows. Next, the PFA and DFI of feature selection were performed. The selected dominant features were then used as the input to build non-dynamic and dynamic tool wear prediction models. The prediction accuracies with different approaches were analyzed and the results are shown in this section.

The resulting singular values are shown in Table 4.10 and plotted in Figure 4.17. It can be seen that the fourth and subsequent singular values for both force and AE signals are very small as compared to the first three singular values. As such, it appears that retaining the first three singular values would be sufficient to capture the relevant trend in tool wear prediction for this experiment. However, it is not clear which of the sixteen features are required for tool wear prediction. We therefore use the DFI methodology to select the most important features.

TABLE 4.10
Principal Components and Singular Values

No	Singular Values (Force)	Singular Values (AE)
1	3.874509	11.08428
2	0.260766	0.228871
3	0.011573	0.020575
4	6.69×10^{-4}	5.56×10^{-4}
5	7.42×10^{-6}	5.37×10^{-5}
6	5.88×10^{-7}	1.31×10^{-6}
7	8.23×10^{-8}	2.21×10^{-10}
8	7.53×10^{-9}	1.27×10^{-11}
9	4.02×10^{-9}	3.86×10^{-14}
10	1.40×10^{-9}	4.48×10^{-15}
11	1.25×10^{-10}	1.66×10^{-15}
12	6.30×10^{-12}	9.01×10^{-17}
13	1.09×10^{-12}	6.08×10^{-18}
14	8.32×10^{-14}	6.91×10^{-19}
15	1.82×10^{-16}	2.13×10^{-19}
16	1.39×10^{-18}	1.40×10^{-19}

In current literature, tool wear is usually predicted with standard non-dynamic linear prediction models, e.g., in [165][166]. The simplest and most commonly used is the linear MRM (4.38). For non-intrusive and online prediction, the RLS algorithm depicted earlier in (4.43) was also used for our experiment.

MRMs using RLS were identified to predict the baseline measured tool wear using all of the original sixteen force and AE features as described in the previous section. An MRE of 7.15% for all sixteen force features and 10.53% for all sixteen AE features are observed as shown in Figure 4.18. A better prediction performance is obtained using the force MRM when compared to the AE MRM. The MRE of the AE MRM is above 10%, and is unacceptable in TCM applications.

Using the ELS techniques presented in the previous section, dynamic ARMAX models in (4.45) were identified to predict the baseline measured tool wear using all sixteen original force and AE features. The actual measured tool wear and the predicted tool wear using ARMAX models as functions of time are shown in Figure 4.19. Clearly, the predictions are good. An MRE of 1.13% using force features and 3.22% using AE features are observed in Figure 4.19, and represent the best possible prediction of tool wear using these two sets of sixteen features. Both force and AE with dynamic ARMAX models using ELS perform better than those using the same features with non-dynamic MRM using RLS. We also observed that using AE with ARMAX and ELS obtained better performance than using force with MRM and

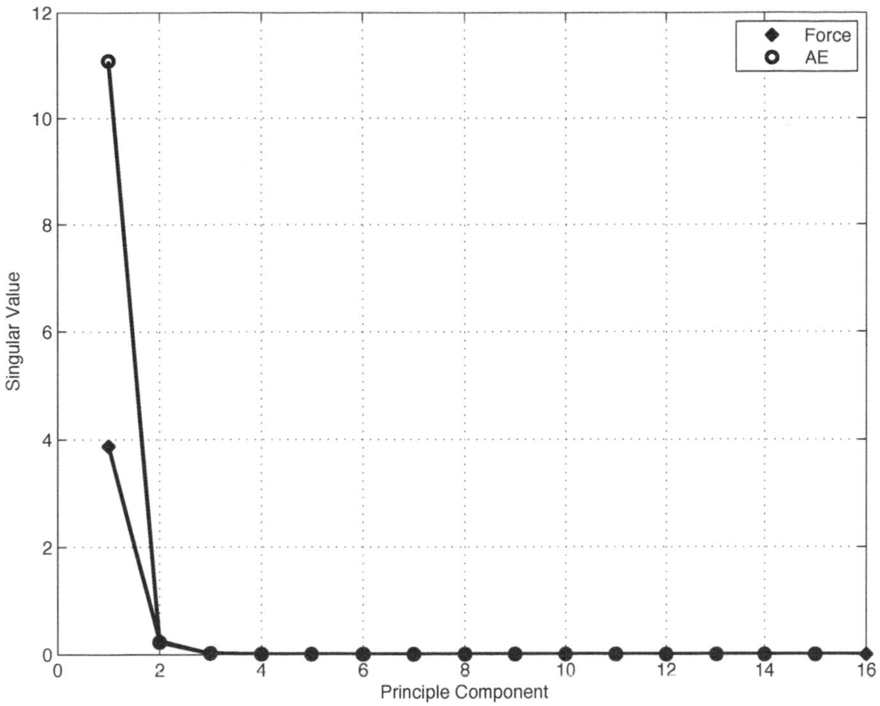

FIGURE 4.17 Plot of principal components vs. singular values.

RLS. This justifies the proposed usage of AE sensors in TCM, as effective tool wear prediction can be realized with improvement in system identification techniques.

Here, we consider a single cutting condition to demonstrate the effectiveness of our combination of dominant feature identification and dynamic ARMAX modeling for tool wear prediction for simplicity, but without loss of generality. It is seen that the prediction of tool wear using these techniques is extremely good. In our future work, we will examine tool wear prediction under different cutting conditions.

4.5.4 COMPARISON OF PROPOSED ARMAX MODEL USING ELS WITH DFI, MRM USING RLS WITH DFI, AND MRM USING RLS WITH PRINCIPAL FEATURE ANALYSIS (PFA)

It is desired to use fewer features to predict the tool wear. Next, we use the DFI method [37] and PFA method [129][130] in current literature as the feature selection techniques to reduce the number of features in the feature space and select the important features. The selected features obtained using both DFI and PFA methods are then used as inputs, and the ARMAX models using ELS and MRMs using RLS techniques are used to build the prediction models to forecast tool wear.

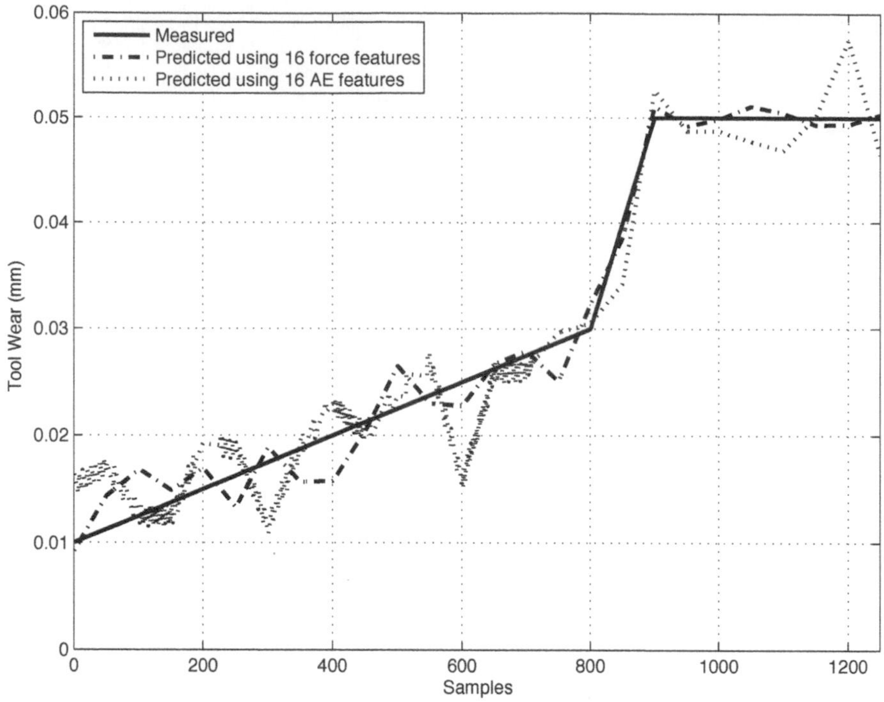

FIGURE 4.18 RLS using all sixteen force and AE features.

To showcase the effectiveness of our proposed new DFI methodology using AE sensors and ARMAX models with ELS, only AE features are used in this comparison for simplicity, but without loss of generality. Here, four dominant features (namely kurtosis, count, area, and duration) were selected using DFI, and four dominant features (namely kurtosis, delta, count, and duration) were selected by using PFA. Using ELS and RLS to compute the ARMAX model and MRM, respectively, mathematical models were built to predict tool wear as shown in Figure 4.20. The actual measured tool wear, tool wear prediction using ARMAX model with ELS and DFI (new DFI), tool wear prediction using MRM with RLS and DFI, and tool wear prediction using MRM with RLS and PFA, with all predictions using the best four features with DFI and PFA, are shown in Figure 4.20. The MREs of the different models are shown in Table 4.11. A reduction in 16.83% in MRE is observed when the proposed new DFI methodology is used, as compared to other methods proposed in the literature.

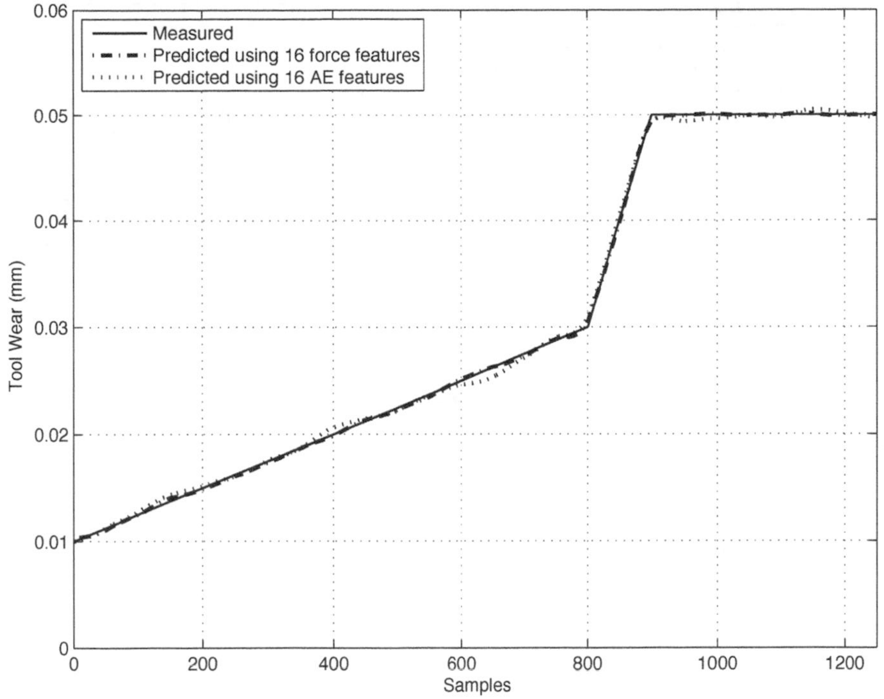

FIGURE 4.19 ARMAX models using all sixteen force and AE features.

TABLE 4.11
Comparison of Model Accuracies

Model type	MRE(%)
ARMAX using ELS with DFI	7.19
MRM using RLS with DFI	17.52
MRM using RLS with PFA	24.02

4.5.5 EFFECTS OF DIFFERENT NUMBERS OF RETAINED SINGULAR VALUES AND FEATURES SELECTED

Also, comparison studies were carried out using increasing numbers q retained singular values and p selected features using DFI and PFA. Comparison results using three retained singular values with DFI using ARMAX models and ELS (new DFI), DFI using MRMs with RLS, and PFA using MRMs with RLS are shown in Tables 4.12, 4.13, and 4.14, respectively. Similarly, the results using four retained singular values with

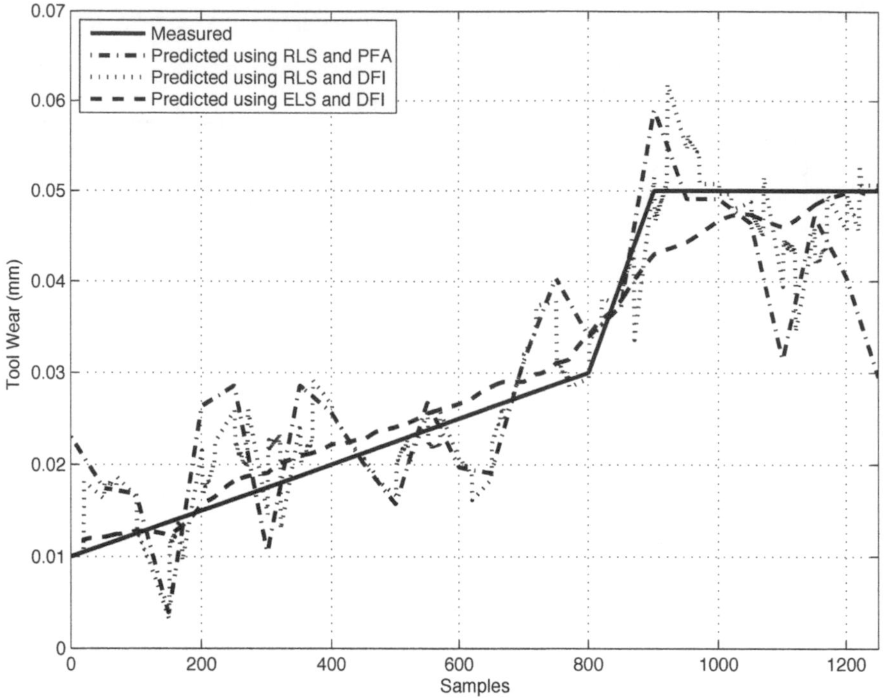

FIGURE 4.20 Examples of tool wear prediction using four dominant features and three principal components.

DFI using ARMAX models and ELS (new DFI), DFI using MRMs with RLS, and PFA using MRMs with RLS are shown in Tables 4.15, 4.16, and 4.17, respectively. In these tables, the first column denotes the number of features selected using DFI or PFA to be used for tool wear prediction, and the second column displays the selected features. The MSE and MRE obtained using the corresponding prediction models for tool wear forecast are displayed in columns three and four, respectively.

The observations from these tables are as follows:

1. The dominant features are consistent when using the DFI methodology, even when different numbers of features are used. However, the features are slightly different using the PFA method when different numbers of p are used, especially when the numbers of features chosen are four and five;
2. Besides the consistency in selection of features, the features chosen are also different using DFI and PFA. Using the same MRM with RLS, DFI gives a smaller MSE and MRE when compared to that using PFA, and hence provides a better accuracy than PFA in tool wear prediction. This is especially obvious when the numbers of features chosen are four and five.

TABLE 4.12
Results of DFI and ARMAX Model with ELS (New DFI) Using Three Retained Singular Values

No. of selected features	Features selected	MSE (mm^2)	MRE (%)
4	AEkts, AEc, AEca, AEd	0.26	7.19
5	AEskew, AEkts, AEr, AEca, AEd	0.25	6.25
6	AEskew, AEkts, AEc, AEr, AEca, AEd	0.24	5.11
7	AEcrest, AEkts, AEstd, AEc, AEca, AEd	0.17	3.77
8	AEcrest, AEkts, AEca, AEstdb, AEc, AEr, AEca, AEd	0.14	4.42
9	AEskew, AEcrest, AEkts, AEa, AEstdb, AEc, AEr, AEca, AEd	0.12	3.49

TABLE 4.13
Results of DFI and MRM with RLS Using Three Retained Singular Values

No. of selected features	Features selected	MSE (mm^2)	MRE (%)
4	AEkts, AEc, AEca, AEd	0.72	24.02
5	AEskew, AEkts, AEr, AEca, AEd	0.52	17.34
6	AEskew, AEkts, AEc, AEr, AEca, AEd	0.44	15.67
7	AEcrest, AEkts, AEstd, AEc, AEca, AEd	0.43	15.28
8	AEcrest, AEkts, AEca, AEstdb, AEc, AEr, AEca, AEd	0.40	13.03
9	AEskew, AEcrest, AEkts, AEa, AEstdb, AEc, AEr, AEca, AEd	0.37	12.67

TABLE 4.14

Results of PFA and MRM with RLS Using Three Retained Singular Values

No. of selected features	Features selected	MSE (mm^2)	MRE (%)
4	AEkts, AEc, AEca, AEd	1.262	25.12
5	AEskew, AEkts, AEr, AEca, AEd	1.202	19.11
6	AEskew, AEkts, AEc, AEr, AEca, AEd	1.111	15.64
7	AEcrest, AEkts, AEstd, AEc, AEca, AEd	1.111	15.33
8	AEcrest, AEkts, AEca, AEstdb, AEc, AEr, AEca, AEd	0.946	13.03
9	AEskew, AEcrest, AEkts, AEa, AEstdb, AEc, AEr, AEca, AEd	0.946	13.03

TABLE 4.15

Results of DFI and ARMAX Model with ELS (New DFI) Using Four Retained Singular Values

No. of selected features	Features selected	MSE (mm^2)	MRE (%)
5	AEskew, AEkts, AEr, AEca, AEd	0.25	6.25
6	AEskew, AEkts, AEc, AEr, AEca, AEd	0.24	5.11
7	AEcrest, AEkts, AEstd, AEc, AEca, AEd	0.17	3.77
8	AEcrest, AEkts, AEca, AEstdb, AEc, AEr, AEca, AEd	0.14	3.42
9	AEskew, AEcrest, AEkts, AEa, AEstdb, AEc, AEr, AEca, AEd	0.12	3.49

TABLE 4.16
Results of DFI and MRM RLS Using Four Retained Singular Values

No. of selected features	Features selected	MSE (mm^2)	MRE (%)
5	AEskew, AEkts, AEr, AEca, AEd	0.52	17.34
6	AEskew, AEkts, AEc, AEr, AEca, AEd	0.44	15.67
7	AEcrest, AEkts, AEstd, AEc, AEca, AEd	0.43	15.28
8	AEcrest, AEkts, AEca, AEstdb, AEc, AEr, AEca, AEd	0.40	13.03
9	AEskew, AEcrest, AEkts, AEa, AEstdb, AEc, AEr, AEca, AEd	0.37	12.67

TABLE 4.17
Results of PFA and MRM RLS Using Four Retained Singular Values

No. of selected features	Features selected	MSE (mm^2)	MRE (%)
5	AEskew, AEkts, AEr, AEca, AEd	0.54	19.15
6	AEskew, AEkts, AEc, AEr, AEca, AEd	0.52	17.34
7	AEcrest, AEkts, AEstd, AEc, AEca, AEd	0.44	15.61
8	AEcrest, AEkts, AEca, AEstdb, AEc, AEr, AEca, AEd	0.44	15.27
9	AEskew, AEcrest, AEkts, AEa, AEstdb, AEc, AEr, AEca, AEd	0.40	12.67

When using a small number of selected features, the computational time for features processing is also greatly reduced. Our experiments also found that the computational time required to build MRMs using the original sixteen features was about five times longer, when compared to that using four selected features only; and

3. When using different system identification techniques using non-dynamic or dynamic models with the same feature sets selected by DFI, ARMAX models using ELS always provide better predictions when compared to MRMs using RLS.

As such with the improvement in feature selection and modeling techniques, the AE sensors can replace force sensors in real-time tool wear prediction. This justifies the effectiveness of our proposed new DFI methodology in prediction of tool wear using dynamic ARMAX models with ELS for TCM applications.

4.5.6 COMPARISON OF TOOL WEAR PREDICTION USING AE MEASUREMENTS AND FORCE MEASUREMENTS

Next, we compare the tool wear prediction when using force and AE sensors using four dominant features selected by DFI only. An ARMAX model with ELS is built with the four selected AE dominant features (new DFI), and an MRM model is built with the four selected force dominant features. The actual measured tool wear and prediction results from the constructed models are shown in Figure 4.21. An MRE of 7.19% when using the four AE features and 13.56% when using the four force features are observed. The ARMAX model with four AE dominant features performs better than MRM model with four force dominant features [37], and further justifies the use of AE sensors with DFI and dynamic ARMAX models using ELS for effective TCM.

The feature selection of DFI was performed offline in this study. In the online monitoring, only the selected features were extracted in order to save the computational effort and also for fast data processing. The ARMAX model was then updated dynamically online with the new feature points. This approach is different from the normal tool-wear monitoring system in which static tool reference models [165][166] are necessary. Since the nature of the cutting conditions is always changing in the real production, building a generic static model suitable for all cutting conditions is very challenging and almost impossible. On the other hand, different static tool-wear models are built for different cutting conditions when a generic model is lacking, which is very time consuming. The proposed approach in this section updates the dynamic ARMAX model on-line with new data points that makes it suitable for different cutting conditions, as no static models are needed. The dynamic ARMAX model that is updated on-line also gives a more accurate tool wear prediction.

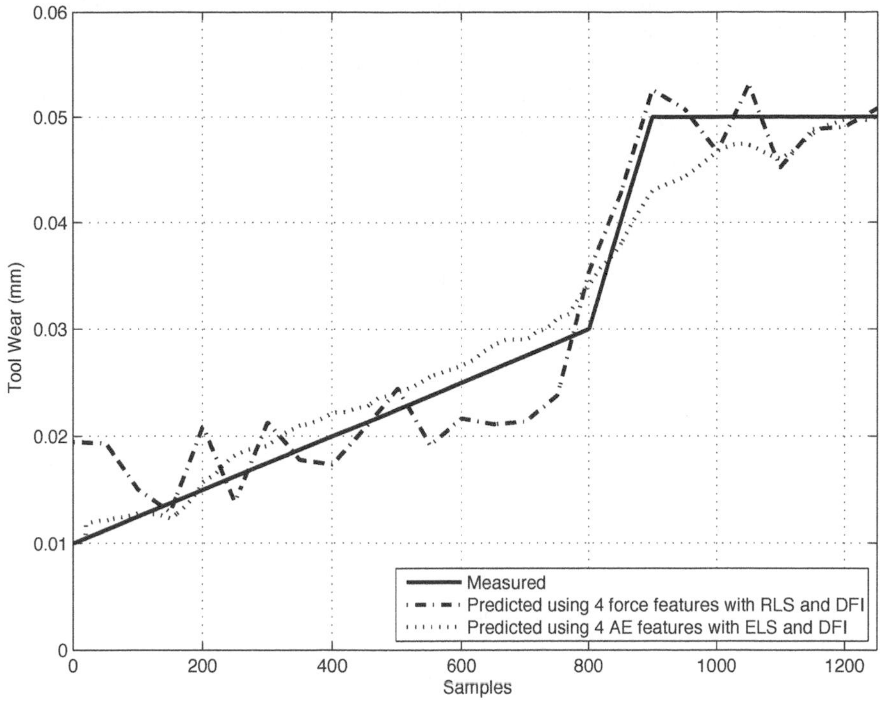

FIGURE 4.21 Examples of prediction using four dominant features and three principal components.

4.6 DFI FOR INDUSTRIAL FAULT DETECTION AND ISOLATION (FDI)

In this section, we extend the earlier proposed Dominant Feature Identification (DFI) method for industrial Fault Detection and Isolation (FDI) using Neural Networks (NNs) to classify faults based on a reduced feature set. We propose two methods of using DFI for industrial FDI applications. We show how to apply DFI to fault detection by two methods that seek to identify the important features in a given set of faults. Then, based on the determined reduced feature set, an NN is used for online fault classification and multiple fault prediction in such a proposed two-stage framework. We note that any NN structure and activation function can be used for the second stage. Our experimental results on a machine fault simulator reduce the number of features from 120 to 13 and sensors from 8 to 4 at an accuracy of 99.4%.

Suppose there is available measured time series data of length m samples from several fault conditions. These may be signals measured from sensors such as accelerometers, force sensors, electric current, and Hall Effect, etc. For each fault condition k, n features are computed (rms value, energy, skew, etc.) from the measured data time series. For the kth fault condition, form feature matrix $X_k \in \mathbb{R}^{m\times n}$ by arranging the n features as its columns, each one a time series of length m.

Based on the feature matrices X_k for fault conditions k, we propose two methods of industrial FDI based on DFI; namely an Augmented DFI (ADFI) approach and Decentralized DFI (DDFI) approach. In both cases, the dominant features identified will be used for training an NN for fault classification as detailed below.

4.6.1 AUGMENTED DOMINANT FEATURE IDENTIFICATION (ADFI)

The Augmented DFI method performs DFI on an augmented X matrix that concatenates all the sensor (or processed feature) data simultaneously for different fault conditions k. We call this method Augmented DFI (ADFI).

Suppose there are K fault conditions. The feature matrices $X_k \in \mathbb{R}^{m\times n}$ for fault conditions k are concatenated into an overall feature matrix

$$X = \begin{bmatrix} X_1^T & X_1^T & \cdots & X_k^T \end{bmatrix}, \tag{4.54}$$

where $X \in \mathbb{R}^{mK\times n}$. We now perform DFI on this augmented composite fault feature matrix X to identify the p dominant features for FDI across all faults simultaneously. This procedure allows us to select p dominant features for describing all the fault conditions, and will be used for fault classification using a NN as described subsequently.

ADFI uses an approach similar to that used in facial recognition [167], whereby the singular values that best describe all of the set of faces together are retained and used for classification. The features resulting after ADFI on the concatenated matrix X in (4.54) are those that best describe the complete set of fault conditions *as a whole*.

4.6.2 DECENTRALIZED DOMINANT FEATURE IDENTIFICATION (DDFI)

The second method performs DFI on the fault feature matrices $X_k \in \mathbb{R}^{m\times n}$ individually for each fault case k. We call this method Decentralized DFI (DDFI).

Thus, we perform DFI on each individual fault feature matrix $X_k \in \mathbb{R}^{m\times n}$ and extract the dominant features for each fault condition k. The dominant features for FDI across all the faults are then given by the intersection of the sets of features for each fault condition. These features will be used for fault classification using a NN in the next section.

4.6.3 FAULT CLASSIFICATION WITH NEURAL NETWORKS

With the dominant features selected using either ADFI and DDFI as described, Neural Networks (NNs) are trained to classify the faults that have occurred based on the reduced set of p dominant features.

In general, a two-layer NN is described by

$$y_i = \sigma\left[\sum_{l=1}^{L} w_{il}\,\sigma\left(\sum_{j=1}^{n} v_{lj}x_j + v_{l0}\right) + w_{i0}\right], \tag{4.55}$$

for $i = 1, 2, \cdots, m$, where x_j and y_i are the inputs and outputs of the NN, respectively. v_{lj} and w_{il} are the weights of the hidden and output layer, respectively. n, L, and m are the orders of the input, hidden layer, and output, respectively, and $\sigma(.)$ is the activation function [132]. The details of the two-layer NN are shown in Figure 4.22.

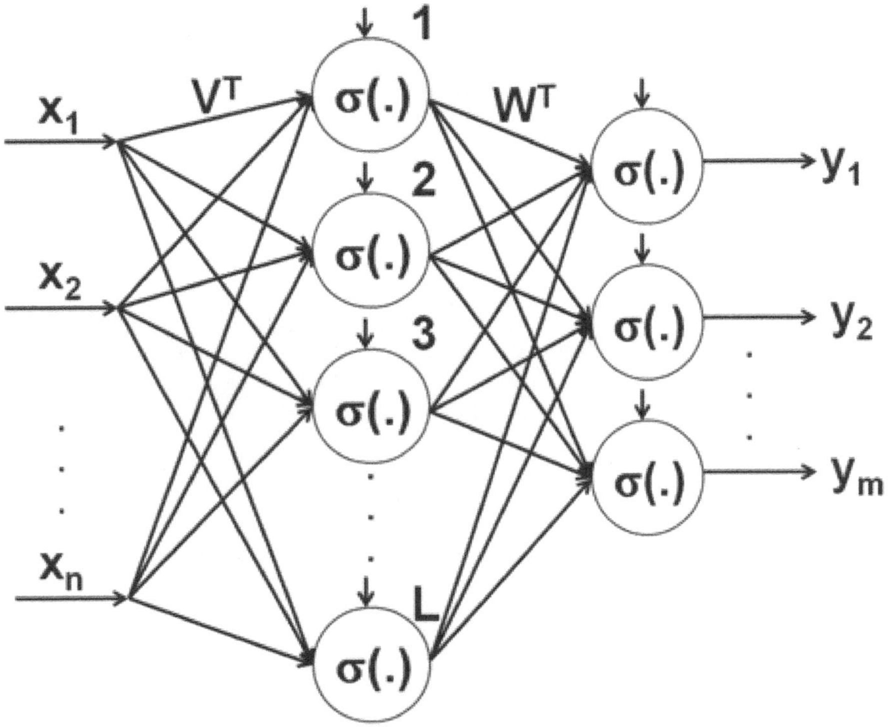

FIGURE 4.22 Two-layer NN trained using backpropagation for industrial fault classification after proposed ADFI and DDFI.

The inputs to the NN are the p dominant features' time series, each of length m. These are packed into a matrix of length mp as inputs to the NN. Such a "feature-time signature" matrix corresponds to each fault condition k.

We can now use the dominant features collected via ADFI or DDFI to classify or even predict fault occurrences in typical FDI applications using the NN. The NN is trained using these feature time signatures of length mp of each of the K faults. For the training phase, the backpropagation algorithm is used with Gaussian Radial Basis Functions (RBFs) as the activation functions. RBFs are chosen because they provide good approximations to non-linear functions and are fast in convergence. More importantly, the number of hidden neurons to-be used in the hidden layer can be automatically determined during the NN-training phase for an optimized architecture [168]. It is worth noting that any NN structure and activation function can be used for the second stage, and we have chosen the above-mentioned to verify the performance of the proposed new method in FDI based on DFI without loss of generality.

The NN is trained iteratively via batch processing until the inaccuracies (or errors) of fault classifications are sufficiently small. Upon completion of training and validation, the NN is put online, and presented with the *mp* feature time signature matrix as observed from the operating machine. The details of using the NN for industrial fault classification are shown in the next section.

4.6.4 EXPERIMENTAL SETUP

For our experiment, the proposed methodologies are evaluated on a machine fault simulator as shown in Figure 4.23.

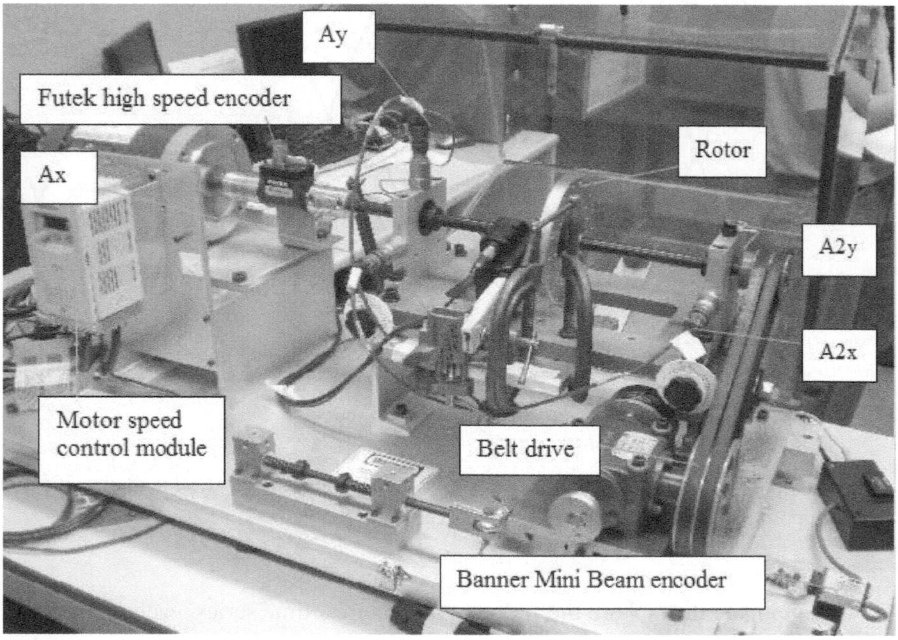

FIGURE 4.23 Machine fault simulator and the eight sensors.

The machine fault simulator by SpectraQuest is designed to study the signature of common machinery faults, such as unbalance, alignment, resonance, bearing and belt drive, etc., as depicted in [169]. Eight sensors are installed for fault detection: namely three current clamp meters to measure the motor line current, one Futek torque sensor placed on the shaft between the motor and first bearing to measure the torque generated, and four Kistler vibration sensors are placed near the two bearings to measure the vibration levels generated. The system overview is shown in Figure 4.23.

It should be noted that the positions of the clamp meters are hidden under the motor and are not shown in Figure 4.23. Each sensor has a corresponding channel on the data acquisition module.

The names of the channels and the corresponding sensors attached are shown in Table 4.18.

TABLE 4.18
Nomenclature and Corresponding Sensors

Name	Sensor
AI0	Clamp meter C1, white cable of motor current input (neutral)
AI1	Clamp meter C2, black cable of motor current input (line 1)
AI2	Clamp meter C3, red cable of motor current input (line 2)
AI3	Futek high speed encoder, torque signals
AI4	Sensor Ax, acceleration signals
AI5	Sensor Ay, acceleration signals
AI6	Sensor A2x, acceleration signals
AI7	Sensor A2y, acceleration signals
PFI4	Banner mini beam encoder, speed signals
PFI8	Futek high speed encoder, speed signals

In our experiment, the sensitivity of the accelerometer is 100 mV/g and that of the clamp meter is 100 mV/A. The zoomed-in view of the four vibration sensors, Ax, Ay, A2x, and A2y, are shown in Figure 4.24. The current clamp meters and Futek torque sensors are shown in Figure 4.25.

The procedure of computation of corresponding features is detailed here. Firstly, signal processing is performed using envelope method in [149] to extract the demodulated signals. Next, *statistical* features calculation from time domain are used for current, torque, and vibration signal analysis [170][171][172][173]. The statistical features that are used in this chapter are

$$\text{Min} = \min(\{S_i\}_{i\in[a,b]}), \tag{4.56}$$

$$\text{Max} = \max(\{S_i\}_{i\in[a,b]}), \tag{4.57}$$

$$\text{Average}(\mu) = \frac{\sum\limits_{i\in[a,b]} x_i}{b-a}, \tag{4.58}$$

$$\text{Root Mean Square (RMS)} = \sqrt{\frac{\sum\limits_{i\in[a,b]} (x_i-\bar{x})^2}{b-a}}, \tag{4.59}$$

$$\text{Standard Deviation}(\sigma) = \sqrt{\frac{1}{N}\sum_{i=1}^{N} (x_i-\mu)^2}, \tag{4.60}$$

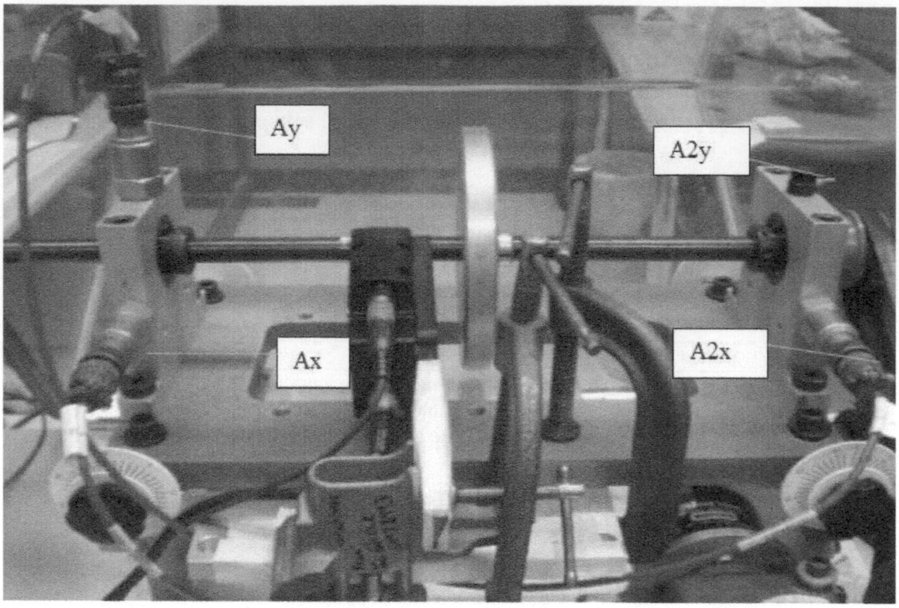

FIGURE 4.24 Zoomed-in view of the four vibration sensors Ax, Ay, A2x, A2y, with the corresponding data cables attached.

$$\begin{aligned} \text{Skewness} &= \frac{\mu_3}{\sigma^3} \\ &= \frac{E[(x-\mu)^3]}{E[(x-\mu)^2]^{3/2}}, \end{aligned} \tag{4.61}$$

$$\text{Kurtosis} = \frac{\mu_4}{\sigma^4}, \tag{4.62}$$

$$\text{Crest Factor} = \frac{|x|_{peak}}{x_{rms}}. \tag{4.63}$$

Feature calculations are also carried out in the *frequency* domain. The features used for evaluating the rotating imbalance include the first, second, and third harmonics of the fundamental frequency [174][175][176]. If f_0 is the fundamental frequency, then

$$f_{1\text{st}} = f_0, \tag{4.64}$$

$$f_{2\text{nd}} = 2f_0, \tag{4.65}$$

$$f_{3\text{rd}} = 3f_0. \tag{4.66}$$

The features used for evaluating the ball bearing faults include the first, second, and third harmonic of ball defect frequency [176]. A typical schematic of a ball bearing is shown in Figure 4.26.

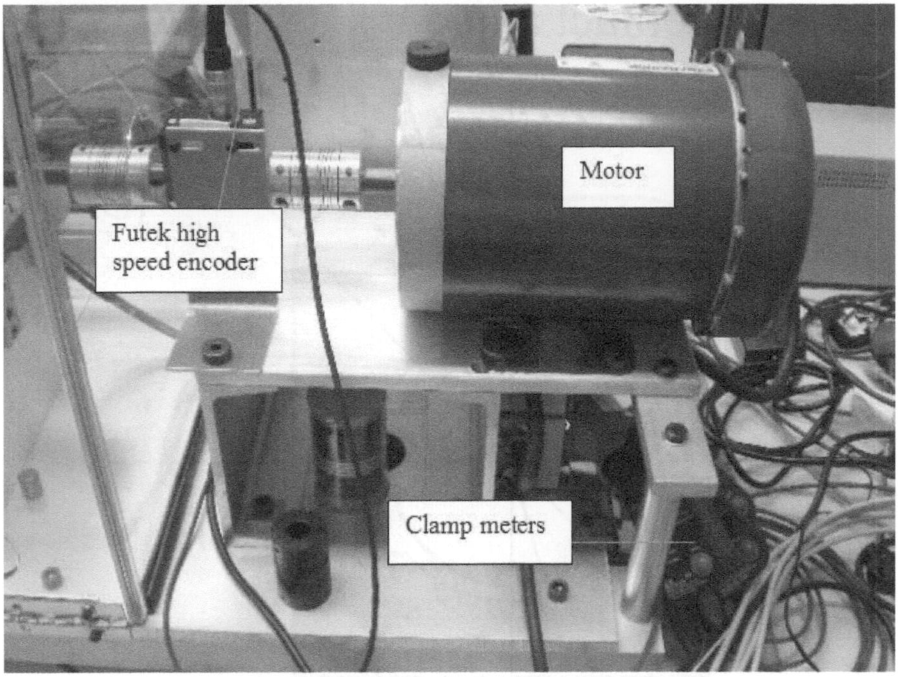

FIGURE 4.25 Current clamp meters and Futek torque sensor.

For a ball bearing fault in ball defect, the following features (in Hz) are calculated,

$$f_{\text{BFF}} = \frac{PD}{BD} f_r \left[1 - (\frac{BD}{PD} \cos\beta)^2 \right], \quad (4.67)$$

$$f_{\text{1st\_BFF}} = f_{\text{BFF}}, \quad (4.68)$$

$$f_{\text{2nd\_BFF}} = 2 f_{\text{BFF}}, \quad (4.69)$$

$$f_{\text{3rd\_BFF}} = 3 f_{\text{BFF}}, \quad (4.70)$$

where n is the number of balls or rollers, and f_r is the relative revolutions per second between inner and outer races. Similarly, the features used for evaluating the bearing outer race faults include the first, second, and third harmonic of outer race defect frequency [176] as

$$f_{\text{BPFO}} = \frac{n}{2} f_r \left(1 - \frac{BD}{PD} \cos\beta \right), \quad (4.71)$$

$$f_{\text{1st\_BPFO}} = f_{\text{BPFO}}, \quad (4.72)$$

$$f_{\text{2nd\_BPFO}} = 2 f_{\text{BPFO}}, \quad (4.73)$$

$$f_{\text{3rd\_BPFO}} = 3 f_{\text{BPFO}}. \quad (4.74)$$

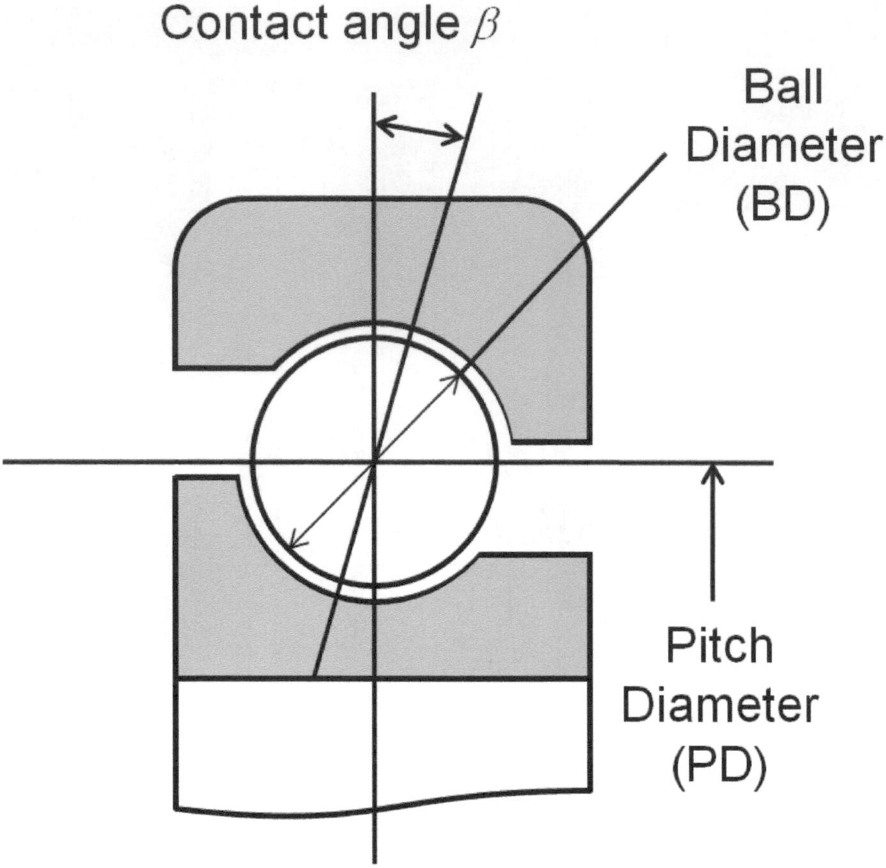

FIGURE 4.26 Schematic of a typical ball bearing.

The computed features from the corresponding eight sensors are shown in Tables 4.19–4.21.

Neural Networks (NNs) are used to build the fault classification model for FDI in estimating the machine status and health condition. The obtained features are used as the input of the NNs. The machine statuses are digitalized as the output of the NN and are defined in Table 4.22.

Our experiments are conducted for 10 seconds at a 2 kHz sampling rate. The 12 features in Table 4.19 are extracted from each of current clamp meter (AI0-AI2). Similarly, 12 features in Table 4.20 are extracted from torque (AI3), and 18 features in Table 4.21 are extracted from each of accelerometers (AI4-AI7). A total of 120 features are thus extracted from the 8 sensors in the machine fault simulator in this study. From the 120 principle components, the most dominant 25 (of largest magnitude in singular values) and their corresponding singular values are shown in Table 4.23

TABLE 4.19
(AI0-AI2) Motor Current Features and Nomenclature

No.	Features	References
1	Minimum	[170]
2	Maximum	[170]
3	Peak to Peak	[170]
4	Average	[170]
5	Root Mean Square (RMS)	[170]
6	Standard deviation	[170]
7	Skewness (3rd movement)	[170]
8	Kurtosis (4th movement)	[170]
9	Crest factor	[170]
10	Amplitude of 1st harmonic of rotational frequency	[174]
11	Amplitude of 2nd harmonic of rotational frequency	[174]
12	Amplitude of 3rd harmonic of rotational frequency	[174]

TABLE 4.20
AI3 Torque Features and Nomenclature

No.	Features	References
1	Minimum	[171]
2	Maximum	[171]
3	Peak to Peak	[171]
4	Average	[171]
5	Root Mean Square (RMS)	[171]
6	Standard deviation	[171]
7	Skewness (3rd movement)	[171]
8	Kurtosis (4th movement)	[171]
9	Crest factor	[22]
10	Amplitude of 1st harmonic of rotational frequency	[175]
11	Amplitude of 2nd harmonic of rotational frequency	[175]
12	Amplitude of 3rd harmonic of rotational frequency	[175]

and Figure 4.27.

It can be seen that the 14th and subsequent singular values are very small as compared to the first 13 singular values. As such, we retain the first 13 singular values which will be sufficient to capture the relevant trend in fault detection for our experiments. However, since singular values are not directly associated with features, it is not clear which of the 120 features are required for fault detection. We therefore use the DFI methodology to select the most important features.

TABLE 4.21
(AI4-AI7) Acceleration Features and Nomenclature

No.	Features	References
1	Minimum	[172]
2	Maximum	[172]
	Peak to Peak	[172]
4	Average	[172]
5	Root Mean Square (RMS)	[172]
6	Standard deviation	[172]
7	Skewness (3rd movement)	[173]
8	Kurtosis (4th movement)	[173]
9	Crest factor	[173]
10	Amplitude of 1st harmonic of rotational frequency	[176]
11	Amplitude of 2nd harmonic of rotational frequency	[176]
12	Amplitude of 3rd harmonic of rotational frequency	[176]
13	Amplitude of 1st harmonic of cage frequency	[176]
14	Amplitude of 2nd harmonic of cage frequency	[176]
15	Amplitude of 3rd harmonic of cage frequency	[176]
16	Amplitude of 1st harmonic of outer race frequency	[176]
17	Amplitude of 2nd harmonic of outer race frequency	[176]
18	Amplitude of 3rd harmonic of outer race frequency	[176]

TABLE 4.22
Machine Status and Representation

Machine Status	Output of NN
Normal	0
Bearing ball fault	1
Imbalance	2
Loose belt	3
Bearing outer race fault	4

TABLE 4.23
The 25 Most Dominant Principal Components and Singular Values

No	Singular Values
1	29.95252
2	8.533333
3	2.886794
4	2.410079
5	1.382239
6	0.972716
7	0.36262
8	0.310402
9	0.28623
10	0.193081
11	0.175312
12	0.140064
13	0.119166
14	0.090004
15	0.086305
16	0.064154
17	0.060056
18	0.05459
19	0.053143
20	0.044893
21	0.037535
22	0.035038
23	0.033172
24	0.031379
25	0.028036

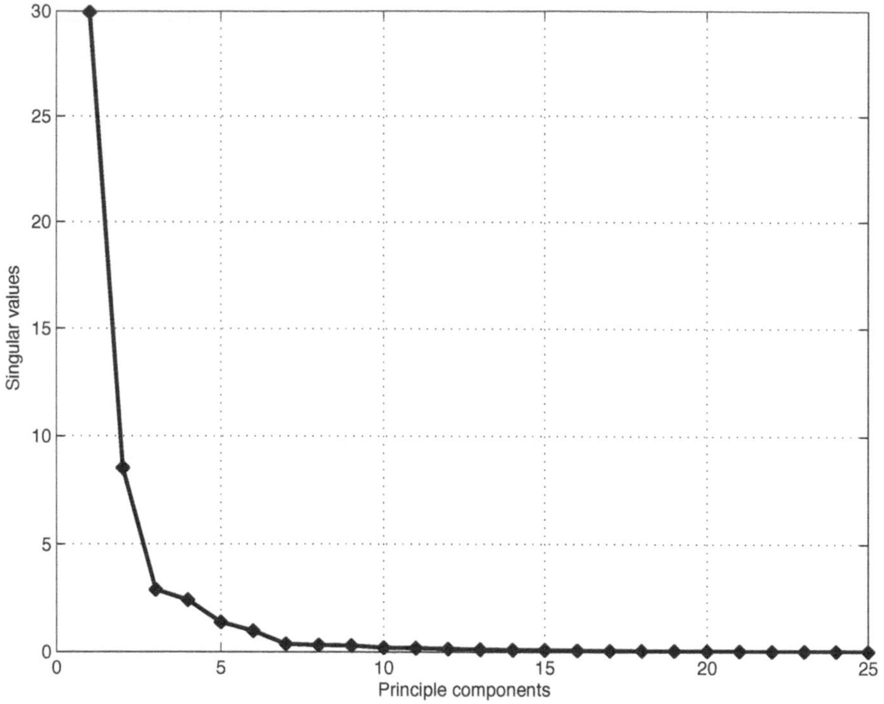

FIGURE 4.27 Plot of principal components vs. singular values.

4.6.5 FAULT DETECTION USING 120 FEATURES

In this section, we conduct a baseline study that includes all the measured features for comparison with results using only reduced feature sets. In our experiment, we use one set of data which consists of measurements from the 8 sensors at five different machine conditions, namely normal, bearing ball fault, imbalance, loose belt, and bearing outer race fault. A total of 120 features are extracted from the signals collected from the 8 sensors. We use the extracted features to conduct fault detection. The first set of data uses all the 120 features as the input, along with the output status defined earlier in Table 4.22 for training the NN. A one hidden layer feedforward NN with five neurons is created and trained. The input and target samples are automatically divided into training and validation sets. The backpropagation algorithm is used to train the NN. The mean squared error goal setting is 0.001, and the best validation performance is 0.00096283 and happens after seven epochs.

The trained NN can now be tested with the testing samples. We conduct another set of experiments under the same five machine conditions. Now, one hundred points are collected for each machine status. The second set of data is used for testing. The testing data is not used in training in any ways and hence provides an uncorrelated

TABLE 4.24
Machine Status and Fault Detection

Machine Status	Estimation	
	Correct	Wrong
Normal	100	0
Bearing ball fault	99	1
Imbalance	100	0
Loose belt	100	0
Bearing outer race fault	100	0

"out-of-sample" data set to validate the NN. This allows us to understand how well the NN will perform when tested with realistic industrial data. The results using the testing data with our trained NN for fault estimation are shown in Table 4.24. In Table 4.24, the first column shows the real machine status, while the second and third columns show the estimation results using the NN with all 120 features from 8 eight sensors.

The overall estimation yields an accuracy of four hundred and ninety nine cases out of the total five hundred. The overall accuracy of the fault detection is hence established as 99.8%. These data are used for comparison with subsequent fault classification tests using reduced feature sets.

4.6.6 AUGMENTED DOMINANT FEATURE IDENTIFICATION (ADFI) AND NN FOR FAULT DETECTION

It is now desired to use fewer sensors and features to detect the machine faults. An augmented X matrix is constructed using all the original 120 features arising from all the 8 sensors. The data are concatenated from five different machine conditions. The proposed ADFI is used to select the dominant features from the original 120 features arising from all the 8 sensors across five different machine statuses simultaneously. The selected 13 dominant features are shown in Table 4.25.

Our results reduced the number of features from 120 to 13! These 13 features come from only 4 sensors, and so the number of required sensors has been pruned from 8 to 4. This represents a significant amount of savings from sensor to signal processing costs in industrial manufacturing.

We now use the selected features in Table 4.25 to conduct the fault detection tests. Similarly, the first set of data is used for training the one hidden layer feedforward NN with five neurons. By setting the MSE goal to 0.001, the best achievable performance is 0.00052268 after thirty epochs.

The trained NN is then validated with testing samples from a second set of uncorrelated data. The results of the testing are shown in Table 4.26.

The overall correct estimation is four hundred and ninety seven cases out of five hundred. As such, the overall accuracy of the fault detection is 99.4% for ADFI.

TABLE 4.25

Features Selection Using ADFI

Sensor	Features
C2	Average
C2	Maximun
C2	Crest factor
C2	Amplitude of 1st harmonic of rotational frequency
C2	Amplitude of 2nd harmonic of rotational frequency
Torque	Minimum
Torque	Standard deviation
Ax1	Minimum
Ax1	RMS
Ax1	Amplitude of 2nd harmonic of outer race frequency
Ax2	Maximum
Ax2	Minimum
Ax2	Amplitude of 2nd harmonic of cage frequency

TABLE 4.26

Machine Status and Fault Detection Using Reduced Number of Sensors and Features from ADFI and NN

Machine Status	Estimation	
	Correct	Wrong
Normal	99	1
Bearing ball fault	99	1
Imbalance	100	0
Loose belt	100	0
Bearing outer race fault	99	1

4.6.7 DECENTRALIZED DOMINANT FEATURE IDENTIFICATION (DDFI) AND NN FOR FAULT DETECTION

Using the proposed DDFI method, we perform DFI on each individual fault feature matrix X_k and extract dominant features for each of the five machine conditions. The selected dominant features for each condition are shown in Tables 4.27–4.31.

TABLE 4.27
Feature Selection Using Convention DFI–Normal

Sensor	Features
C1	mean
C1	std
C1	skew
Torque	min
Torque	mean
Torque	peak_peak
Torque	skew
Ax1	min
Ax1	mean
Ax1	cage_third_harmonic
Ax1	cage_first_harmonic
Ay2	min
Ay2	mean
Ay2	rRms
Ay2	outer_third_harmonic

The combination features are decided using a cross selection of all the features for each fault condition. The combination of the twenty-eight dominant features for DDFI is shown in Table 4.32.

Similarly, we now use the selected features in Table 4.32 to conduct the fault detection. The first set of data is used for training the one hidden layer feedforward NN with five neurons. The MSE goal setting is 0.001, and the best performance is 0.496 after twelve epochs.

A second set of uncorrelated experimental data is used for validating the NN, and the results of the testing are shown in Table 4.33. The overall correct estimation is four hundred and twenty-two cases out of five hundred. This gives an overall accuracy of fault detection of 84.4%.

In all, the fault detection accuracy is 99.8% when using all the original 8 sensors and 120 features followed by a one hidden layer NN (five neurons) for fault detection. However, we can achieve a fault detection accuracy of 99.4% when using proposed ADFI with 13 features or 84.4% when using DDFI with 28 features. Obviously, the performance of ADFI is better than DDFI both in features and sensors number reduction as well as fault detection accuracy. As such, the proposed ADFI with NN for

TABLE 4.28
Feature Selection Using Convention DFI–Bearing

Sensor	Features
C1	mean
C1	crest
C1	skew
Torque	min
Torque	mean
Torque	rRms
Ax1	min
Ax1	mean
Ax1	skew
Ax1	cage_first_harmonic
Ax1	outer_third_harmonic
Ay2	min
Ay2	mean
Ay2	crest
Ay2	second_harmonic

TABLE 4.29
Feature Selection Using Convention DFI–Imbalance

Sensor	Features
C1	mean
C1	std
C1	skew
Torque	min
Torque	mean
Torque	rRms
Torque	skew
Ax1	min
Ax1	mean
Ay2	min
Ay2	std
Ay2	skew
Ay2	outer_third_harmonic

TABLE 4.30
Feature Selection Using Convention DFI–Loose Belt

Sensor	Features
C1	skew
C1	first_harmonic
Torque	min
Torque	mean
Torque	skew
Torque	crest
Ax1	min
Ax1	mean
Ax1	crest
Ax1	outer_second_harmonic
Ay2	min
Ay2	crest
Ay2	first_harmonic
Ay2	outer_third_harmonic

TABLE 4.31
Feature Selection Using Convention DFI–Bearing Outer Race Fault

Sensor	Features
C1	mean
C1	std
C1	skew
Torque	min
Torque	mean
Torque	rRms
Torque	skew
Ax1	min
Ax1	mean
Ax1	rRms
Ay2	min
Ay2	std
Ay2	skew
Ay2	outer_third_harmonic

TABLE 4.32
Feature Selection Using Proposed DDFI

Sensor	Features
C1	Mean
C1	Std
C1	Skew
C1	Crest
C1	first_harmonic
Torque	Min
Torque	Mean
Torque	peak_peak
Torque	rRms
Torque	Crest
Torque	Skew
Ax1	Min
Ax1	Mean
Ax1	Crest
Ax1	Skew
Ax1	outer_second_harmonic
Ax1	outer_third_harmonic
Ax1	cage_first_harmonic
Ax1	cage_third_harmonic
Ay2	Min
Ay2	Mean
Ay2	rRms
Ay2	Std
Ay2	Skew
Ay2	Crest
Ay2	first_harmonic
Ay2	second_harmonic
Ay2	outer_third_harmonic

TABLE 4.33
Machine Status and Fault Detection Using Reduced Number of Sensors and Features from DDFI and NN

Machine Status	Estimation	
	Correct	Wrong
Normal	95	5
Bearing ball fault	42	58
Imbalance	94	6
Loose belt	97	3
Bearing outer race fault	94	6

FDI is a promising technology which saves both hardware costs and computational time in signal processing.

4.7 CONCLUSION

In this chapter, the Dominant Feature Identification (DFI) methodology using Singular Value Decomposition (SVD) of collected tool wear data from force sensors and Acoustic Emission (AE) is proposed for prediction of times series of deterioration of an industrial cutting tool using both static and dynamic models, respectively. The proposed DFI uses SVD which operates on the inner product matrix at a lower dimension, and reduces the Least Squares Error (LSE) induced when selecting the principal components, and clustering to identify the dominant features. Our experimental results show Mean Squares Errors (MSEs) values from 0.946 mm^2 to 1.262 mm^2 and Mean Relative Error (MRE) values from 8.86% to 11.61% between the actual measured tool wear to that predicted from the Multiple Regression Models (MRMs) using RLS with static models. AE sensors also show significant reduction in both MSEs and MRE when a dynamic ARMAX model with Extended Least Squares (ELS) technique is employed, which is promising in replacing force sensors and conventional non-dynamic models for effective online Tool Condition Monitoring (TCM) and tool wear prediction.

We also combine DFI with Neural Networks (NNs) for Fault Detection and Identification (FDI) applications. The number of sensors used is reduced from 8 to 4, and the required number of features is significantly reduced from 120 to 13 when using proposed Augmented DFI (ADFI) with NN for fault detection on a machine fault simulator. The FDI accuracy is 99.4% in this study, which translates to reduction in both hardware and computational time from signal processing without compromising the FDI accuracy. Our future works include rigorous theoretical derivations of the errors induced during DFI and using NNs for clustering to form the dominant features, and improving signal-to-noise ratio of the AE signals, as well as their corresponding signal processing and spectral analysis.

In the following chapter, we propose a probabilistic small signal stability assessment method for analyzing stability of large-scale interconnected systems under parametric variations. The proposed framework is tested on a New England 39-Bus Test System, and the effectiveness of our methodology is evaluated with extensive simulations on non-deterministic parametric as well as load variations.

5 Probabilistic Small Signal Stability Assessment

With the deregulation of the power industry in many countries, the traditionally vertically integrated power systems have been experiencing dramatic changes leading to competitive electricity markets. Power system planning in such an environment is now facing increasing requirements and challenges because of the deregulation, and deregulations (coupled with market practices in power industry) have brought great challenges to the system planning area. In particular, they introduce a variety of uncertainties to system planning, and these changes introduce a variety of uncertainties to system planners. As such, new system planning methods are required to deal with these uncertainties, to ensure robust stability.

The traditional deterministic power system analysis techniques have, in many cases, limited capabilities in revealing the increasing uncertainties in today's power systems. As a promising approach, probabilistic methods are attracting more and more attention by system planners. The power system operation and planning are demonstrating probabilistic characteristics which require emphasis on probabilistic techniques. A key probabilistic power system analysis technique is the probabilistic power system small signal stability assessment technique. In small signal stability analysis, generation control parameters play an important role in determining the stability margin. With the many factors such as demand uncertainties, market price elasticities, and unexpected system congestion, it is more appropriate to have probabilistic power system stability assessment results rather than deterministic one, especially for the sake of risk management in a competitive electricity market.

Since small signal stability is the ability of power systems to maintain synchronism under small disturbances, the disturbances are considered sufficiently small for the linearization of system equations to be permissible for applying linear state-space techniques. In state-space domain, the system stability is determined by the eigenvalues of its state matrix, and the small signal stability problem is usually due to insufficient damping of system oscillations. There are many uncertainty factors affecting system small signal stability, e.g., dynamic parameter deviation of a generator and its control parameters, forecast error of load, and measurement error of transmission network parameters, etc. The system uncertainty also comes from uncertainties in system parameters (e.g., forecasted loads). It is therefore critical to investigate whether the system is stable under normal operating conditions, especially after the restructure of the power industry.

Also, ongoing power industry changes are creating a need for more advanced transmission planning methods and associated tools. New and improved methods are required to perform transmission planning in the competitive environment, taking maximum advantage of existing transmission assets. Investigation of probabilistic small signal stability is the second step (following the development of probabilistic load flow program) of applying probabilistic methodology to the system planning area. Similarly, load modeling also plays an important role in power system dynamic stability assessment. One of the widely used methods in assessing load model impact on system dynamic response is parametric sensitivity analysis. A composite load model-based load sensitivity analysis framework is also proposed. This enables comprehensive investigation into load modeling impacts on system stability considering the dynamic interactions between load and system dynamics. The effect of the location of individual as well as patches of composite loads in the vicinity on the sensitivity of the oscillatory modes can then be investigated, along with the impact of load composition on the overall sensitivity of the load.

The objective of this chapter is thus to perform a theoretical study on probabilistic small signal stability assessment techniques so that the study results can be used to help RTOs and ISOs perform planning studies under the open access environment. By using the methodology developed in this chapter, system planners can model the system uncertainties when performing small signal stability analysis. Based on the results obtained, system planners have better understanding of system stability boundaries and margins. The power system has been modeled in state space and the critical eigenvalues of the system have been identified. Both analytical and numerical methods have been developed and compared for computing the sensitivity factors of eigenvalues to non-deterministic parameters. The probabilistic density function associated with each critical eigenvalue is computed based on the assumption that the change of non-deterministic parameters follows normal distribution. The overall system probabilistic small signal stability is computed based on Tetrachoric series, and the impact of induction motor load and their parameters in terms of mobility of critical modes are also discussed, using the New England test system is used as the test case.

5.1 INTRODUCTION

Power system stability is defined as the ability of a power system to remain in a state of operating equilibrium (i.e., synchronicity) under normal operating conditions and to regain an acceptable state of equilibrium after being subjected to external disturbances. A disturbance in the power system is a sudden change or a sequence of changes in one or more of the parameters of the system or in one or more of the operating quantities. The system is presumed to be in a static equilibrium state when the disturbance is initiated. The disturbance causes a dynamic motion of the system state variables to either a new equilibrium condition or an unstable condition, and the latter is regarded as a system failure in engineering sense.

Power system disturbances are generally classified as small or large depending on whether a linear approximation of the system model is valid for analysis. If the equations that describe the dynamics of the power system can be linearized, the disturbance is called a *small* disturbance; otherwise, it is called a *large* disturbance. According to the size of disturbance, two different types of stabilities are usually defined. A power system is steady-state stable for particular steady-state operation conditions if, following any small disturbance, it reaches a steady-state operating condition which is *identical or similar* to the pre-disturbance operating conditions. This is also known as *small signal stability* of a power system.

Power system stability analysis is presently conducted in a deterministic framework using large-scale computer program to solve the system differential and algebraic equations as a function of time. System engineers study time-domain results and compare the results of many runs to determine the adequacy of a given system design with respect to a given set of disturbances. Usually, no detailed knowledge is available concerning the probability of the given disturbance for which stability is in question, although the engineer will usually recognize that all disturbances are not equally likely. This is factored into judgmental decisions concerning the adequacies of the system. As such, stability analysis is basically a probabilistic rather than a deterministic problem. The factors initiating system disturbances coupled with the conditions of the system at that time are probabilistic in nature, and therefore it appears logical to attempt to examine them from that point of view.

Recently, power systems have experienced more and more uncertainties, especially under an open access deregulated environment. The system uncertainties may come from various sources, but the main contribution is from uncertainties in system parameters and forecasted loads. Because of deregulation, the ISO or RTO planners have no access to the Independent Power Producer (IPP) facilities in many cases, and therefore cannot perform tests in order to measure the *real* system parameters. Consequently, uncertainties are inevitably introduced into the ISO or RTO's planning process. This has resulted in challenges for system planners in an open access electricity market. In order to have a comprehensive picture of the system stability in planning, probabilistic stability assessment is attracting more and more attention over the traditional deterministic approach. Sensitivity analysis is the first step for probabilistic small signal stability studies. Sensitivity analysis has been investigated in various aspects in [177][178][179][180][181][182][183], and will be discussed in detail in Section 5.4. However, these previous works did not investigate the computational efficiencies of analytical and numerical approaches in sensitivity computation, which we will be discussing in this chapter.

After comparing the results of both approaches, we propose guidelines for selecting the parameter perturbation sizes in sensitivity analysis. We also identify the parameters that have great impact on system stability. Using the sensitivity analysis results, the planners need to model those parameters as random variables when performing small signal stability assessment. Given the fact that a power system is a non-linear, complex, and interconnected large-scale networked system, the parametric sensitivities are coupled into the system state matrix in a very complex manner. Sensitivity of some of the parameters that have direct entry to the system state matrix can be computed analytically by studying their contribution in the state matrix. However, for those parameters which do not have direct entry to the state matrix, e.g., active power components of load, sensitivity computations can be very inefficient and computationally expensive.

In terms of planning, power system planning has been traditionally the least cost based approach using the $n-1$ criteria for reliability assessment. With the deregulation of the power sectors in many countries worldwide, more and more attention has been drawn to the uncertainties involved in power system planning. It has been identified that the reliability assessment based on traditional creditable contingency screening is now facing challenges from more comprehensive probabilistic approaches [184][185], and hence Electric Power Research Institute (EPRI) proposed *probabilistic power system planning*, which involves probabilistic reliability assessment, probabilistic power flow, and probabilistic power system stability assessment. As a matter of fact, Billinton et al. [186][187] had developed rather sophisticated probabilistic-based reliability assessment techniques as early as the 1970s.

Using parametric variation approaches, the stochastic properties of the eigenvalues obtained are identified as random variables with unknown Probability Density Functions (PDFs). To test the stability of eigenvalues under stochastic load perturbations, the probability of the real part of the critical mode will be of interest [177][178]. In order to ensure the mode's stability, the probability will be a feasible measure. As such, the mean value of the real part of the critical eigenvalues should be located on the left half s-plane. The PDF of the real part of the eigenvalues are assumed to be of normal distribution and the probability of the eigenvalues in the region (corresponding to 99.72% confidence interval) to the right tail of the PDF is essential for drawing stability conclusions. To study the stability of a power system under small signal stability, the multivariate normal PDF will be used to analyze the *joint* probability of the system eigenvalues (hence modes and mode shapes). This PDF allows the probability and statistics of the system eigenvalues under stochastic non-deterministic parameter perturbations to be studied simultaneously.

We have covered the necessity of probabilistic small signal stability for large-scale interconnected power systems, considering system uncertainties and parametric variations from the physical system or planning side. However, it is worth noting that load modeling also plays an essential role in power system stability assessment and enhancement. Different load models may result in totally different system stability assessment outcomes, which in turn affect grid operation and planning appreciably. Load modeling in power systems attracted the attention of scientists more than a decade ago [188]. Different types of load models have been proposed based on field measurements. Voltage- and frequency-dependent static load and dynamic Induction Motor (IM)-type load models have been well reported [189][190]. Various studies show that the recent blackout events on August 10 1996 and August 4 2000 in USA could not be reproduced by the existing static load models. The importance of including dynamic IM models in order to obtain similar simulation and actual response has been emphasized in [191]. Based on energy function analysis and detailed simulations, it has been further illustrated that load model characteristics have a significant impact on system transient stability properties [192]. With the development of technology and growth in consumer purchase power, electricity loads are exhibiting more and more diversities and uncertainties in their characteristics. These require more advanced analytical techniques to reliably handle power stability analysis involving load modeling, and to clearly identify the impacts of different load models on system dynamic behavior.

After all, small signal stability is the ability of the power system to maintain synchronism under disturbances. To investigate the small signal stability of a power system, we need to model the dynamic components (e.g., generators) and their control systems (such as excitation control system, and speed governor systems, etc.) in detail. The accuracy of power system stability analysis depends on the accuracy of the models used. Using more accurate models could result in increases of overall power system transfer capability and associated economic benefits. Under open access environment, the planners may not be able to obtain this information as accurately as they used to. It is therefore important to attempt mathematically modeling and analyzing these parameters probabilistically; therefore, the planner can gain a better understanding of the system stability margin.

With this motivation in mind, a probabilistic power system small signal stability assessment framework for system planning in an open access environment is developed, and the rest of the chapter is organized as follows. Section 5.2 presents the detailed power system modeling methods to include the major system components such as generator, excitation controller, system constraints, and other dynamics as differential equations. Section 5.3 derives the essential stator equations and network admittance matrices (along with their reduction) as algebraic equations. Based on the system models represented by a set of differential algebraic equations, the system state matrix is then obtained in Section 5.4 which can be used for small signal stability analysis. Both analytical and numerical approaches to compute the sensitivity of the system state matrix to certain parametric variations (including control system parameters and system loads) are described in Section 5.5. These identified parameter variations need to be investigated with highest priority. Computation of probabilistic density function of each critical eigenvalue to non-deterministic parameters is investigated following the modeling and sensitivity analysis, and the overall system probabilistic index is then calculated based on Tetrachoric series. Section 5.6 discusses the impact of induction motor load and its parameters in terms of mobility of critical modes. More discussions are found in Section 5.7. Our conclusion and future works are summarized in Section 5.8.

5.2 POWER SYSTEM MODELING: DIFFERENTIAL EQUATIONS

For small signal stability analysis, power systems should be represented in linear state space form to characterize system small signal behavior as a Linear Time Invariant (LTI) system around an equilibrium point represented as a steady-state operating condition. Power systems can be treated as dynamic devices, such as generators and their controllers, High Voltage Direct Current (HVDC) and Flexible Alternating Current Transmission Systems (FACTS), interacting through the transmission network. To illustrate this, each dynamic device is considered as a subsystem coupled to the transmission network. Therefore, the complete model of an interconnected power system consists of [193][194][195][196][197][198][199][200] [201][202][203][204]:

1. A set of differential equations describing the local dynamics of each sub-

system; and

2. A set of algebraic equations, involving variables at all nodes adjacent to each subsystem, describing the coupling characteristics between the dynamic device and the transmission network.

In this section, the differential equations for the synchronous machines, IEEE Type 1 exciter, Automatic Voltage Regulator (AVR), speed governor, steam turbine (prime mover), as well as their interactions between synchronous machines and control systems are modeled and derived rigorously.

5.2.1 SYNCHRONOUS MACHINES

Synchronous machines are usually modeled by the third order flux decay model or the fourth-order two-axis model depending on specifications for each synchronous machine. In our study, the fourth-order model is used and the following basic assumptions are made [201]:

1. Damping power is proportional to frequency variation;
2. Machine rotor speed does not vary much from synchronous speed ω_o;
3. Machine rotational power losses due to windage and friction are ignored; and
4. Mechanical shaft power is smooth, i.e., the shaft power is constant except for the results of speed governor action.

The "swing" equations are thus given by

$$\dot{\delta} = \omega_o \omega, \tag{5.1}$$

$$\dot{\omega} = -\frac{D}{2H}\omega + \frac{1}{2H}p_m - \frac{1}{2H}p_e, \tag{5.2}$$

where δ is the rotor angle, ω_o is the electrical synchronous speed, and ω is the rotor angular speed. D is the damping constant, H is the generalized lumped inertia constant (generally the ratio of stored kinetic energy at synchronous speed and generator MVA rating), p_m is the mechanical power, and p_e is the electrical power.

To account for faster changes in initial conditions external to the synchronous machines, the following transient equations are included in machine modeling. New fictitious transient voltages E'_d and E'_q, representing the flux linkages in the rotor windings, are created to exist behind their transient reactances X'_d and X'_q. In the case with damper windings, other circuits exist in the rotor causing subtransients to exist but are neglected in the following model. Together, the electrical equations modeling the direct axis (d-axis) and quadrature axis (q-axis) are

$$\dot{E}_q = -\frac{1}{T'_{d0}}E'_q - \frac{(X_d - X'_d)}{T'_{d0}}I_d + \frac{1}{T'_{d0}}E_{fd}, \tag{5.3}$$

$$\dot{E}_d = -\frac{1}{T'_{q0}}E'_d - \frac{(X_q - X'_q)}{T'_{q0}}I_q, \tag{5.4}$$

where X_d and X_q are the direct axis reactance and quadrature axis reactance, respectively. I_d and I_q are the current in the direct axis and quadrature axis, respectively. T'_{d0} is the open circuit direct axis subtransient time constant and E_{fd} is the field voltage.

Linearizing (5.1), (5.2), (5.3), and (5.4) about an operating (equilibrium) point and arranging the electrical equations first, we get

$$\Delta\dot{E}_q = -\frac{1}{T'_{d0}}\Delta E'_q - \frac{(X_d - X'_d)}{T'_{d0}}\Delta I_d + \frac{1}{T'_{d0}}\Delta E_{fd}, \tag{5.5}$$

$$\Delta\dot{E}_d = -\frac{1}{T'_{q0}}\Delta E'_d - \frac{(X_q - X'_q)}{T'_{q0}}\Delta I_q, \tag{5.6}$$

$$\Delta\dot{\delta} = \omega_o \Delta\omega, \tag{5.7}$$

$$\Delta\dot{\omega} = -\frac{D}{2H}\Delta\omega + \frac{1}{2H}\Delta p_m - \frac{1}{2H}\Delta p_e. \tag{5.8}$$

Now the stator *algebraic* equations are

$$0 = E'_q - V_q - R_a I_q - X'_d I_d, \tag{5.9}$$

$$0 = E'_d - V_d - R_a I_d - X'_q I_q, \tag{5.10}$$

where V_d and V_q are the voltages in the direct axis and quadrature axis, respectively, and R_a is the armature resistance. (5.9) and (5.10) are linearized about their operating points to give

$$0 = \Delta E'_q - \Delta V_q - R_a \Delta I_q - X'_d \Delta I_d, \tag{5.11}$$

$$0 = \Delta E'_d - \Delta V_d - R_a \Delta I_d - X'_q \Delta I_q. \tag{5.12}$$

Now (5.11) and (5.12) are usually packed into the following matrix form

$$\begin{aligned} \begin{bmatrix} \Delta I_d \\ \Delta I_q \end{bmatrix} &= \frac{1}{R_a^2 + X'_d X'_q} \begin{bmatrix} R_a & X'_q \\ X'_d & R_a \end{bmatrix} \begin{bmatrix} \Delta E'_d - \Delta V_d \\ \Delta E'_q - \Delta V_q \end{bmatrix} \\ &= Y_g \begin{bmatrix} R_a & X'_q \\ X'_d & R_a \end{bmatrix} \begin{bmatrix} \Delta E'_d - \Delta V_d \\ \Delta E'_q - \Delta V_q \end{bmatrix}, \end{aligned} \tag{5.13}$$

where $Y_g = \frac{1}{R_a^2 + X'_d X'_q}$ obviously. Substituting the linearized stator equation matrix (5.13) into (5.5) and (5.6), we get

$$\begin{aligned} \Delta\dot{E}_q &= -\frac{1}{T'_{d0}}\Delta E'_q - \frac{(X_d - X'_d)Y_g}{T'_{d0}}(R_a\Delta E'_d - R_a\Delta V'_d + X'_q\Delta E'_q - X'_q\Delta V'_q) + \frac{1}{T'_{d0}}\Delta E_{fd} \\ &= -\frac{1}{T'_{d0}}\Delta E'_q - \frac{C_1}{T'_{d0}}(R_a\Delta E'_d - R_a\Delta V'_d + X'_q\Delta E'_q - X'_q\Delta V'_q) + \frac{1}{T'_{d0}}\Delta E_{fd} \\ &= -\frac{1 + C_1 X'_q}{T'_{d0}}\Delta E'_q - \frac{C_1}{T'_{d0}}R_a\Delta E'_d + \frac{C_1}{T'_{d0}}R_a\Delta V'_d - \frac{C_1}{T'_{d0}}X'_q\Delta V'_q + \frac{1}{T'_{d0}}\Delta E_{fd}, \end{aligned} \tag{5.14}$$

$$\begin{aligned}
\Delta\dot{E}_d &= -\frac{1}{T'_{q0}}\Delta E'_d - \frac{(X_q - X'_q)Y_g}{T'_{q0}}(-X'_d\Delta E'_d + X'_d\Delta V'_d)R_a\Delta E'_q - R_a\Delta V'_q \\
&= -\frac{1}{T'_{q0}}\Delta E'_d - \frac{C_2}{T'_{q0}}(-X'_d\Delta E'_d + X'_d\Delta V'_d)R_a\Delta E'_q - R_a\Delta V'_q \\
&= -\frac{C_2}{T'_{q0}}R_a\Delta E'_q - \frac{1 + C_2 X'_d}{T'_{q0}}\Delta E'_d + \frac{C_2}{T'_{q0}}X'_d\Delta V'_d - \frac{C_2}{T'_{q0}}R_a\Delta V_q, \qquad (5.15)
\end{aligned}$$

where $C_1 = (X_d - X'_d)Y_g$ and $C_2 = (X_q - X'_q)Y_g$. The dynamics of the transient voltages E_q and E_d are now expressed in terms of the direct axis and quadrature axis voltages V_q and V_d.

For representing the synchronous machines in *network* notation, the synchronous machines equations are usually rewritten in the rotating frame (with respect to the rotor). The following transformation is usually used

$$V_d = \cos\delta V_x + \sin\delta V_y, \qquad (5.16)$$

$$V_q = -\sin\delta V_x + \cos\delta V_y, \qquad (5.17)$$

or packed into matrix form as

$$\begin{bmatrix} V_d \\ V_q \end{bmatrix} = \begin{bmatrix} \cos\delta & \sin\delta \\ -\sin\delta & \cos\delta \end{bmatrix} \begin{bmatrix} V_x \\ V_y \end{bmatrix}. \qquad (5.18)$$

This transformation is equally valid for currents as well

$$\begin{bmatrix} I_d \\ I_q \end{bmatrix} = \begin{bmatrix} \cos\delta & \sin\delta \\ -\sin\delta & \cos\delta \end{bmatrix} \begin{bmatrix} I_x \\ I_y \end{bmatrix}, \qquad (5.19)$$

or conversely

$$\begin{bmatrix} I_x \\ I_y \end{bmatrix} = \begin{bmatrix} \cos\delta & -\sin\delta \\ \sin\delta & \cos\delta \end{bmatrix} \begin{bmatrix} I_d \\ I_q \end{bmatrix}. \qquad (5.20)$$

Linearizing (5.18) about its operating point $(V_d^0, V_q^0, \delta^0, V_x^0, V_y^0)$, we obtain

$$\begin{aligned}
\begin{bmatrix} \Delta V_d \\ \Delta V_q \end{bmatrix} &= \begin{bmatrix} \cos\delta & \sin\delta \\ -\sin\delta & \cos\delta \end{bmatrix}_{\delta=\delta^0} \begin{bmatrix} V_x \\ V_y \end{bmatrix}_{V_x=V_x^0, V_y=V_y^0} \Delta\delta \\
&\quad + \begin{bmatrix} \cos\delta & \sin\delta \\ -\sin\delta & \cos\delta \end{bmatrix}_{\delta=\delta^0} \begin{bmatrix} \Delta V_x \\ \Delta V_y \end{bmatrix} \\
&= \begin{bmatrix} -V_x^0 \sin\delta^0 + V_y^0 \cos\delta^0 \\ -V_x^0 \cos\delta^0 - V_y^0 \sin\delta^0 \end{bmatrix} \Delta\delta \\
&\quad + \begin{bmatrix} \cos\delta^0 & \sin\delta^0 \\ -\sin\delta^0 & \cos\delta^0 \end{bmatrix} \begin{bmatrix} \Delta V_x \\ \Delta V_y \end{bmatrix} \\
&= \begin{bmatrix} V_q^0 \\ -V_d^0 \end{bmatrix} \Delta\delta + \begin{bmatrix} \cos\delta^0 & \sin\delta^0 \\ -\sin\delta^0 & \cos\delta^0 \end{bmatrix} \begin{bmatrix} \Delta V_x \\ \Delta V_y \end{bmatrix}. \qquad (5.21)
\end{aligned}$$

The same can be achieved by linearizing (5.20) about its operating point $(I_x^0, I_y^0, \delta^0, I_d^0, I_q^0)$ to give

$$\begin{bmatrix} \Delta I_x \\ \Delta I_y \end{bmatrix} = \begin{bmatrix} -I_y^0 \\ I_x^0 \end{bmatrix} \Delta\delta + \begin{bmatrix} \cos\delta^0 & -\sin\delta^0 \\ \sin\delta^0 & \cos\delta^0 \end{bmatrix} \begin{bmatrix} \Delta I_d \\ \Delta I_q \end{bmatrix}. \tag{5.22}$$

As such, (5.14) and (5.15) can be re-expressed as

$$\begin{aligned} \Delta\dot{E}'_q &= -\frac{1+C_1X'_q}{T'_{d0}}\Delta E'_q - \frac{C_1R_a}{T'_{d0}}\Delta E'_d + \frac{C_1R_a}{T'_{d0}}(V_q^0\Delta\delta + \cos\delta^0\Delta V_x + \sin\delta^0\Delta V_y) \\ &\quad + \frac{C_1X'_q}{T'_{d0}}(-V_d^0\Delta\delta - \sin\delta^0\Delta V_x + \cos\delta^0\Delta V_y) + \frac{1}{T'_{d0}}\Delta E_{fd} \\ &= -\frac{1+C_1X'_q}{T'_{d0}}\Delta E'_q - \frac{C_1R_a}{T'_{d0}}\Delta E'_d + \frac{C_1(R_aV_q^0 - X'_qV_d^0)}{T'_{d0}}\Delta\delta \\ &\quad + \frac{C_1(R_a\cos\delta^0 - X'_q\sin\delta^0)}{T'_{d0}}\Delta V_x + \frac{C_1(R_a\sin\delta^0 + X'_q\cos\delta^0)}{T'_{d0}}\Delta V_y \\ &\quad + \frac{1}{T'_{d0}}\Delta E_{fd} \\ &= A_g^{(1,1)}\Delta E'_q + A_g^{(1,2)}\Delta E'_d + A_g^{(1,3)}\Delta\delta + C_g^{(1,1)}\Delta V_x + C_g^{(1,2)}\Delta V_y + \frac{1}{T'_{d0}}\Delta E_{fd}, \end{aligned} \tag{5.23}$$

where

$$A_g^{(1,1)} = -\frac{1+C_1X'_q}{T'_{d0}}, \tag{5.24}$$

$$A_g^{(1,2)} = \frac{C_1R_a}{T'_{d0}}, \tag{5.25}$$

$$A_g^{(1,3)} = \frac{C_1(R_aV_q^0 - X'_qV_d^0)}{T'_{d0}}, \tag{5.26}$$

$$C_g^{(1,1)} = \frac{C_1(R_a\sin\delta^0 + X'_q\cos\delta^0)}{T'_{d0}}, \tag{5.27}$$

$$C_g^{(1,2)} = \frac{C_1(R_a\sin\delta^0 + X'_q\cos\delta^0)}{T'_{d0}}, \tag{5.28}$$

and

$$\begin{aligned}
\Delta\dot{E}'_d &= -\frac{C_2R_a}{T'_{q0}}\Delta E'_q - \frac{1+C_2X'_d}{T'_{q0}}\Delta E'_d + \frac{C_2X'_d}{T'_{q0}}(V_q^0\Delta\delta + \cos\delta^0\Delta V_x + \sin\delta^0\Delta V_y) \\
&\quad - \frac{C_2R_a}{T'_{q0}}(-V_d^0\Delta\delta - \sin\delta^0\Delta V_x + \cos\delta^0\Delta V_y) \\
&= \frac{C_2R'_a}{T'_{q0}}\Delta E'_q - \frac{1+C_2X'_d}{T'_{q0}}\Delta E'_d + \frac{C_2(X'_dV_q^0 + R_aV_d^0)}{T'_{q0}}\Delta\delta \\
&\quad + \frac{C_2(X'_q\cos\delta^0 + R_a\sin\delta^0)}{T'_{q0}}\Delta V_x + \frac{C_2(X'_d\sin\delta^0 - R_a\cos\delta^0)}{T'_{q0}}\Delta V \\
&= A_g^{(2,1)}\Delta E'_q + A_g^{(2,2)}\Delta E'_d + A_g^{(2,3)}\Delta\delta + C_g^{(2,1)}\Delta V_x + C_g^{(2,2)}\Delta V_y, \qquad (5.29)
\end{aligned}$$

where

$$A_g^{(2,1)} = \frac{C_2R'_a}{T'_{q0}}, \qquad (5.30)$$

$$A_g^{(2,2)} = -\frac{1+C_2X'_d}{T'_{q0}}, \qquad (5.31)$$

$$A_g^{(2,3)} = \frac{C_2(X'_dV_q^0 + R_aV_d^0)}{T'_{q0}}, \qquad (5.32)$$

$$C_g^{(2,1)} = \frac{C_2(X'_q\cos\delta^0 + R_a\sin\delta^0)}{T'_{q0}}, \qquad (5.33)$$

$$C_g^{(2,2)} = \frac{C_2(X'_d\sin\delta^0 - R_a\cos\delta^0)}{T'_{q0}}. \qquad (5.34)$$

The instantaneous electrical power P_e is given by

$$P_e = E'_qI_q + E'_dI_d. \qquad (5.35)$$

Linearizing (5.35) about its operating points ($E_q'^0$, $E_d'^0$, I_q^0, I_d^0), we get

$$\Delta P_e = I_q^0\Delta E'_q + E_q'^0\Delta I_q + I_d^0\Delta E'_d + E_d'^0\Delta I_d. \qquad (5.36)$$

For the New England System operating at 60 Hz, the electrical synchronous speed ω_o is approximately 377 rad/s. Together with (5.36), the swing equations (5.7) and (5.8) can also be written as

$$\Delta\dot{\delta} = 377\Delta\omega, \qquad (5.37)$$

$$\Delta\dot{\omega} = -\frac{D}{2H}\Delta\omega + \frac{1}{2H}\Delta P_m - \frac{1}{2H}(I_q^0\Delta E_q' + E_q'^0\Delta I_q + I_d^0\Delta E_d' + E_d'^0\Delta I_d). \tag{5.38}$$

Now recall from (5.13) that

$$\Delta I_d = Y_g(R_a\Delta E_d' - R_a\Delta V_d + X_q'\Delta E_q' - X_q'\Delta V_q), \tag{5.39}$$

$$\Delta I_q = Y_g(-X_d'\Delta E_d' + X_d'\Delta V_d + R_a,\Delta E_q' - R_a\Delta V_q), \tag{5.40}$$

the equilibrium points $E_q'^0$ and $E_d'^0$ can be found from substituting (5.9) and (5.10) into (5.3) and (5.4) and then setting the derivatives to zero. We will then have the following

$$\begin{aligned}
\Delta\dot{\omega} &= -\frac{D}{2H}\Delta\omega + \frac{1}{2H}\Delta P_m - \frac{1}{2H}(I_q^0\Delta E_q') - \frac{1}{2H}(I_d^0\Delta E_d') \\
&\quad -\frac{E_q'^0 Y_g}{2H}(-X_d'\Delta E_d' + X_d'\Delta V_d + R_a\Delta E_q' - R_a\Delta V_q) \\
&\quad -\frac{E_d'^0 Y_g}{2H}(R_a\Delta E_d' - -R_a\Delta V_d + X_q'\Delta E_q' - X_q'\Delta E_q') \\
&= -\frac{I_q^0 + E_q'^0 Y_g R_a + E_d'^0 Y_g X_q'}{2H}\Delta E_q' - \frac{I_d^0 - E_q'^0 Y_g X_d' + E_d'^0 Y_g R_a}{2H}\Delta E_d' \\
&\quad -\frac{E_q'^0 Y_g X_d' - E_d'^0 Y_g R_a}{2H}\Delta V_d - \frac{-E_q'^0 Y_g R_a - E_d'^0 Y_g X_q'}{2H}\Delta V_q - \frac{D}{2H}\Delta\omega + \frac{1}{2H}\Delta P_m \\
&= -\frac{I_q^0 + P_2 R_a + P_1 X_q'}{2H}\Delta E_q' - \frac{I_d^0 - P_2 X_d' + P_1 R_a}{2H}\Delta E_d' - \frac{P_2 X_d' - P_1 R_a}{2H}\Delta V_d \\
&\quad -\frac{-P_2 R_a - P_1 X_q'}{2H}\Delta V_q - \frac{D}{2H}\Delta\omega + \frac{1}{2H}\Delta P_m,
\end{aligned} \tag{5.41}$$

where $P_1 = E_d'^0 Y_g$ and $P_2 = E_q'^0 Y_g$.

Using (5.22) to remove ΔV_d and ΔV_q, we get

$$
\begin{aligned}
\Delta\dot{\omega} &= -\frac{I_q^0 + P_2R_a + P_1X_q'}{2H}\Delta E_q' - \frac{I_d^0 - P_2X_d' + P_1R_a}{2H}\Delta E_d' \\
&\quad -\frac{P_2X_d' - P_1R_a}{2H}(V_q^0\Delta\delta + \cos\delta^0\Delta V_x + \sin\delta^0\Delta V_y) \\
&\quad -\frac{P_2R_a - P_1X_q'}{2H}(-V_q^0\Delta\delta - \sin\delta^0\Delta V_x + \cos\delta^0\Delta V_y) - \frac{D}{2H}\Delta\omega + \frac{1}{2H}\Delta P_m \\
&= -\frac{I_q^0 + P_2R_a + P_1X_q'}{2H}\Delta E_q' - \frac{I_d^0 - P_2X_d' + P_1R_a}{2H}\Delta E_d' \\
&\quad -\frac{P_1(X_q'V_d^0 - R_aV_q^0) + P_2(R_aV_d^0 + X_q'V_q^0)}{2H}\Delta\delta \\
&\quad -\frac{P_2X_d'\cos\delta^0 - P_1R_a\sin\delta^0 + P_2R_a\sin\delta^0 + P_1X_q'\sin\delta^0}{2H}\Delta V_x \\
&\quad -\frac{P_2X_d'\sin\delta^0 - P_1R_a\sin\delta^0 - P_2R_a\cos\delta^0 - P_1X_q'\cos\delta^0}{2H}\Delta V_y \\
&\quad -\frac{D}{2H}\Delta\omega + \frac{1}{2H}\Delta P_m \\
&= -\frac{I_q^0 + P_2R_a + P_1X_q'}{2H}\Delta E_q' - \frac{I_d^0 - P_2X_d' + P_1R_a}{2H}\Delta E_d' - \frac{P_2P_{22} - P_1P_{11}}{2H}\Delta\delta \\
&\quad -\frac{P_2X_d'\cos\delta^0 - P_1R_a\sin\delta^0 + P_2R_a\sin\delta^0 + P_1X_q'\sin\delta^0}{2H}\Delta V_x \\
&\quad -\frac{P_2X_d'\sin\delta^0 - P_1R_a\sin\delta^0 - P_2R_a\cos\delta^0 - P_1X_q'\cos\delta^0}{2H}\Delta V_y \\
&\quad -\frac{D}{2H}\Delta\omega + \frac{1}{2H}\Delta P_m \\
&= A_g^{(4,1)}\Delta E_q' + A_g^{(4,2)}\Delta E_d' + A_g^{(4,3)}\Delta\delta + A_g^{(4,4)}\Delta\omega + C_g^{(4,1)}\Delta V_x + C_g^{(4,2)}\Delta V_y \\
&\quad + \frac{1}{2H}\Delta P_m, \qquad (5.42)
\end{aligned}
$$

where

$$P_{11} = X_q'V_d^0 - R_aV_q^0, \qquad (5.43)$$

$$P_{22} = R_aV_d^0 + X_q'V_q^0, \qquad (5.44)$$

$$A_g^{(4,1)} = -\frac{I_q^0 + P_2R_a + P_1X_q'}{2H}, \qquad (5.45)$$

$$A_g^{(4,2)} = -\frac{I_d^0 - P_2X_d' + P_1R_a}{2H}, \qquad (5.46)$$

$$A_g^{(4,3)} = -\frac{P_2P_{22} - P_1P_{11}}{2H}, \tag{5.47}$$

$$A_g^{(4,4)} = -\frac{D}{2H}, \tag{5.48}$$

$$C_g^{(4,1)} - \frac{P_2X_d' \cos\delta^0 - P_1R_a \sin\delta^0 + P_2R_a \sin\delta^0 + P_1X_q' \sin\delta^0}{2H}, \tag{5.49}$$

$$C_g^{(4,2)} = -\frac{P_2X_d' \sin\delta^0 - P_1R_a \sin\delta^0 - P_2R_a \cos\delta^0 - P_1X_q' \cos\delta^0}{2H}. \tag{5.50}$$

Now letting $\mathbf{\Delta X_G} = \begin{bmatrix} \Delta E_q' & \Delta E_d' & \Delta\delta & \Delta\omega \end{bmatrix}^T$ and $\mathbf{\Delta V_G} = \begin{bmatrix} \Delta V_x & \Delta V_y \end{bmatrix}^T$, we can arrange (5.23), (5.29), and (5.42) into the following matrices $\mathbf{A_G}$ and $\mathbf{C_G}$ as

$$\begin{aligned}
\mathbf{\Delta \dot{X}_G} &= \begin{bmatrix} A_g^{(1,1)} & A_g^{(1,2)} & A_g^{(1,3)} & 0 \\ A_g^{(2,1)} & A_g^{(2,2)} & A_g^{(2,3)} & 0 \\ 0 & 0 & 0 & 377 \\ A_g^{(4,1)} & A_g^{(4,2)} & A_g^{(4,3)} & A_g^{(4,4)} \end{bmatrix} \mathbf{\Delta X_G} + \begin{bmatrix} C_g^{(1,1)} & C_g^{(1,2)} \\ C_g^{(2,1)} & C_g^{(2,2)} \\ 0 & 0 \\ C_g^{(4,1)} & C_g^{(4,2)} \end{bmatrix} \mathbf{\Delta V_G} \\
&\quad + \begin{bmatrix} \frac{1}{T_{d0}'} \\ 0 \\ 0 \\ 0 \end{bmatrix} \Delta E_{fd} + \begin{bmatrix} 0 \\ 0 \\ 0 \\ \frac{1}{2H} \end{bmatrix} \Delta P_m \\
&= \mathbf{A_G \Delta X_G} + \mathbf{C_G \Delta V_G} + \begin{bmatrix} \frac{1}{T_{d0}'} \\ 0 \\ 0 \\ 0 \end{bmatrix} \Delta E_{fd} + \begin{bmatrix} 0 \\ 0 \\ 0 \\ \frac{1}{2H} \end{bmatrix} \Delta P_m.
\end{aligned} \tag{5.51}$$

These equations are the dynamical equations for the synchronous machine.

For dynamic power system transient considerations of more than one second, it is essential and important that the effects of machine controllers are included in the modeling process. This is especially true with more power systems operating at (or close to) their limits with near critical fault clearing times.

The two principal controllers of the turbine-generator are the AVR and the speed governor. In the next two subsections, their simulation model equations will be derived.

5.2.2 EXCITER AND AUTOMATIC VOLTAGE REGULATOR (AVR)

The Automatic Voltage Regulator (AVR) basically consists of voltage sensing equipment, voltage and speed comparators, as well as amplifiers controlling the synchronous machines. Many different AVR models have been developed to represent

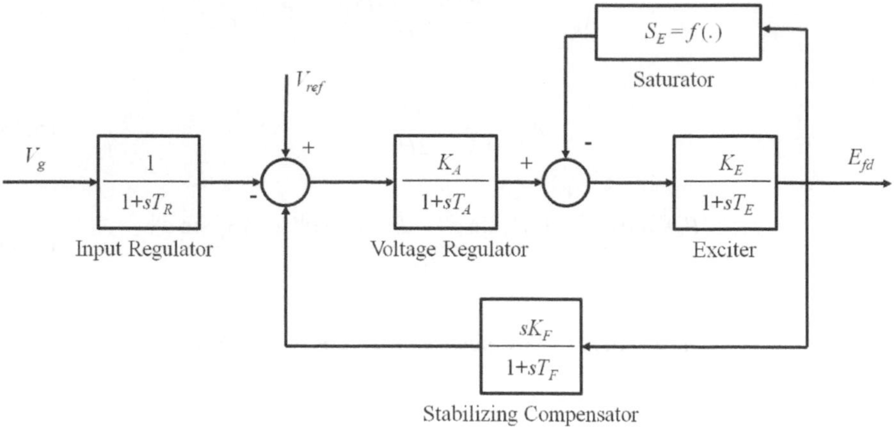

FIGURE 5.1 Block diagram of IEEE Type 1 Exciter and AVR system.

the various types of AVR used in a power system. In this section, only the IEEE Type 1 Exciter and AVR model from [198] as shown in Figure 5.1 is considered. The Type 1 excitation system is made up of a continuously acting regulator and exciter. The stabilizing compensator is included to improve dynamic response of the exciter system and at the same time to improve system's damping for less field voltage E_{fd} overshoot.

For simplicity but without loss of generality in our stability analysis on the New England Power system, the following simplifications and assumptions are made:

1. The Power System Stabilizer (PSS) is assumed to be disabled;
2. The simple time constant of the input regulator T_R is very small and is assumed to be zero; and
3. The effects of non-linear saturation function S_E are pre-calculated and modified into the exciter constant K_E to make K_E'.

As such, the simplified exciter and AVR model with its internal states is shown in Figure 5.2. The dynamics of the AVR and exciter system are given by

$$\dot{V}_{ex1} = -\frac{1}{T_A}V_{ex1} - \frac{K_A}{T_A}V_{ex3} + \frac{K_A}{T_A}V_{ref} - \frac{K_A}{T_A}\vec{V}_G, \tag{5.52}$$

$$\dot{V}_{ex2} = \frac{1}{T_E}V_{ex1} - \frac{K_E'}{T_A}V_{ex2}, \tag{5.53}$$

$$\dot{V}_{ex3} = -\frac{K_F}{T_E T_F}V_{ex1} - \frac{K_E' K_F}{T_E T_F}V_{ex2} - \frac{1}{T_F}V_{ex3}, \tag{5.54}$$

and $\vec{V}_G = \vec{V}_x + \vec{V}_y$ is the generated voltage vector.

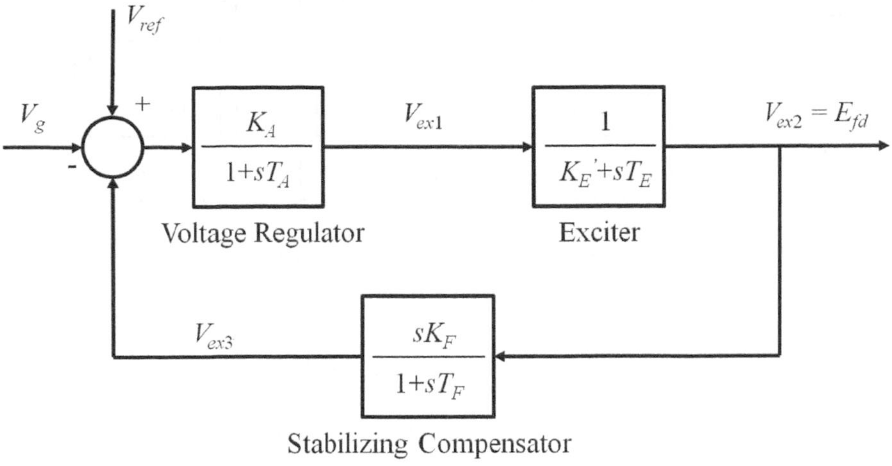

FIGURE 5.2 Block diagram of simplified exciter and AVR system.

The above dynamic equations (5.52), (5.53), and (5.54) can be linearized about their operating points V_{ex1}^0, V_{ex2}^0, and V_{ex3}^0, respectively, to give

$$\Delta\dot{V}_{ex1} = -\frac{1}{T_A}\Delta V_{ex1} - \frac{K_A}{T_A}\Delta V_{ex3} + \frac{K_A|V_x|}{T_A|V_G|}\Delta V_x - \frac{K_A|V_x|}{T_A|V_G|}\Delta V_y, \tag{5.55}$$

$$\Delta\dot{V}_{ex2} = \frac{1}{T_E}\Delta V_{ex1} - \frac{K_E'}{T_A}\Delta V_{ex2}, \tag{5.56}$$

$$\Delta\dot{V}_{ex3} = -\frac{K_F}{T_E T_F}\Delta V_{ex1} - \frac{K_E' K_F}{T_E T_F}\Delta V_{ex2} - \frac{1}{T_F}\Delta V_{ex3}. \tag{5.57}$$

Similarly, letting $\mathbf{\Delta X_E} = \begin{bmatrix} \Delta V_{ex1} & \Delta V_{ex2} & \Delta V_{ex3} \end{bmatrix}^T$ and $\mathbf{\Delta V_G} = \begin{bmatrix} \Delta V_x & \Delta V_y \end{bmatrix}^T$, we can arrange (5.55) to (5.57) into the following matrices $\mathbf{A_E}$ and $\mathbf{C_E}$ as

$$\begin{aligned} \mathbf{\Delta\dot{X}_E} &= \begin{bmatrix} -\frac{1}{T_A} & 0 & -\frac{K_A}{T_A} \\ \frac{1}{T_E} & -\frac{K_E'}{T_E} & 0 \\ \frac{K_F}{T_E T_F} & -\frac{K_E' K_F}{T_E T_F} & -\frac{1}{T_F} \end{bmatrix} \mathbf{\Delta X_E} + \begin{bmatrix} -\frac{K_A|V_x|}{T_A|V_G|} & -\frac{K_A|V_y|}{T_A|V_G|} \\ 0 & 0 \\ 0 & 0 \end{bmatrix} \mathbf{\Delta V_G} \\ &= \mathbf{A_E \Delta X_E} + \mathbf{C_E \Delta V_G}. \end{aligned} \tag{5.58}$$

5.2.3 SPEED GOVERNOR AND STEAM TURBINE

A typical speed governing system consists of a speed governor, speed relay, hydraulic servomotor and a set of governor controlled valves [199]. For the newer generation of speed governing systems, electronic circuits are used in place of the mechanical components. In this section, a simplified speed governor model as shown in Figure 5.3

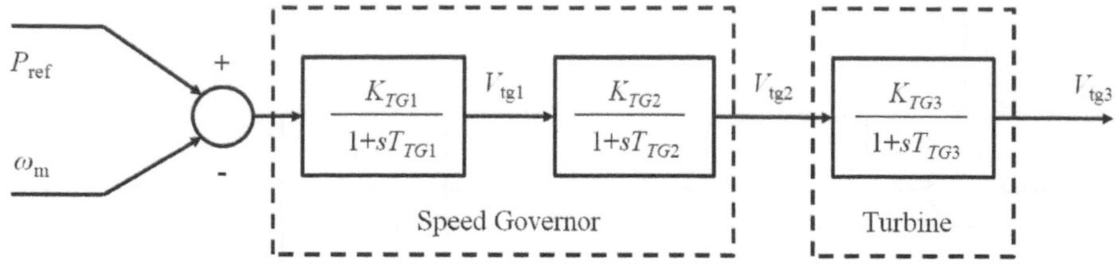

FIGURE 5.3 Simplified block diagram of speed governor system.

is used. This model is valid for thermal governor and valve systems as well as the hydro governor and valve systems.

A low order steam turbine system is modeled and considered. All steam turbine systems utilize governor-controlled valves at the inlet to the high pressure turbine to control steam flow [199]. This model can be used for both a mechanical-hydraulic system as well as an electro-hydraulic system.

For the steam turbine, the steam chest and inlet piping to the first turbine cylinder and reheaters, and crossover piping downstream, all introduce delays between valve movement and changes in steam flow. As such, the fundamental principal in modeling the steam system for stability studies is to account for these delays (phase lags), and a first order model usually suffices. Flows into and out of any steam vessel are related by a simple time constant. The diagram of a steam turbine model to be considered is also shown in Figure 5.3.

The dynamics equations of the simplified speed governor and steam turbine system are

$$\dot{V}_{tg1} = \frac{K_{TG1}}{T_{TG4}}(P_{ref} - \omega) - \frac{1}{T_{TG4}} V_{tg1}, \tag{5.59}$$

$$\dot{V}_{tg2} = \frac{K_{TG2}}{K_{TG45}} V_{tg1} - \frac{1}{T_{TG5}} V_{tg2}, \tag{5.60}$$

$$\dot{V}_{tg3} = \frac{K_{TG3}}{T_{TG6}} V_{tg2} - \frac{1}{T_{TG6}} V_{tg3}, \tag{5.61}$$

with the mechanical power output P_m as

$$P_m = (1 - K_{TG2})V_{tg1} + (1 - K_{TG3})V_{tg2} + V_{tg3}. \tag{5.62}$$

Linearizing the speed governor and steam turbine system about their operating points V_{tg1}^0, V_{tg2}^0, and V_{tg3}^0, respectively, we can get

$$\Delta\dot{V}_{tg1} = -\frac{1}{T_{TG4}} \Delta V_{tg1} - \frac{K_{TG1}}{T_{TG4}} \Delta\omega, \tag{5.63}$$

$$\Delta\dot{V}_{tg2} = \frac{K_{TG2}}{K_{TG45}} \Delta V_{tg1} - \frac{1}{T_{TG5}} \Delta V_{tg2}, \tag{5.64}$$

$$\Delta\dot{V}_{tg3} = \frac{K_{TG3}}{T_{TG6}}\Delta V_{tg2} - \frac{1}{T_{TG6}}\Delta V_{tg3}, \tag{5.65}$$

and also (5.62) becomes

$$\Delta P_m = (1 - K_{TG2})\Delta V_{tg1} + (1 - K_{TG3})\Delta V_{tg2} + \Delta V_{tg3}. \tag{5.66}$$

As above, letting $\mathbf{\Delta X_T} = \begin{bmatrix} \Delta V_{tg1} & \Delta V_{tg2} & \Delta V_{tg3} \end{bmatrix}^T$ and $\mathbf{\Delta V_G} = \begin{bmatrix} \Delta V_x & \Delta V_y \end{bmatrix}^T$, we can arrange (5.63), (5.64), and (5.65) into the following matrices $\mathbf{A_T}$ and $\mathbf{C_T}$ as

$$\begin{aligned}
\mathbf{\Delta\dot{X}_T} &= \begin{bmatrix} -\frac{1}{T_{TG4}} & 0 & 0 \\ \frac{K_{TG2}}{T_{TG5}} & -\frac{1}{T_{TG5}} & 0 \\ 0 & -\frac{K_{TG3}}{K_{TG6}} & -\frac{1}{T_{TG6}} \end{bmatrix} \mathbf{\Delta X_T} + \begin{bmatrix} 0 & 0 \\ 0 & 0 \\ 0 & 0 \end{bmatrix} \mathbf{\Delta V_G} \\
&\quad + \begin{bmatrix} -\frac{K_{TG1}}{T_{TG4}} \\ 0 \\ 0 \end{bmatrix} \Delta\omega \\
&= \mathbf{A_T\Delta X_T} + \mathbf{C_T\Delta V_G} + \begin{bmatrix} -\frac{K_{TG1}}{T_{TG4}} \\ 0 \\ 0 \end{bmatrix} \Delta\omega,
\end{aligned} \tag{5.67}$$

and

$$\Delta P_m = \begin{bmatrix} (1 - K_{TG2}) & (1 - K_{TG3}) & 1 \end{bmatrix} \mathbf{\Delta X_T} + \begin{bmatrix} 0 & 0 \end{bmatrix} \mathbf{\Delta V_G}. \tag{5.68}$$

Now that the dynamical equations for the individual system are derived, we shall now investigate the coupling and interactions between them.

5.2.4 INTERACTION BETWEEN A SYNCHRONOUS MACHINE AND ITS CONTROL SYSTEMS

Since ΔE_{fd} is the field voltage deviation of the synchronous machine (and also the state V_{ex2} of the excitation system) and ΔP_m is the mechanical power deviation (which is output of the speed governor and turbine system), we can proceed to augment the interactions between the synchronous machine and its excitation systems into a differential matrix equation.

Recall from (5.51), (5.58), (5.67), and (5.68) that

$$\mathbf{\Delta\dot{X}_G} = \mathbf{A_G\Delta X_G} + \mathbf{C_G\Delta V_G} + \begin{bmatrix} \frac{1}{T'_{d0}} \\ 0 \\ 0 \\ 0 \end{bmatrix} \Delta E_{fd} + \begin{bmatrix} 0 \\ 0 \\ 0 \\ \frac{1}{2H} \end{bmatrix} \Delta P_m, \tag{5.69}$$

$$\mathbf{\Delta\dot{X}_E} = \mathbf{A_E\Delta X_E} + \mathbf{C_E\Delta V_G}, \tag{5.70}$$

$$\mathbf{\Delta\dot{X}_T} = \mathbf{A_T\Delta X_T} + \mathbf{C_T\Delta V_G} + \begin{bmatrix} -\frac{K_{TG1}}{T_{TG4}} \\ 0 \\ 0 \end{bmatrix} \Delta\omega, \tag{5.71}$$

$$\Delta P_m = \begin{bmatrix} (1-K_{TG2}) & (1-K_{TG3}) & 1 \end{bmatrix} \mathbf{\Delta X_T} + \begin{bmatrix} 0 & 0 \end{bmatrix} \mathbf{\Delta V_G}. \tag{5.72}$$

Now with $\Delta E_{fd} = \Delta V_{ex2} = \begin{bmatrix} 0 & 1 & 0 \end{bmatrix} \mathbf{\Delta X_E}$ and $\Delta\omega = \begin{bmatrix} 0 & 0 & 0 & 1 \end{bmatrix} \mathbf{\Delta X_G}$, we can rewrite the above equations as

$$\begin{aligned} \mathbf{\Delta \dot{X}_G} &= \mathbf{A_G \Delta X_G} + \mathbf{C_G \Delta V_G} \\ &+ \begin{bmatrix} 0 & \frac{1}{T'_{d0}} & 0 \\ 0 & 0 & 0 \\ 0 & 0 & 0 \\ 0 & 0 & 0 \end{bmatrix} \mathbf{\Delta X_E} + \begin{bmatrix} 0 & 0 & 0 \\ 0 & 0 & 0 \\ 0 & \frac{1-K_{TG3}}{2H} & \frac{1}{2H} \end{bmatrix} \mathbf{\Delta X_T} \\ &= \mathbf{A_G \Delta X_G} + \mathbf{A_{GE} \Delta X_E} + \mathbf{A_{GT} \Delta X_T} + \mathbf{C_G \Delta V_G}, \end{aligned} \tag{5.73}$$

$$\mathbf{\Delta \dot{X}_E} = \mathbf{A_E \Delta X_E} + \mathbf{C_E \Delta V_G}, \tag{5.74}$$

$$\begin{aligned} \mathbf{\Delta \dot{X}_T} &= \mathbf{A_T \Delta X_T} + \begin{bmatrix} 0 & 0 & 0 & -\frac{K_{TG1}}{T_{TG4}} \\ 0 & 0 & 0 & 0 \\ 0 & 0 & 0 & 0 \end{bmatrix} \mathbf{\Delta X_G} + \mathbf{C_T \Delta V_G} \\ &= \mathbf{A_{TG} \Delta X_G} + \mathbf{A_T \Delta X_T} + \mathbf{C_T \Delta V_G}. \end{aligned} \tag{5.75}$$

Finally, let $\mathbf{\Delta X_M} = \begin{bmatrix} \mathbf{\Delta X_G} & \mathbf{\Delta X_E} & \mathbf{\Delta X_T} \end{bmatrix}^T$. The interactions as well as the dynamic machine and control equations can be augmented into matrices $\mathbf{A_M}$ and $\mathbf{C_M}$ as

$$\begin{aligned} \mathbf{\Delta \dot{X}_M} &= \begin{bmatrix} \mathbf{A_G} & \mathbf{A_{GE}} & \mathbf{A_{GT}} \\ \mathbf{0} & \mathbf{A_E} & \mathbf{0} \\ \mathbf{A_{TG}} & \mathbf{0} & \mathbf{A_T} \end{bmatrix} \mathbf{\Delta X_M} + \begin{bmatrix} \mathbf{C_G} \\ \mathbf{C_E} \\ \mathbf{C_T} \end{bmatrix} \mathbf{\Delta V_G} \\ &= \mathbf{A_M \Delta X_M} + \mathbf{C_M \Delta V_G}. \end{aligned} \tag{5.76}$$

This completes the dynamic power system modeling in differential equations.

5.3 POWER SYSTEM MODELING: ALGEBRAIC EQUATIONS

In this section, the algebraic equations (fast dynamics) of the power system components mentioned above will be derived. The algebraic equations are obtained from the stator equations, network admittance matrices, and the reduced admittance matrices.

5.3.1 STATOR EQUATIONS

Recall from (5.22) that

$$\begin{bmatrix} \Delta I_x \\ \Delta I_y \end{bmatrix} = \begin{bmatrix} -I_y^0 \\ I_x^0 \end{bmatrix} \Delta\delta + \begin{bmatrix} \cos\delta^0 & -\sin\delta^0 \\ \sin\delta^0 & \cos\delta^0 \end{bmatrix} \begin{bmatrix} \Delta I_d \\ \Delta I_q \end{bmatrix}. \tag{5.77}$$

Substituting the algebraic stator equations (matrix form) in (5.13), we get

$$\begin{aligned}
\begin{bmatrix} \Delta I_x \\ \Delta I_y \end{bmatrix} &= \begin{bmatrix} -I_y^0 \\ I_x^0 \end{bmatrix}\Delta\delta \\
&\quad + Y_g \begin{bmatrix} \cos\delta^0 & -\sin\delta^0 \\ \sin\delta^0 & \cos\delta^0 \end{bmatrix}\begin{bmatrix} R_a & X_q' \\ -X_d' & R_a \end{bmatrix}\begin{bmatrix} \Delta E_d' - \Delta V_d \\ \Delta E_q' - \Delta V_q \end{bmatrix} \\
&= \begin{bmatrix} -I_y^0 \\ I_x^0 \end{bmatrix}\Delta\delta \\
&\quad + Y_g \begin{bmatrix} R_a\cos\delta^0 + X_d'\sin\delta^0 & X_q'\cos\delta^0 - R_a\sin\delta^0 \\ R_a\sin\delta^0 - X_d'\cos\delta^0 & R_a\cos\delta^0 + X_q'\sin\delta^0 \end{bmatrix}\begin{bmatrix} \Delta E_d' \\ \Delta E_q' \end{bmatrix} \\
&\quad - Y_g \begin{bmatrix} R_a\cos\delta^0 + X_d'\sin\delta^0 & X_q'\cos\delta^0 - R_a\sin\delta^0 \\ R_a\sin\delta^0 - X_d'\cos\delta^0 & R_a\cos\delta^0 + X_q'\sin\delta^0 \end{bmatrix}\begin{bmatrix} \Delta V_d \\ \Delta V_q \end{bmatrix}.
\end{aligned} \tag{5.78}$$

Removing ΔV_d and ΔV_q using (5.23) to express in terms of $\mathbf{X_M}$ and $\mathbf{V_G}$, we obtain

$$\begin{aligned}
\begin{bmatrix} \Delta I_x \\ \Delta I_y \end{bmatrix} &= Y_g \begin{bmatrix} R_a\cos\delta^0 + X_d'\sin\delta^0 & X_q'\cos\delta^0 - R_a\sin\delta^0 \\ R_a\sin\delta^0 - X_d'\cos\delta^0 & R_a\cos\delta^0 + X_q'\sin\delta^0 \end{bmatrix}\begin{bmatrix} \Delta E_d' \\ \Delta E_q' \end{bmatrix} \\
&\quad + \begin{bmatrix} -I_y^0 \\ I_x^0 \end{bmatrix}\Delta\delta \\
&\quad - Y_g \begin{bmatrix} R_a\cos\delta^0 + X_d'\sin\delta^0 & X_q'\cos\delta^0 - R_a\sin\delta^0 \\ R_a\sin\delta^0 - X_d'\cos\delta^0 & R_a\cos\delta^0 + X_q'\sin\delta^0 \end{bmatrix}\begin{bmatrix} V_q^0 \\ V_d^0 \end{bmatrix}\Delta\delta \\
&\quad - Y_g \begin{bmatrix} R_a\cos\delta^0 + X_d'\sin\delta^0 & X_q'\cos\delta^0 - R_a\sin\delta^0 \\ R_a\sin\delta^0 - X_d'\cos\delta^0 & R_a\cos\delta^0 + X_q'\sin\delta^0 \end{bmatrix}\cdots \\
&\quad \times \begin{bmatrix} \cos\delta^0 & \sin\delta^0 \\ -\sin\delta^0 & \cos\delta^0 \end{bmatrix}\begin{bmatrix} \Delta V_x \\ \Delta V_y \end{bmatrix} \\
&= \begin{bmatrix} Y_g(R_a\cos\delta^0 + X_d'\sin\delta^0) & Y_g(X_q'\cos\delta^0 - R_a\sin\delta^0) \\ Y_g(R_a\sin\delta^0 - X_d'\cos\delta^0) & Y_g(R_a\cos\delta^0 + X_q'\sin\delta^0) \end{bmatrix}\begin{bmatrix} \Delta E_d' \\ \Delta E_q' \end{bmatrix} \\
&\quad + \begin{bmatrix} -I_y^0 \\ I_x^0 \end{bmatrix}\Delta\delta \\
&\quad - Y_g \begin{bmatrix} V_q^0 R_a\cos\delta^0 + V_q^0 X_d'\sin\delta^0 - V_d^0 X_q'\cos\delta^0 + V_d^0 R_a\sin\delta^0 \\ V_q^0 R_a\sin\delta^0 - V_q^0 X_d'\cos\delta^0 - V_d^0 R_a\cos\delta^0 - V_d^0 X_q'\sin\delta^0 \end{bmatrix}\Delta\delta \\
&\quad - Y_g \left[\begin{matrix} R_a\cos^2\delta^0 + X_d'\sin\delta^0\cos\delta^0 + R_a\sin^2\delta^0 - X_q'\sin\delta^0\cos\delta^0 \\ R_a\sin\delta^0\cos\delta^0 - X_d'\cos^2\delta^0 - X_q'\sin^2\delta^0 - R_a\sin\delta^0\cos\delta^0 \end{matrix} \right| \\
&\quad \left| \begin{matrix} R_a\cos^2\delta^0 + X_d'\sin\delta^0\cos\delta^0 + R_a\sin^2\delta^0 - X_q'\sin\delta^0\cos\delta^0 \\ R_a\sin\delta^0\cos\delta^0 - X_d'\cos^2\delta^0 - X_q'\sin^2\delta^0 - R_a\sin\delta^0\cos\delta^0 \end{matrix} \right]\cdots \\
&\quad \times \begin{bmatrix} \Delta V_x \\ \Delta V_y \end{bmatrix}.
\end{aligned} \tag{5.79}$$

This can be simplified using identity $\sin^2\delta^0+\cos^2\delta^0=1$ to give

$$\begin{aligned}
\begin{bmatrix}\Delta I_x\\ \Delta I_y\end{bmatrix} &= \begin{bmatrix} Y_g(R_a\cos\delta^0+X'_d\sin\delta^0) & Y_g(X'_q\cos\delta^0-R_a\sin\delta^0)\\ Y_g(R_a\sin\delta^0-X'_d\cos\delta^0) & Y_g(R_a\cos\delta^0+X'_q\sin\delta^0)\end{bmatrix}\begin{bmatrix}\Delta E'_d\\ \Delta E'_q\end{bmatrix}\\
&\quad+\begin{bmatrix}-I_y^0\\ I_x^0\end{bmatrix}\Delta\delta\\
&\quad-\begin{bmatrix} Y_g(R_aV_y^0-V_d^0X'_d\cos\delta^0+V_q^0V'_d\sin\delta^0)\\ -Y_g(R_aV_x^0+V_q^0X'_d\cos\delta^0+V_d^0X'_q\sin\delta^0)\end{bmatrix}\Delta\delta\\
&\quad-Y_g\begin{bmatrix} R_a+(X'_d-X'_q)\sin\delta^0\cos\delta^0 & X'_d\sin^2\delta^0+X'_q\cos^2\delta^0\\ -(X'_d\cos^2\delta^0+X'_q\sin^2\delta^0) & R_a+(X'_d-X'_q)\sin\delta^0\cos\delta^0\end{bmatrix}\cdots\\
&\quad\times\begin{bmatrix}\Delta V_x\\ \Delta V_y\end{bmatrix}
\end{aligned}$$

$$\begin{aligned}
&= \begin{bmatrix} Y_g(R_a\cos\delta^0+X'_d\sin\delta^0) & Y_g(X'_q\cos\delta^0-R_a\sin\delta^0)\\ Y_g(R_a\sin\delta^0-X'_d\cos\delta^0) & Y_g(R_a\cos\delta^0+X'_q\sin\delta^0)\end{bmatrix}\begin{bmatrix}\Delta E'_d\\ \Delta E'_q\end{bmatrix}\\
&\quad+\begin{bmatrix}-I_y^0\\ I_x^0\end{bmatrix}\Delta\delta-\begin{bmatrix}P_{13}\\ P_{23}\end{bmatrix}\Delta\delta\\
&\quad+\left[\begin{matrix} Y_g[-R_a+(X'_q-X'_d)\sin\delta^0\cos\delta^0]\\ Y_g(X'_d\cos^2\delta^0+X'_q\sin^2\delta^0)\end{matrix}\right|\\
&\quad\left|\begin{matrix} Y_g(-X'_d\sin^2\delta^0-X'_q\cos^2\delta^0)\\ Y_g[-R_a-(X'_d-X'_q)\sin\delta^0\cos\delta^0]\end{matrix}\right]\begin{bmatrix}\Delta V_x\\ \Delta V_y\end{bmatrix}\\
&= \begin{bmatrix} Y_g(R_a\cos\delta^0+X'_d\sin\delta^0) & Y_g(X'_q\cos\delta^0-R_a\sin\delta^0)\\ Y_g(R_a\sin\delta^0-X'_d\cos\delta^0) & Y_g(R_a\cos\delta^0+X'_q\sin\delta^0)\end{bmatrix}\begin{bmatrix}\Delta E'_d\\ \Delta E'_q\end{bmatrix}\\
&\quad+\begin{bmatrix}-(P_{13}+I_y^0)\\ -(P_{23}-I_x^0)\end{bmatrix}\Delta\delta\\
&\quad+\left[\begin{matrix} Y_g[-R_a+(X'_q-X'_d)\sin\delta^0\cos\delta^0]\\ Y_g(X'_d\cos^2\delta^0+X'_q\sin^2\delta^0)\end{matrix}\right|\\
&\quad\left|\begin{matrix} Y_g(-X'_d\sin^2\delta^0-X'_q\cos^2\delta^0)\\ Y_g[-R_a-(X'_d-X'_q)\sin\delta^0\cos\delta^0]\end{matrix}\right]\begin{bmatrix}\Delta V_x\\ \Delta V_y\end{bmatrix},
\end{aligned} \tag{5.80}$$

with

$$P_{13}=Y_g(R_aV_y^0-V_d^0X'_d\cos\delta^0+V_q^0V'_d\sin\delta^0), \tag{5.81}$$

$$P_{23}=Y_g(R_aV_x^0+V_q^0X'_d\cos\delta^0+V_d^0X'_q\sin\delta^0). \tag{5.82}$$

Rewriting (5.80) with $\mathbf{\Delta I_M}=\begin{bmatrix}\Delta I_x & \Delta I_y\end{bmatrix}^T$ in terms of state variables $\mathbf{\Delta X_M}$

and $\mathbf{\Delta V_G}$, we get

$$\begin{aligned}
\mathbf{\Delta I_M} &= \begin{bmatrix} Y_g(R_a\cos\delta^0 + X_d'\sin\delta^0) & Y_g(X_q'\cos\delta^0 - R_a\sin\delta^0) \\ Y_g(R_a\sin\delta^0 - X_d'\cos\delta^0) & Y_g(R_a\cos\delta^0 + X_q'\sin\delta^0) \end{bmatrix. \\
&\quad \left. \begin{matrix} -(P_{13}+I_y^0) & 0 & \cdots & 0 \\ -(P_{23}-I_x^0) & 0 & \cdots & 0 \end{matrix} \right] \mathbf{\Delta X_M} \\
&\quad + \left[\begin{matrix} Y_g[-R_a + (X_q' - X_d')\sin\delta^0\cos\delta^0] \\ Y_g(X_d'\cos^2\delta^0 + X_q'\sin^2\delta^0) \end{matrix} \right. \\
&\quad \left. \begin{matrix} Y_g(-X_d'\sin^2\delta^0 - X_q'\cos^2\delta^0) \\ Y_g[-R_a - (X_d' - X_q')\sin\delta^0\cos\delta^0] \end{matrix} \right] \mathbf{\Delta V_G} \\
&= \mathbf{W_G \Delta X_M} + \mathbf{N_G \Delta V_G}. \qquad (5.83)
\end{aligned}$$

5.3.2 NETWORK ADMITTANCE MATRIX $\mathbf{Y_N}$

From the previous subsection, the injected currents are derived in terms of the power system dynamic state variables and the generated terminal voltages. To include probabilistic loads into the algebraic equations, the load flow equations for perturbed loads are augmented into the transmission network admittance matrix $\mathbf{Y_T}$ to form an overall network admittance matrix $\mathbf{Y_N}$.

To calculate the overall load flow solution, a Norton equivalent circuit of the overall transmission network will then be needed. This transmission network admittance matrix $\mathbf{Y_T}$ can be obtained by sparse matrix analysis algorithms using the bus and line data. Probabilistic load (real power demand) can be augmented into a load admittance matrix $\mathbf{Y_L}$ using a similar approach analogous to constant impedance loads. In this case, the steady-state voltage, phase, and complex power obtained from load flow solutions are used to derive a new steady-state equivalent admittance matrix $\mathbf{Y_L}$ which is included in $\mathbf{Y_T}$.

In the course of each execution of load-flow solution loop, the perturbed loads are solved individually and sequentially along with the generators to obtain

$$\mathbf{Y_L} = \begin{bmatrix} \frac{S_1^*}{|V_1|^2} & 0 & \cdots & 0 \\ 0 & \frac{S_2^*}{|V_2|^2} & 0 & 0 \\ \vdots & 0 & \ddots & \vdots \\ 0 & 0 & \cdots & \frac{S_m^*}{|V_m|^2} \end{bmatrix}, \qquad (5.84)$$

where m is the number of buses and superscript * indicates complex conjugate. The injected current now compensates for the change in network and equivalent load impedances. Together, the network admittance matrix $\mathbf{Y_B}$ considering the transmission network and probabilistic loads can be augmented via

$$\mathbf{Y_B} = \mathbf{Y_T} + \mathbf{Y_L}. \qquad (5.85)$$

5.3.3 REDUCED ADMITTANCE MATRIX $\mathbf{Y_R}$

The overall network admittance matrix $\mathbf{Y_B}$ derived in the previous subsection represents the admittance of the synchronous machine (generator) buses and load buses. To enable small signal stability studies, $\mathbf{Y_B}$ will have to be reduced so as to formulate the overall system state matrix $\mathbf{A}$.

Now let the injected current vector be $\mathbf{\Delta I_S} = \left[\begin{array}{cc} \mathbf{\Delta I_G} & \mathbf{\Delta I_L} \end{array}\right]^T$ where $\mathbf{\Delta I_G}$ is the vector of synchronous machine (generator) currents and $\mathbf{\Delta I_L}$ is the vector of load currents. Similarly, let $\mathbf{\Delta V_S} = \left[\begin{array}{cc} \mathbf{\Delta V_G} & \mathbf{\Delta V_L} \end{array}\right]^T$ where $\mathbf{\Delta V_G}$ is the vector of synchronous machine (generator) terminal voltages as defined earlier, and $\mathbf{\Delta V_L}$ is the vector of load bus voltages. We can then partition $\mathbf{Y_B}$ in (5.85) into the following

$$\mathbf{\Delta I_S} = \left[\begin{array}{cc} \mathbf{Y_{GG}} & \mathbf{Y_{GL}} \\ \mathbf{Y_{LG}} & \mathbf{Y_{LL}} \end{array}\right] \mathbf{\Delta V_S}. \tag{5.86}$$

For the New England 39-Bus Test System, $\mathbf{Y_{GG}}$, $\mathbf{Y_{GL}}$, $\mathbf{Y_{LG}}$, and $\mathbf{Y_{LL}}$ are matrices of sizes 10×10, 10×29, 29×10, and 29×29, respectively. The above equation when expanded yields

$$\mathbf{\Delta I_G} = \mathbf{Y_{GG}} \mathbf{\Delta V_G} + \mathbf{Y_{GL}} \mathbf{\Delta V_L}, \tag{5.87}$$

$$\mathbf{\Delta I_L} = \mathbf{Y_{LG}} \mathbf{\Delta V_G} + \mathbf{Y_{LL}} \mathbf{\Delta V_L}. \tag{5.88}$$

At steady-state for small signal stability studies, $\mathbf{\Delta I_L} = 0$ because the injection currents to other nodes other than generator internal nodes are zero. Hence, $\mathbf{\Delta V_L} = -\mathbf{Y_{LL}}^{-1} \mathbf{Y_{LG}} \mathbf{\Delta V_G}$, and we can substitute this into (5.87) to get

$$\begin{aligned} \mathbf{\Delta I_G} &= (\mathbf{Y_{GG}} - \mathbf{Y_{GL}} \mathbf{Y_{LL}}^{-1} \mathbf{Y_{LG}}) \mathbf{\Delta V_G} \\ &= \mathbf{Y_R} \mathbf{\Delta V_G}, \end{aligned} \tag{5.89}$$

where $\mathbf{Y_R}$ is the *reduced admittance matrix*. This method is only applicable when the loads are treated as constant impedance loads as assumed earlier. If not, these load nodes have to be kept in the original $\mathbf{Y_N}$ matrix.

Since the relationship between nodal injected current vector $\mathbf{\Delta I_M}$ and generator terminal voltages $\mathbf{\Delta V_G}$ can be given by the derived reduced admittance matrix as

$$\begin{aligned} \mathbf{\Delta I_M} &= \left[\begin{array}{cc} \Re\{\mathbf{Y_R}^{(1,1)}\} & \Im\{\mathbf{Y_R}^{(1,2)}\} \\ \Im\{\mathbf{Y_R}^{(2,1)}\} & \Re\{\mathbf{Y_R}^{(2,2)}\} \end{array}\right] \mathbf{\Delta V_G} \\ &= \mathbf{Y_{TXY}} \mathbf{\Delta V_G}, \end{aligned} \tag{5.90}$$

where $\Re\{.\}$ and $\Im\{.\}$ denote the real and imaginary parts of the contents in the brackets, respectively. We can rewrite (5.83) as

$$\mathbf{Y_{TXY}} \mathbf{\Delta V_G} = \mathbf{W_G} \mathbf{\Delta X_M} + \mathbf{N_G} \mathbf{\Delta V_G}, \tag{5.91}$$

which can be rearranged to give

$$\begin{aligned} \mathbf{0} &= \mathbf{W_G} \mathbf{\Delta X_M} + (\mathbf{N_G} - \mathbf{Y_{TXY}}) \mathbf{\Delta V_G} \\ &= \mathbf{A_S} \mathbf{\Delta X_M} + \mathbf{C_S} \mathbf{\Delta V_G}, \end{aligned} \tag{5.92}$$

and the above equation is the algebraic equation for the New England system under consideration.

5.4 STATE MATRIX AND CRITICAL MODES

From *Lyapunov's first method*, the set of non-linear power system dynamic differential and algebraic equations from the previous sections are linearized about their equilibrium operating points using Taylor's expansions. The use of Lyapunov's first method consists of the following procedures [205]:

1. linearization of the original system;
2. elimination of the algebraic variables to form the reduced dynamic state matrix **A**;
3. computation of the eigenvalues and eigenvectors of the state matrix **A**;
4. stability study of the system [206];
   a. If eigenvalues of the state matrix are located in the left hand side of the complex plane, then the system is said to be small-signal stable at the studied equilibrium point;
   b. If the most right eigenvalue is zero, the system is on the edge of small-signal aperiodic instability;
   c. If the most right complex conjugate pair of eigenvalues has zero real part and non-zero imaginary part, the system on the edge of oscillatory instability depending on the transversality condition [207];
   d. If the system has eigenvalue with positive real parts, the system is not stable;
   e. For the stable case, analyze several characteristics including damping and frequencies for all modes, eigenvalue sensitivities to the system parameters, excitability, observability, and controllability of the modes.

As such, the synchronous machines and their control systems when subjected to small signal disturbance perturbations can be represented by a set of linear time-invariant dynamic and algebraic state-space equations, respectively, as

$$\mathbf{\Delta\dot{X}_M} = \mathbf{A_M \Delta X_M} + \mathbf{C_M \Delta V_G}, \tag{5.93}$$

$$\mathbf{0} = \mathbf{A_S \Delta X_M} + \mathbf{C_S \Delta V_G}, \tag{5.94}$$

where $\mathbf{\Delta X_M}$ is a vector of dynamic machine variables and $\mathbf{\Delta V_G}$ is a vector of generator terminal voltages in x and y-axis. Matrices $\mathbf{A_M}$, $\mathbf{C_M}$, $\mathbf{A_S}$, and $\mathbf{C_S}$ are in essence the *Jacobians* of the non-linear machine and stator equations, and were derived in the previous section.

The New England test system shown in Figure 5.4 consists of ten interconnected synchronous machines.

Now augmenting all ten machine states together simultaneously to derive the overall state matrix **A** with $\mathbf{X} = \begin{bmatrix} \mathbf{\Delta X_M}^{(1)} & \mathbf{\Delta X_M}^{(2)} & \cdots & \mathbf{\Delta X_M}^{(10)} \end{bmatrix}^T$ and the generated terminal voltages together with $\mathbf{V} = \begin{bmatrix} \mathbf{\Delta V_G}^{(1)} & \mathbf{\Delta V_G}^{(2)} & \cdots & \mathbf{\Delta V_G}^{(10)} \end{bmatrix}^T$, we can combine the dynamic and algebraic equations together to form matri-

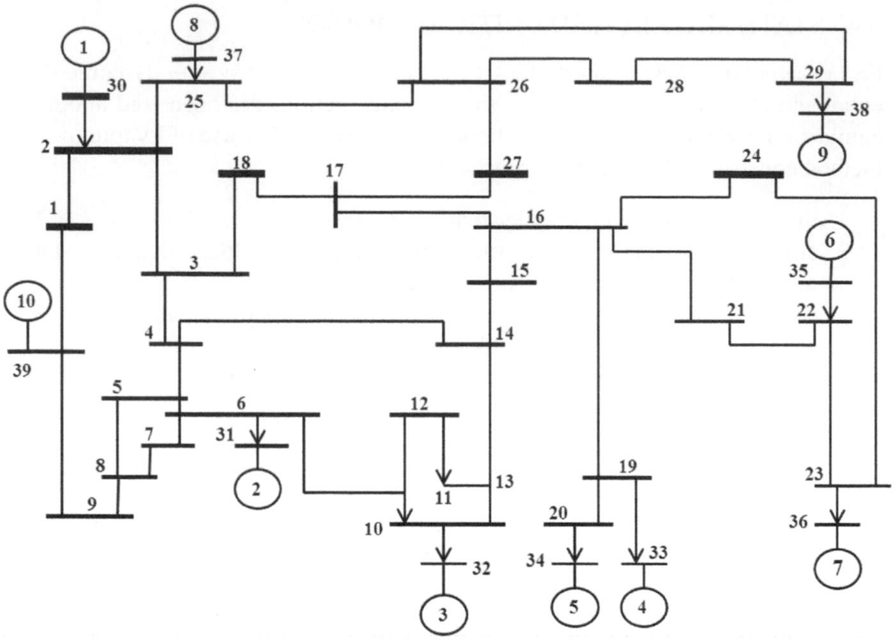

FIGURE 5.4 New England test system.

ces $\mathbf{A_D}$, $\mathbf{C_D}$, $\mathbf{A_A}$, and $\mathbf{C_A}$ as

$$\begin{aligned}\dot{\mathbf{X}} &= \begin{bmatrix} \mathbf{A_M}^{(1)} & \mathbf{0} & \mathbf{0} & \mathbf{0} \\ \mathbf{0} & \mathbf{A_M}^{(2)} & \mathbf{0} & \mathbf{0} \\ \vdots & \mathbf{0} & \ddots & \vdots \\ \mathbf{0} & \mathbf{0} & \cdots & \mathbf{A_M}^{(10)} \end{bmatrix} \mathbf{X} \\ &+ \begin{bmatrix} \mathbf{C_M}^{(1)} & \mathbf{0} & \mathbf{0} & \mathbf{0} \\ \mathbf{0} & \mathbf{C_M}^{(1)} & \mathbf{0} & \mathbf{0} \\ \vdots & \mathbf{0} & \ddots & \vdots \\ \mathbf{0} & \mathbf{0} & \cdots & \mathbf{C_M}^{(10)} \end{bmatrix} \mathbf{V} \\ &= \mathbf{A_D X} + \mathbf{C_D V}, \end{aligned} \tag{5.95}$$

$$\begin{aligned}\mathbf{0} &= \begin{bmatrix} \mathbf{A_S}^{(1)} & \mathbf{0} & \mathbf{0} & \mathbf{0} \\ \mathbf{0} & \mathbf{A_S}^{(2)} & \mathbf{0} & \mathbf{0} \\ \vdots & \mathbf{0} & \ddots & \vdots \\ \mathbf{0} & \mathbf{0} & \cdots & \mathbf{A_S}^{(10)} \end{bmatrix} \mathbf{X} \\ &+ \begin{bmatrix} \mathbf{C_S}^{(1)} & \mathbf{0} & \mathbf{0} & \mathbf{0} \\ \mathbf{0} & \mathbf{C_S}^{(1)} & \mathbf{0} & \mathbf{0} \\ \vdots & \mathbf{0} & \ddots & \vdots \\ \mathbf{0} & \mathbf{0} & \cdots & \mathbf{C_S}^{(10)} \end{bmatrix} \mathbf{V} \\ &= \mathbf{A_A X} + \mathbf{C_A V}, \end{aligned} \tag{5.96}$$

where the superscripts indicate the machine number. For the New England Power System, $\mathbf{A_D}$ is of size 94×94 as bus number 39 is an inertia bus serving as the slack (swing) bus. The sizes of $\mathbf{C_D}$, $\mathbf{A_A}$, and $\mathbf{C_A}$ are hence of sizes 94×20, 20×94, and 20×20, respectively. The augmented algebraic equation (5.96) can also be rewritten as

$$\mathbf{V} = -\mathbf{C_A}^{-1}\mathbf{A_A X} \tag{5.97}$$

if $\mathbf{C_A}$ is non-singular.

Power system modeling is essential to power system small signal stability assessment. Small signal stability is based on eigenvalue analysis which is sensitive to system model variations. A small change in the system states may have significant impact on the system eigenvalue properties. Consequently, a reliable and accurate system model including both the system dynamics and algebraic equations is required for small signal stability analysis in the following sections.

Bifurcation theory is often used to study the different types of unstable equilibria of the system. Bifurcations occur where, by slowly varying certain system parameters in some direction, the system properties change qualitatively or quantitatively at a certain point. Local bifurcations can be detected by monitoring the behavior of eigenvalues of the systems operation point. In some direction of parameter variation, the system may become unstable due to the singularity of the system dynamic state matrix associated with zero eigenvalue, or because of a pair of complex conjugate eigenvalues crossing the imaginary axes of the complex plane. These two phenomena are known as the *saddle-node* and *Hopf* bifurcations, respectively. Other conditions which may drive the system state into instability may also occur, such as *singularity-induced* bifurcations, *cyclic fold*, *period doubling*, *blue sky* bifurcations, or even *chaos* [208][209][210].

For example, saddle-node bifurcation occurs when the state matrix **A** by substituting (5.97) into (5.95) in

$$\begin{aligned}\dot{\mathbf{X}} &= (\mathbf{A_D} - \mathbf{C_D}\mathbf{C_A}^{-1}\mathbf{A_A})\mathbf{X} \\ &= \mathbf{AX} \qquad (5.98)\end{aligned}$$

becomes singular with a *negative real eigenvalue crossing into the open right half complex plane*. A typical physical phenomenon associated with this type of bifurcation is a static voltage collapse or aperiodic type instability.

In order to study small signal stability for power systems, the state matrix will have to be formulated to address the different physical phenomena and mathematical descriptions. The dynamics of the system can be described by the linearized differential equations. The stability of the system is therefore determined by the eigenvalues of the state matrix **A**. Based on the small signal stability theorem and system dynamics, critical eigenvalues of the system can then be identified based on their mode of oscillations. With these critical eigenvalues of a power system, the small signal stability properties can be obtained at the particular equilibrium only. In order to assess the system small signal stability region over a range of operating points, repeated computations are required so that the system state matrix **A** and corresponding critical eigenvalues can be computed at each operating point to obtain an overall picture of the small signal stability property. Given the complexity of a power system, the total number of possible parameter variations can be huge, and makes this approach of computing critical eigenvalues computationally inefficient and even impractical in most cases. This motivates us to investigate this probabilistic small signal stability study.

Using the equations derived above, a case study is performed on the New England test system as shown in Figure 5.4. The New England test system under consideration consists of ten synchronous machines, thirty-nine buses, thirty-four transmission lines, nine IEEE Type 1 AVR and exciters, nine speed governors and turbines, as well as one PSS (Power System Stabilizer). Bus 39 is a slack bus (inertia bus). Each linearized synchronous machine is modeled as a fourth order system. The IEEE Type 1 AVR and exciter as well as speed governors and turbines are modeled as third order systems. As such, these machines (and control) together form a tenth order system. Augmenting all the ten machines together and noting that there is no IEEE Type 1 AVR and exciter, governor or speed governor and turbine for the slack bus, the dynamic behavior of the power system can be described by a state matrix **A** as described in (5.98). **A** is a square matrix of size 94×94. Stability analysis can then be done on **A** by considering its eigenvalues and eigenvectors (mode shapes).

It is well known that small signal stability is based on eigenvalue analysis of the state matrix. The system state matrix **A** obtained earlier includes the dynamic and static properties of a power system. Such characteristics can be revealed and investigated by eigenvalue analysis of the system state matrix. By far, the detailed power system models have been presented and the state matrix of a power system is derived. These have laid out the foundation for further probabilistic small signal stability analysis. In the following sections, the New England test system is used for all case studies.

In this section, the eigenvalue analysis of New England test system is performed. Critical eigenvalues will be identified for probabilistic small signal stability analysis. In general, the number of eigenvalues or eigenvectors equals the order of the dynamic system [201][205]. For a system with m differential equations, there will be m eigenvalues and m eigenvectors. However, not all of the system eigenvalues need to be studied for stability, because many of them remain in the left complex plane under small signal disturbances. There are only a limited number of eigenvalues which could be pushed to cross the imaginary axis leading to instability under small disturbances. These eigenvalues are critical eigenvalues which correspond to low frequency oscillation modes with poor damping factor among all system eigenvalues. The critical eigenvalues will be identified here. General eigenvalue and eigenvector analysis for identification of electromechanical oscillation is first introduced based on the state matrix **A**.

The eigenvalue of **A** are the roots of the characteristic equation corresponding to the state equations and are defined as the values that satisfy

$$\det(\lambda_i \mathbf{I} - \mathbf{A}) = 0. \tag{5.99}$$

The total number of eigenvalues is equal to the number of state variables and each eigenvalue represents a natural mode. Eigenvectors corresponding to the *mode shapes* can then be identified from

$$\mathbf{A}\mathbf{v}_i = \lambda_i \mathbf{v}_i \tag{5.100}$$

$$\mathbf{w}_i^T \mathbf{A} = \lambda_i \mathbf{w}_i^T, \tag{5.101}$$

where

$$\begin{aligned} \mathbf{v}_i \cdot \mathbf{w}i^T &= \mathbf{w}_i^T \mathbf{v}_i \\ &= 1. \end{aligned} \tag{5.102}$$

$\mathbf{v}_i$ and $\mathbf{w}_i^T$ are known as the *right* and *left* eigenvectors, respectively, for the same eigenvalue λ_i. The above-mentioned notions have also been covered earlier in Chapter 2.

We define critical modes as the modes whose eigenvalues are associated with electromechanical oscillations. According to the nature of power system, there are two major electromechanical modes:

1. *Local plant mode*: Local plant mode is associated with the torque angles of the generating units swinging with respect to the rest of the power system.

This problem is more pronounced with the inclusion of fast response excitation system. It typically has natural frequency in the range of 0.8 to 2 Hz; and

2. *Inter-area mode*: Inter-area mode is associated with many generators in one part of the system swinging against generators in other parts. It is caused by two or more groups of closely coupled generators which are interconnected by weak tie lines between groups. The natural frequency of this type of mode is typically in the range of 0.1 to 0.8 Hz.

The critical eigenvalues of the New England test system at 830 MW are given in Table 5.1.

TABLE 5.1
Critical Eigenvalues and Their Locations at 830MW Level

No.	Location	Critical Frequency	Frequency	Damping Ratio
1	11	$-0.3676 \pm 8.7533j$	1.39	0.0420
2	13	$-0.4021 \pm 8.6764j$	1.38	0.0464
3	15	$-0.3131 \pm 8.4747j$	1.35	0.0370
4	21	$-0.2740 \pm 7.4602j$	1.19	0.0368
5	23	$-0.0001 \pm 6.9946j$	1.11	-0.0002
6	25	$-0.2475 \pm 6.9946j$	1.11	0.0356
7	27	$-0.2501 \pm 6.3504j$	1.01	0.0394
8	29	$-0.2609 \pm 5.9670j$	0.95	0.0433
9	33	$-0.2796 \pm 3.7787j$	0.61	0.0725

The critical eigenvalues are then identified and their corresponding normalized left and right eigenvectors are obtained. With these critical modes, we can proceed to calculate the sensitivity matrix **S** to non-deterministic system parameters in the following section.

5.5 EIGENVALUE SENSITIVITY MATRIX

In this section, we introduce the concepts of eigenvalue sensitivity analysis techniques. When a power system is subject to small signal disturbances and perturbations, the state matrix **A** contains non-deterministic variables. As such, these random variables will cause eigenvalues of **A** to be non-deterministic. These non-deterministic phenomena are results of system uncertainties, un-modeled dynamics, or measurement noises, etc.

We present the development of algorithms to calculate the sensitivity matrix of the system eigenvalues to non-deterministic random variables of the system [179][180][205][211][212][213][214]. Both analytical and numerical methods are investigated and compared. The effectiveness of our proposed schemes are verified with case studies based on the New England test system.

Recall from (2.81) in Chapter 2 that

$$\frac{\partial \lambda_i}{\partial \mathbf{K}_j} = \frac{\mathbf{w}_i^T \frac{\partial \mathbf{A}}{\partial \mathbf{K}_j} \mathbf{v}_i}{\mathbf{w}_i^T \mathbf{v}_i}, \tag{5.103}$$

and

$$\Delta\mathbf{\Lambda} = \mathbf{S}\Delta\mathbf{\Gamma}, \tag{5.104}$$

where $\Delta\mathbf{\Gamma} = \begin{bmatrix} \Delta\mathbf{K}_1 & \Delta\mathbf{K}_2 & \cdots & \Delta\mathbf{K}_j \end{bmatrix}^T$ is a vector of non-deterministic system parameters and $\Delta\mathbf{\Lambda} = \begin{bmatrix} \Delta\lambda_i & \Delta\lambda_i & \cdots & \Delta\lambda_i \end{bmatrix}^T$ is a vector of critical eigenvalues of interest. As such, $\mathbf{S} \in \mathbb{R}^{i \times j}$ is the eigenvalue sensitivity matrix, and the main obstacle in computing $\mathbf{S}$ is the term $\frac{\partial \mathbf{A}}{\partial \mathbf{K}_j}$ as can be observed in (5.103).

In order to evaluate the sensitivity matrix of the system with respect to its non-deterministic parameters, the partial derivative of the matrix $\frac{\partial \mathbf{A}}{\partial \mathbf{K}_j}$ will have to be calculated first. The base values of $\mathbf{K}_j$ are obtained from conventional Newton-Raphson load flow solutions.

Moreover, if the non-deterministic system parameters are states of the state matrix **A**, then the matrix $\frac{\partial \mathbf{A}}{\partial \mathbf{K}_j}$ can be obtained from an analytical or a direct approximation method as shown in the sequel. This approach is made possible as the matrix **A** can be expressed explicitly in terms of the state variables, hence the required system parameters.

However, if the non-deterministic parameters are not states of the system, an analytical solution (though possible) is proven to be computational intensive and numerically inefficient. As such, a partial finite difference approach is recommended and used. For our application, the perturbed exciter voltage regulator gain K_A is chosen as the parameter of study for illustration purposes but without loss of generality.

If the required perturbed parameters appear explicitly in state matrix **A**, e.g., the voltage regulator gain K_A in the IEEE Type 1 Exciter as shown in Figure 5.1, then $\frac{\partial \mathbf{A}}{\partial K_A}$ can be obtained by direct differentiation of elements in **A**, i.e.,

$$\frac{\partial \mathbf{A}}{\partial K_A} = \frac{\partial}{\partial K_A}\{a_{ij}\}, \tag{5.105}$$

where a_{ij} are entries in **A**. The sensitivity of the state matrix **A** to the exciter voltage regulator gain K_A at the ith machine is obtained by

$$\begin{bmatrix} \Delta\dot{V}_{ex1}^{(i)} \\ \Delta\dot{V}_{ex2}^{(i)} \\ \Delta\dot{V}_{ex3}^{(i)} \\ \Delta\dot{E}_{fd}^{(i)} \end{bmatrix} = \begin{bmatrix} -\frac{1}{T_R^{(i)}} & 0 & 0 & 0 \\ -\frac{K_A^{(i)}}{T_A^{(i)}} & -\frac{1}{T_A^{(i)}} & \frac{K_A^{(i)}T_A^{(i)}}{K_F^{(i)}T_F^{(i)}} & -\frac{K_A^{(i)}T_A^{(i)}}{K_F^{(i)}T_F^{(i)}} \\ 0 & 0 & -\frac{1}{T_F^{(i)}} & \frac{1}{T_F^{(i)}} \\ 0 & \frac{1}{T_E^{(i)}} & 0 & -\frac{K_E^{(i)}+S_E^{(i)}}{T_E^{(i)}} \end{bmatrix} \begin{bmatrix} \Delta V_{ex1}^{(i)} \\ \Delta V_{ex2}^{(i)} \\ \Delta V_{ex3}^{(i)} \\ \Delta E_{fd}^{(i)} \end{bmatrix}$$

$$+ \begin{bmatrix} \frac{V_{gx}^{(i)}}{T_R^{(i)}V_g^{(i)}} & \frac{V_{gy}^{(i)}}{T_R^{(i)}V_g^{(i)}} \\ 0 & 0 \\ 0 & 0 \\ 0 & 0 \end{bmatrix} \begin{bmatrix} \Delta V_{gx}^{(i)} \\ \Delta V_{gy}^{(i)} \end{bmatrix}, \tag{5.106}$$

$$\therefore \frac{\partial \mathbf{A}}{\partial K_A^{(i)}} = \begin{bmatrix} \cdots & \cdots & \cdots & \cdots & \cdots & \cdots \\ \cdots & -\frac{1}{T_A^{(i)}} & 0 & \frac{T_A^{(i)}}{K_F^{(i)}T_F^{(i)}} & -\frac{T_A^{(i)}}{K_F^{(i)}T_F^{(i)}} & \cdots \\ \cdots & \cdots & \cdots & \cdots & \cdots & \cdots \\ \cdots & \cdots & \cdots & \cdots & \cdots & \cdots \end{bmatrix}. \tag{5.107}$$

We now provide another example where the perturbed parameters appear explicitly in state matrix $\mathbf{A}$. Normally, speed governor systems have slow responses and do not significantly affect local modes with frequencies greater than 1 Hz. However, they may have negative damping effects on inter-area mode oscillations in an interconnected power system. A simple governor model is used here for sensitivity analysis [199][201] as shown in Figure 5.3. The dynamic equations of a typical speed governor system installed at the ith generator are given by

$$\Delta\dot{\mathbf{X}}_T = \begin{bmatrix} \frac{-1}{T_{TG4}} & 0 & 0 \\ \frac{K_{TG2}}{T_{TG5}} & \frac{-1}{T_{TG5}} & 0 \\ 0 & \frac{K_{TG3}}{T_{TG6}} & \frac{-1}{T_{TG6}} \end{bmatrix} \Delta\mathbf{X}_T + \begin{bmatrix} 0 & 0 \\ 0 & 0 \\ 0 & 0 \end{bmatrix} \Delta\mathbf{V}_G + \begin{bmatrix} \frac{-K_{TG1}}{T_{TG4}} \\ 0 \\ 0 \end{bmatrix} \Delta w, \tag{5.108}$$

where $\Delta\mathbf{X}_T = \begin{bmatrix} \Delta V_{tg1} & \Delta V_{tg2} & \Delta V_{tg3} \end{bmatrix}^T$ and $\Delta\mathbf{V}_G = \begin{bmatrix} \Delta V_x & \Delta V_y \end{bmatrix}^T$. The interactions between the synchronous machines and the governor system to be considered are then

$$\Delta\dot{\mathbf{X}}_{sg(i)} = \begin{bmatrix} \cdots & \cdots & \cdots & \cdots \\ 0 & 0 & \frac{1-K_{TG2(i)}}{2H_{(i)}} & \frac{1-K_{TG3(i)}}{2H_{(i)}} \end{bmatrix} \Delta\mathbf{X}_{sg(i)}. \tag{5.109}$$

Similar to that of the excitation system parameters, the speed governor system gains K_{TG1}, K_{TG2}, and K_{TG3}'s individual contribution to the state matrix sensitivity are then given as

$$\frac{\partial \mathbf{A}}{\partial K_{TG2}^{(i)}} = \begin{bmatrix} 0 & 0 & 0 \\ \frac{1}{T_{TG3}} & 0 & 0 \\ 0 & 0 & 0 \end{bmatrix}, \tag{5.110}$$

$$\frac{\partial \mathbf{A}}{\partial K_{TG3}^{(i)}} = \begin{bmatrix} 0 & 0 & 0 \\ 0 & 0 & 0 \\ 0 & \frac{1}{T_{TG6}} & 0 \end{bmatrix}, \tag{5.111}$$

$$\frac{\partial \mathbf{A}}{\partial K_{TG1}^{(i)}} = \begin{bmatrix} \frac{-1}{T_{TG4}} \\ 0 \\ 0 \end{bmatrix}, \tag{5.112}$$

$$\frac{\partial \mathbf{A}}{\partial K_{TG2(i)}} = \begin{bmatrix} \cdots & \cdots & \cdots & \cdots \\ 0 & 0 & \frac{-1}{2H_{(i)}} & 0 \end{bmatrix}, \tag{5.113}$$

$$\frac{\partial \mathbf{A}}{\partial K_{TG3(i)}} = \begin{bmatrix} \cdots & \cdots & \cdots & \cdots \\ 0 & 0 & 0 & \frac{-1}{2H_{(i)}} \end{bmatrix}. \tag{5.114}$$

For a governor system, the time constants T_{TGi} may have some deviation from their rated values over the time as well. Because these time constants are often denominators in the state matrix, they may have more impact than the system gains K_{TGi}. Their contribution to the partial derivative of the state matrix can be obtained in a similar fashion to that of K_{TGi}.

So far, the analytical approach only applies to parameters which have direct entry into the system state matrix such that an analytical form of the sensitivity can be accurately computed. For example, it is well known that when the load changes, the system voltages and voltage angles will also change to meet the required power flow conditions. However, many parameters, e.g., rotor angles and load powers, etc., do not have a direct entry into the state matrix but do have an impact on the system state matrix. To find out the analytical form of sensitivity with respect to such parameters, e.g., $\frac{\partial \mathbf{A}}{\partial \mathbf{P}_i}$, is a very complex task. For large scale systems, it may even become impractical or even impossible to derive an analytical form. A numerical approach is hence more appropriate in view of computational efficiency and practicality. We shall introduce the numerical approach to computing sensitivity next.

The numerical method of determining $\frac{\partial \mathbf{A}}{\partial \mathbf{K}_j}$ will be derived. Consider for example, the active load power $\Delta\mathbf{P}$ where $\Delta\mathbf{P}$ does not appear explicitly in the state matrix $\mathbf{A}$. As such, we have to first decouple $\Delta\mathbf{P}_j$ from $\mathbf{A}$ in order to formulate $\frac{\delta \mathbf{A}}{\delta \Delta \mathbf{P}_j}$, and the perturbation of the active load power $\Delta\mathbf{P}_j$ is assumed to be small and normal. In most cases, a 1% perturbation is a good choice, as parameter changes that are too large violates local linearity assumptions, while parameter changes that are too small causes high round-off errors after division [212].

From the definition and first principle of partial derivative, we have

$$\frac{\partial \Lambda}{\partial \mathbf{K}_j} = \lim_{\partial \mathbf{K}_j \to 0} \frac{\Lambda(\mathbf{K}_j + \Delta\mathbf{K}_j) - \Lambda(\mathbf{K}_j)}{\Delta \mathbf{K}_j}, \tag{5.115}$$

where $\mathbf{\Lambda}(\mathbf{K}_j + \partial \mathbf{K}_j)$ and $\mathbf{\Lambda}(\mathbf{K}_j)$ are the eigenvalues of the system matrix $\mathbf{A}$ after and before the parameter perturbation, respectively. Normally, only the critical eigenvalues need to be evaluated subjected to small perturbations. The following assumption is valid

$$\frac{\partial \mathbf{\Lambda}}{\partial \mathbf{K}_j} \approx \frac{\mathbf{\Lambda}(\mathbf{K}_j + \Delta \mathbf{K}_j) - \mathbf{\Lambda}(\mathbf{K}_j)}{\Delta \mathbf{K}_j}, \tag{5.116}$$

if $\Delta \mathbf{K}_j$ is small when compared to the entries in $\mathbf{\Lambda}$.

The numerical approach also applies to computing the sensitivity of a state matrix to perturbed real load power $\frac{\partial \mathbf{A}}{\partial \mathbf{P}_j}$ and to perturbed control system parameter $\frac{\partial \mathbf{A}}{\partial K_A}$.

Recall from (5.104) that

$$\Delta \mathbf{\Lambda} = \mathbf{S} \Delta \mathbf{\Gamma}, \tag{5.117}$$

where $\Delta \mathbf{\Gamma} = \begin{bmatrix} \Delta \mathbf{K}_1 & \Delta \mathbf{K}_2 & \cdots & \Delta \mathbf{K}_j \end{bmatrix}^T$, $\Delta \mathbf{\Lambda} = \begin{bmatrix} \Delta \lambda_1 & \Delta \lambda_2 & \cdots & \Delta \lambda_i \end{bmatrix}^T$, and $\mathbf{S} \in \mathbb{R}^{i \times j}$ is the eigenvalue sensitivity matrix. $\mathbf{S}$ relates how the ith eigenvalue is affected by the jth non-deterministic system parameter $\mathbf{K}_j$, i.e., the rate of change of λ_i with respect to $\mathbf{K}_j$. The first order sensitivity matrix $\mathbf{S}$ is in essence the rate of change of the eigenvalues of interest with respect to the various parameters. It should be noted that $\mathbf{S}$ is a local sensitivity rather than a global sensitivity of the parameter space. Expanding the above equation about operating point $\mathbf{\Lambda}_0$ with Taylor expansion and ignoring higher order terms, we get the following approximation

$$\mathbf{\Lambda} \approx \mathbf{\Lambda}_0 + \mathbf{S} \Delta \mathbf{\Gamma}, \tag{5.118}$$

where $\mathbf{\Lambda}_0$ and $\mathbf{\Lambda}$ are the eigenvalues vectors before and after introducing a small change in $\mathbf{\Gamma}$.

Now, assuming $\mathbf{\Gamma}$ to be multivariate normal (which is justifiable as uncertainties and forecast errors in power systems are assumed to be available from past data and stationary in time internal of study, i.e., ergodicity and independence are not essential), $\mathbf{\Lambda}$ will also be multivariate normal. For conventional dynamic small signal stability studies, we shall only look at the real part of the eigenvalues by

$$\Re\{\mathbf{\Lambda}\} \approx \Re\{\mathbf{\Lambda}_0\} + \Re\{\mathbf{S}\} \Delta \mathbf{\Gamma}. \tag{5.119}$$

It is hence evident that the analytical approach may become too complex for some parameters such as system loads. However, the numerical approach provides a convenient way to approximate the analytical sensitivity results. The amount of perturbation to perform the numerical approach, however, needs to be chosen carefully in order to ensure the robustness of the sensitivity computation. If it is too small, the system may not be able to respond to such perturbation numerically; if it is too large, the system may not be able to maintain smooth change of states. In either case, the final results will not be a true approximation to the actual sensitivities. Subsequent case studies on the New England 39-Bus Test System will indicate a suitable choice of perturbations at realistic and acceptable levels.

5.5.1 SENSITIVITY ANALYSIS OF THE NEW ENGLAND POWER SYSTEM

The New England system shown in Figure 5.4 is used to test the derived algorithms in computing the eigenvalue sensitivities. Both the numerical and analytical approaches are applied to compute the eigenvalue sensitivity factor with respect to the selected power system parameters.

Firstly, the analytical approach of eigenvalue sensitivity computation with respect to variations in K_A is performed as shown in Table 5.2. Both analytical and numerical approaches have been applied to study the eigenvalue sensitivity based on the New England system for different perturbation levels as given in Tables 5.2–5.6, which show that a 1% parametric perturbation using numerical methods can produce reasonably good results compared with their analytical counterparts. This is very important for large-scale system analysis and for sensitivity computations with respect to other parameters not directly appear in the state matrix **A**.

TABLE 5.2
Sensitivity Factor of Critical Eigenvalue to Exciter Gain K_A of Generator at Bus 30 Using the Analytical Approach

No.	Oscillation Mode	Sensitivity Factor ($\times 10^{-3}$)
1	$-0.3678 \pm j8.7547$	0.00078162184764
2	$-0.4031 \pm j8.6747$	0.01106642351696
3	$-0.3139 \pm j8.4773$	−0.00063404657501
4	$-0.2745 \pm j7.4595$	0.00022037834264
5	$0.0013 \pm j6.9647$	−0.00957289052384
6	$-0.2493 \pm j6.9965$	−0.00513654645813
7	$-0.2507 \pm j6.3571$	−0.00071777018732
8	$-0.2600 \pm j5.9958$	−0.00214383955583
9	$-0.2798 \pm j3.8493$	−0.06791209198673

Tables 5.7–5.9 show the sensitivity values with respect to governor time constant T_{TG6}, and Tables 5.10–5.12 show the sensitivity values with respect to governor gain K_{TG2} for all generators and all critical modes.

It is shown in Tables 5.7–5.9 and Tables 5.10–5.12 that T_{TG6} at Bus 30 is the most sensitive one among all T_{TG6} at critical mode $5(0.003414985)$, and K_{TG1} at Bus 34 is most sensitive at critical mode $8(0.404146387)$. It can also be seen that even though the analytical approach does provide an accurate sensitivity factor, it carries with it a very complex numerical analysis, and a different analysis has to be done for different system parameters or even for different systems. The numerical approach is computationally simple and has been proven to be accurate enough as compared with the analytical approach. Generally, a 1% perturbation is able to generate acceptable results, comparable to that of the analytical analysis results, but saves significant analytical time and possibility of errors. Because of the simplicity of the numerical approach, it will be used in sensitivity analysis for other parameters as well, with a 1%

TABLE 5.3
Sensitivity Factor of Critical Eigenvalue to Exciter Gain K_A of Generator at Bus 30 Using the Numerical Approach

No.	Oscillation Mode	Sensitivity Factor($\times 10^{-3}$)
1	$-0.3678 \pm j8.7547$	0.00078163731043
2	$-0.4031 \pm j8.6747$	0.01106188979083
3	$-0.3139 \pm j8.4773$	−0.00063335882650
4	$-0.2745 \pm j7.4595$	0.00022046186654
5	$0.0013 \pm j6.9647$	−0.00959517106325
6	$-0.2493 \pm j6.9965$	−0.00513931605273
7	$-0.2507 \pm j6.3571$	−0.00071194674489
8	$-0.2600 \pm j5.9958$	−0.00208513232680
9	$-0.2798 \pm j3.8493$	−0.06636688800743

TABLE 5.4
Sensitivity Analysis Errors for Different Exciter Gain K_A Perturbation Levels

No.	Error(%) K_A from 5.0 to 6.0	Error(%) K_A from 5.0 to 5.1	Error(%) K_A from 5.0 to 5.01
1	0.008	0.003	0.0004
2	−0.98	−0.09	−0.009
3	−2.10	−0.2	−0.020
4	0.46	0.04	0.004
5	15.41	1.52	0.152
6	0.80	0.08	0.008
7	19.74	1.95	0.194
8	−14.88	−1.45	−0.145
9	−121.60	−11.92	−1.190

perturbation.

Based on the analytical, and more importantly, numerical methods discussed in previous sections, the eigenvalue sensitivity analysis is performed to all the system excitation system gains K_{Ai}, governor gains K_{TGi}, and governor time constant T_{TGi} to find the parameters which have the most impact on system eigenvalue variations, i.e., the parameters with which the system eigenvalue sensitivities are the highest. Tables 5.7–5.12 are selected complete eigenvalue sensitivities for all nine synchronous generators. Similar analysis is performed on other parameters at all machines. Tables 5.13–5.17 summarize the findings and identify the most sensitive parameters, critical eigenvalues, and corresponding generators. By identifying such parameters and their associated machines, the system operator is able to pay more attention to

TABLE 5.5
Comparison of Analytical to Numerical Approaches in Computing the Eigenvalue Sensitivities (Speed Governor Gain K_{TG1} at Bus 30, Sensitivity Values $\times 10^{-3}$)

Mode No.	Numerical	Analytical	Error(%)
1	−0.000272155	−0.000272677	0.191684
2	0.016839917	0.016877105	0.220346
3	−0.00044598	−0.000441561	1.000776
4	0.000301912	0.000303428	0.499824
5	0.498817537	0.498859015	0.008315
6	−0.01390707	−0.013915327	0.059333
7	0.00113737	0.001136542	0.072868
8	0.070758053	0.070710436	0.067340
9	0.043422405	0.043419504	0.006681

TABLE 5.6
Percentage Errors for Eigenvalue Sensitivity Computation between Numerical and Analytical Methods at Bus 30 (Numerical Method Uses 1% Perturbation)

No.	K_A	K_{TG1}	K_{TG2}	K_{TG3}	T_{TG4}	T_{TG5}	T_{TG6}
1	0.002	0.1917	1.0752	0.6984	0.5102	1.6292	0.9819
2	0.041	0.2203	1.356	0.007	1.0787	0.997	1.4562
3	0.1085	1.0008	1.205	0.0087	1.4968	0.9993	1.0484
4	0.0379	0.4998	2.3475	0.0166	0.8038	1.0065	1.4636
5	0.2327	0.0083	0.0371	0.0042	0.7114	0.9841	1.082
6	0.0539	0.0593	1.0034	0.0174	0.2743	0.9695	1.1047
7	0.8113	0.0729	0.5967	0.1796	0.1166	0.7947	0.9689
8	2.7384	0.0673	0.5954	0.1815	0.0649	0.7923	0.9671
9	2.2753	0.0067	0.1604	0.0441	0.3598	0.9286	0.9419

these identified sensitive parameters and machines to manage the system stability more efficiently and effectively.

From Tables 5.7–5.17, we can conclude that:

1. Machine 34 is most sensitive to parameter uncertainties subject to small signal stability;
2. The governor system parameters are most sensitive for Machine 34 to the eighth critical mode of oscillation;
3. The governor time constants have the most impact on eigenvalue sensitivity as compared with other control system gains in a scale of 10^3, and therefore more attention should be paid to these time constants than gains;
4. The excitation system gain K_A has about 10^{-3} times less impact on eigen-

TABLE 5.7
Eigenvalue Sensitivity Analysis of the Governor Time Constant T_{TG6} at Machines 30–32

No.	Machine 30	Machine 31	Machine 32
1	-1.11551×10^{-6}	-3.558×10^{-11}	-4.788×10^{-11}
2	6.46559×10^{-5}	-4.0387×10^{-9}	-1.18767×10^{-9}
3	-1.95761×10^{-6}	-2.57155×10^{-9}	3.168×10^{-11}
4	1.76353×10^{-6}	-2.59354×10^{-6}	$\mathbf{-5.21521 \times 10^{-8}}$
5	**0.003414985**	-1.18294×10^{-7}	4.4259×10^{-9}
6	-8.95896×10^{-5}	-8.1947×10^{-7}	1.08412×10^{-8}
7	8.58656×10^{-6}	$\mathbf{-3.89898 \times 10^{-6}}$	2.06921×10^{-8}
8	0.000572916	-1.26273×10^{-7}	3.49368×10^{-8}
9	0.000838442	-1.08889×10^{-6}	-3.6804×10^{-8}

TABLE 5.8
Eigenvalue Sensitivity Analysis of the Governor Time Constant T_{TG6} at Machines 33–35

No.	Machine 33	Machine 34	Machine 35
1	6.6291×10^{-8}	3.80224×10^{-8}	-1.93559×10^{-7}
2	-3.71654×10^{-6}	-3.09263×10^{-7}	4.77072×10^{-9}
3	**−0.00016377**	-3.88693×10^{-5}	1.12878×10^{-7}
4	-1.55722×10^{-8}	-2.07263×10^{-8}	-1.72079×10^{-8}
5	-3.76821×10^{-8}	1.09418×10^{-5}	-1.76852×10^{-7}
6	-2.99485×10^{-7}	−0.000156474	$\mathbf{4.64143 \times 10^{-7}}$
7	-1.86785×10^{-6}	-5.44971×10^{-5}	3.64671×10^{-9}
8	-2.40172×10^{-5}	**−0.000387368**	-7.10243×10^{-8}
9	-9.65502×10^{-5}	−0.000242267	-3.66778×10^{-7}

value sensitivity to all buses as compared with governor time constants;

5. Machines 30 and 33 are also sensitive to parameter uncertainties. However, this is less than that of Machine 34; and
6. The governor system gains have much more impact on eigenvalue sensitivity than excitation system gains in an approximately 10^3 times more.

These observations indicate that the governor gains and time constants have more impact on eigenvalue sensitivity than excitation system gains, and therefore need more attention to prevent small signal instability. It also indicates that Machine 34 may need more attention in dispatch and maintenance to avoid possible small signal instability.

With these analyses, we can proceed to perform a *probabilistic small signal assess-*

TABLE 5.9

Eigenvalue Sensitivity Analysis of the Governor Time Constant T_{TG6} at Machines 36–38

No.	Machine 36	Machine 37	Machine 38
1	**−0.000139367**	5.07058×10^{-8}	-2.84968×10^{-8}
2	9.64761×10^{-7}	-3.36758×10^{-7}	-1.10567×10^{-5}
3	-1.45484×10^{-5}	1.47502×10^{-7}	-1.57724×10^{-6}
4	-2.90804×10^{-6}	-1.818×10^{-11}	2.34882×10^{-8}
5	-4.33723×10^{-6}	$\mathbf{6.69378 \times 10^{-7}}$	−0.000150108
6	−0.000130285	-2.65358×10^{-8}	-5.22308×10^{-5}
7	5.08404×10^{-8}	1.65754×10^{-9}	−0.000316083
8	-6.30757×10^{-6}	-1.02286×10^{-7}	**−0.000482093**
9	−0.000118408	-5.26866×10^{-7}	−0.000349147

TABLE 5.10

Eigenvalue Sensitivity of the Governor Gain K_{TG2}

No.	Machine 30	Machine 31	Machine 32
1	−0.000219504	2.06935×10^{-6}	4.53992×10^{-7}
2	0.006788523	0.000101141	1.16527×10^{-5}
3	−0.001191229	2.25112×10^{-5}	-1.59018×10^{-7}
4	0.000103514	0.026501181	**0.000309328**
5	**0.336134994**	−0.001874333	-4.24984×10^{-5}
6	−0.009386866	0.013020627	-6.905×10^{-5}
7	0.000899972	**0.050480814**	−0.000180781
8	0.055932815	−0.001087897	−0.00017022
9	0.03323575	0.007897174	5.57848×10^{-5}

ment on the New England 39-Bus Test System. In the following, we will first introduce the concepts of statistical functions which are instrumental to the formulation of the essential probabilistic tool sets.

5.5.2 STATISTICAL FUNCTIONS

Let $F(x)$ be a continuous and non-decreasing function. For any finite or infinite intervals $[a,b]$ in $\mathbb{R}^1$, we have

$$F(a) - F(b) = P(a < X \leq b), \tag{5.120}$$

where X is a random variable. If $F(x)$ also fulfills the following relations

$$F(x) = P(-\infty < X \leq x), \tag{5.121}$$

TABLE 5.11
Eigenvalue Sensitivity of the Governor Gain K_{TG2}

No.	Machine 33	Machine 34	Machine 35
1	0.001545566	-3.82499×10^{-5}	0.002097239
2	0.000753583	0.000157565	-4.18803×10^{-5}
3	**0.181296538**	0.069240347	−0.001011621
4	2.01391×10^{-5}	5.19781×10^{-5}	0.000117185
5	3.27926×10^{-5}	−0.010366876	0.001057575
6	3.94168×10^{-5}	0.208190867	**−0.003035191**
7	0.001762589	0.066927192	-1.87307×10^{-5}
8	0.017075462	**0.404146387**	0.000330395
9	0.037064593	0.138868907	0.000628983

TABLE 5.12
Eigenvalue Sensitivity of the Governor Gain K_{TG2}

No.	Machine 36	Machine 37	Machine 38
1	**0.170009071**	−0.000185777	5.3425×10^{-6}
2	−0.001559526	0.001113672	0.00355742
3	0.022957771	−0.000533284	0.000290418
4	0.003536287	1.79402×10^{-7}	4.75415×10^{-6}
5	0.00768202	**−0.001788257**	0.03501701
6	0.114276293	6.4296×10^{-5}	0.017539022
7	-9.46254×10^{-5}	-9.72278×10^{-6}	0.034078409
8	0.005577833	0.00010949	**0.067435278**
9	0.043878354	0.000250131	0.041316363

$$0 \leq F(x) \leq 1, \tag{5.122}$$

$$F(-\infty) = 0, \tag{5.123}$$

$$F(+\infty) = 1, \tag{5.124}$$

then $F(x)$ is also known as a *Cumulative Distribution Function* (CDF) of X.

Now, consider the *mean density ratio* defined as

$$\begin{aligned} f(x) &= \lim_{h \to 0} \frac{F(x+h) - F(x-h)}{2h}, \quad x-h < X \leq x+h \\ &= F'(x). \end{aligned} \tag{5.125}$$

TABLE 5.13
Identify the Most Sensitive Eigenvalue to Exciter Gain K_A at Each Generator

	K_A	
Machine No.	$\frac{\partial\lambda}{\partial K_A}(\times 10^{-3})$	Critical Mode No.
30	−0.06791209198673	9
31	0.34809738436733	7
32	0.32487814171438	9
33	−0.14168737887021	3
34	0.43863295050830	8
35	0.26266275761013	1
36	0.29353585500314	9
37	0.25569157094522	2
38	0.34868931208412	9
Maximum	0.43863295050830	(34, 8)

TABLE 5.14
Identify the Most Sensitive Eigenvalue to Speed Governor Gain K_{TG1} at Each Generator

	K_{TG1}	
Machine No.	$\frac{\partial\lambda}{\partial K_{TG1}}(\times 10^{-3})$	Critical Mode No.
30	0.49885901529951	5
31	0.09469911688998	4
32	−0.00000534910129	7
33	0.21556520763527	3
34	−0.19964981762726	9
35	−0.00002071690747	6
36	0.49600058177804	1
37	−0.00002595289485	5
38	0.09035773666933	7
Maximum	0.49885901529951	(30, 5)

If the above derivative exists, then $f(x)$ is also known as the *Probability Density Function* (PDF) or *frequency function* of X. The above cases are valid for the scalar set, i.e., in $\mathbb{R}$. In the following sections, we shall extend it to a multivariate case.

Before the multivariate normal PDF for all the real parts of the critical eigenvalues are formulated, we shall first introduce the *Characteristic Function* (CF). The characteristic functions of distributions in $\mathbb{R}^1$ extending to a general multi-dimensional case $\mathbb{R}^n$ will be covered [179][212][215].

TABLE 5.15
Identify the Most Sensitive Eigenvalue to Speed Governor Gain K_{TG3} at Each Generator

Machine No.	K_{TG3}	
	$\frac{\partial\lambda}{\partial K_{TG3}}(\times 1)$	Critical Mode No.
30	−0.00098710467616	5
31	−0.04169354401310	4
32	−0.00021299073944	4
33	−0.11734670906906	3
34	0.07513066243194	9
35	−0.00048199751110	6
36	−0.11193931004559	1
37	−0.00144899144764	2
38	−0.04415122152392	7
Maximum	−0.11734670906906	(33,3)

TABLE 5.16
Identify the Most Sensitive Eigenvalue to Speed Governor Time Constant T_{TG4} at Each Generator

Machine No.	T_{TG4}	
	$\frac{\partial\lambda}{\partial T_{TG4}}(\times 1)$	Critical Mode No.
30	−0.58393197169450	5
31	−0.00992954680167	4
32	−0.00000012095101	4
33	−0.02916175607370	3
34	0.79719080800131	8
35	−0.00000084238796	9
36	0.07860269814019	9
37	0.00000087982427	5
38	0.03049729669530	9
Maximum	0.79719080800131	(34,8)

Let t be a real number in $\mathbb{R}^1$. Now define a function $g(x)$ as

$$\begin{aligned} g(x) &= e^{itx} \\ &= \cos tx + i \sin tx. \end{aligned} \tag{5.126}$$

TABLE 5.17
Identify the Most Sensitive Eigenvalue to Speed Governor Time Constant T_{TG5} at Each Generator

	T_{TG5}	
Machine No.	$\frac{\partial \lambda}{\partial T_{TG4}}(\times 1)$	Critical Mode No.
30	−0.00029901705056	5
31	0.09996306255147	7
32	−0.00000005215206	4
33	0.02636963298285	9
34	0.65079925596713	8
35	0.00000046414332	6
36	0.08358576995736	9
37	0.00000087982427	5
38	0.15184475179912	8
Maximum	0.65079925596713	(34, 8)

Since $|g(x)| = |e^{itx}| = 1$, using *Lebesgue-Stieltjes Integrals* with respect to $F(x)$ gives

$$\begin{aligned} \varphi(t) &= \int_{-\infty}^{\infty} e^{itx} dF(x) \\ &= \int_{-\infty}^{\infty} e^{itx} f(x) dx, \end{aligned} \tag{5.127}$$

and $\varphi(t)$ is also known as the *Characteristic Function* (CF) of the distribution corresponding to $F(x)$ [215]. In general, $\varphi(t)$ is a complex-value function of t. The following properties of $\varphi(t)$ hold

$$\varphi(0) = 1, \tag{5.128}$$

$$|\varphi(t)| \leq \int_{-\infty}^{\infty} dF(x) = 1, \tag{5.129}$$

$$\varphi(-t) = \varphi^*(t), \tag{5.130}$$

where the superscript * denotes the complex conjugate (or Hermitian) operation. Obviously $\varphi(t)$ is continuous for all real t.

Before we relate the CF to the PDF, we introduce the following results without proof

$$\int_{-\infty}^{\infty} e^{-x^2} dx = \sqrt{\pi}. \tag{5.131}$$

By scaling down a factor of $\sqrt{\frac{h}{2}}$ and provided that $h \neq 0$, we get

$$\int_{-\infty}^{\infty} e^{\frac{-hx^2}{2}} dx = \sqrt{\frac{2\pi}{h}}. \tag{5.132}$$

Using Lebesgue-Stieltjes Integrals again, we may differentiate the above integral any number of times with respect to h such that

$$\int_{-\infty}^{\infty} x^{2v} e^{\frac{-hx^2}{2}} dx = \frac{2v!}{2^v v!} \sqrt{2\pi h}^{-v-\frac{1}{2}} \tag{5.133}$$

for all $v \in \{0,1,2,...\}$.

Extending the above formulae to this integral, we obtain

$$\int_{-\infty}^{\infty} e^{itx-\frac{hx^2}{2}} dx = \int_{-\infty}^{\infty} \sum_{0}^{\infty} \frac{(itx)^v}{v!} e^{-\frac{1}{2}hx^2} dx, \tag{5.134}$$

and the partial sums of the series under the last integral are dominated by $e^{|tx|-\frac{1}{2}hx^2}$ which is integrable over the interval $(-\infty,\infty)$. Hence, we may integrate the series term by term to obtain

$$\begin{aligned}
\int_{-\infty}^{\infty} e^{itx-\frac{hx^2}{2}} dx &= \sum_{0}^{\infty} \frac{(it)^v}{v!} \int_{-\infty}^{\infty} x^v e^{-\frac{1}{2}hx^2} dx \\
&= \sum_{0}^{\infty} \frac{it^{2v}}{(2v)!} \frac{(2v)!}{2^v v!} \sqrt{2\pi h}^{-v-\frac{1}{2}} \\
&= \sqrt{\frac{2\pi}{h}} e^{\frac{-t^2}{2h}},
\end{aligned} \tag{5.135}$$

and all the odd terms vanish. With $h = 1$, and introducing the function $\Phi(x)$ as

$$\Phi(x) = \frac{1}{\sqrt{2\pi}} \int_{-\infty}^{x} e^{-\frac{t^2}{2}} dt, \tag{5.136}$$

we can have the following

$$\begin{aligned}
\int_{-\infty}^{\infty} e^{itx} d\Phi(x) &= \frac{1}{\sqrt{2\pi}} \int_{-\infty}^{\infty} e^{itx-\frac{x^2}{2}} dx \\
&= e^{-\frac{t^2}{2}}.
\end{aligned} \tag{5.137}$$

Obviously, $\Phi(x)$ is non-decreasing and continuous with $\Phi(-\infty) = 0$ and $\Phi(+\infty) = 1$. As such, $\Phi(x)$ is a CDF corresponding to CF of $e^{-t^2/2}$. $\Phi(x)$ is also the important *normal* distribution.

Next, we shall extend the above concepts to a multi-dimensional case. Now, let $t = \begin{bmatrix} t_1 & t_2 & \cdots & t_n \end{bmatrix}^T$ and $x = \begin{bmatrix} x_1 & x_2 & \cdots & x_n \end{bmatrix}^T$ be vectors in $\mathbb{R}^n$. Extending (5.127), we get

$$\begin{aligned}
\varphi(t) &= \varphi(t_1, t_2, \cdots, t_n) \\
&= \int_{\mathbb{R}^n} e^{it^T x} dP,
\end{aligned} \tag{5.138}$$

where P is the CDF in $\mathbb{R}^n$. Obviously, the multivariate CF $\varphi(t)$ is a function of n real variables. The following properties of $\varphi(t)$ hold

$$\varphi(0,0,\cdots,0) = 1, \tag{5.139}$$

$$|\varphi(t)| \leq 1, \tag{5.140}$$

$$\varphi(-t) \quad = \quad \varphi^*(t), \tag{5.141}$$

where the superscript * denotes complex conjugate operation. $\varphi(t)$ is continuous everywhere.

If $|\varphi(t)|$ is integrable over $\mathbb{R}^n$ and the following derivative exists and is continuous for all x, we have

$$\begin{aligned} f(x) &= f(x_1,x_2,\cdots,x_n) \\ &= \frac{\partial^n}{\partial x_1,\partial x_2,\cdots,\partial x_n}F \\ &= \frac{1}{(2\pi)^n}\int_{-\infty}^{\infty}\int_{-\infty}^{\infty}\cdots\int_{-\infty}^{\infty} e^{-it^Tx}\varphi(t)\,dt_1,dt_2,\cdots,dt_n, \end{aligned} \tag{5.142}$$

and the above equation is the multivariate joint PDF $f(x)$ corresponding to the CF $\varphi(t)$ [215]

$$\varphi(t) = \int_{-\infty}^{\infty}\int_{-\infty}^{\infty}\cdots\int_{-\infty}^{\infty} e^{-it^Tx}f(x)\,dx_1,dx_2,\cdots,dx_n. \tag{5.143}$$

With these concepts, we are able to represent the real parts of the critical eigenvalues into their respective normal distributions. We shall first consider a single variate case and extend it to multivariate normal PDF in order to calculate the joint probability for all critical eigenvalues lying on the open left half plane.

5.5.3 SINGLE VARIATE NORMAL PDF OF α_I

Assuming the real part of a critical eigenvalue λ_i to be of normal distribution, i.e., $\alpha_i \sim (\bar{\alpha}_i, \sigma^2_{\alpha_i})$ where

$$\bar{\alpha}_i \quad = \quad E[\Re\{\alpha_i\}], \tag{5.144}$$

$$\sigma^2_{\alpha_i} \quad = \quad C_{\Delta\Lambda(i,i)}, \tag{5.145}$$

with $E[\Re\{\alpha_i\}]$ as the *expectation* of the nominal real part of the eigenvalue α_i, and $\mathbf{C}_{\Delta\Lambda(i,i)}$ as the eigenvalues' covariances for $i \neq j$. The normal random variables can then be transformed into standard normal distributions with zero mean and unity variance via

$$\alpha_{Z,i} = \frac{\alpha_i - \bar{\alpha}_i}{\sigma_{\alpha_i}}, \tag{5.146}$$

and the PDF of $\alpha_{Z,i}$ will be $\alpha_{Z,i} \sim N(0,1)$.

Now that the single variate normal PDFs of the real parts of the individual critical eigenvalues are known, we shall proceed to combine them into a joint multivariate normal PDF. The multivariate PDF allows us to calculate the joint probability of dynamic stability, i.e., the probability when *all* the real parts of the critical eigenvalues lie on the open left-half complex plane.

5.5.4 MULTIVARIATE NORMAL PDF

In this section, the multivariate normal PDF will be used to analyze the joint probability of all the system's eigenvalues. The CF of the set of random variables α_i is given by [216]

$$\Phi(\mathbf{\Omega}) = e^{-\frac{1}{2}\mathbf{\Omega}\mathbf{C}\mathbf{\Omega}^T + j\mathbf{\Omega}\mathbf{M}}, \tag{5.147}$$

where $\mathbf{\Omega} = \begin{bmatrix} \omega_1 & \omega_2 & \cdots & \omega_n \end{bmatrix}^T$ is a vector of *Fourier-transformed* variables. $\mathbf{C}$ is the eigenvalue covariance matrix and $\mathbf{M}$ is the mean vector. The random variables α_i are also assumed to be *jointly Gaussian.*

As such, normalization gives the following multivariate PDF $f(\mathbf{X}_\alpha)$

$$f(\mathbf{X}_\alpha) = \frac{1}{\sqrt{(2\pi)^n |C|}} e^{-\frac{(X_\alpha - M)C^{-1}(X_\alpha - M)^T}{2}}, \tag{5.148}$$

where $\mathbf{X}_\alpha = \begin{bmatrix} \alpha_1 & \alpha_2 & \cdots & \alpha_n \end{bmatrix}^T$ is a vector of the real parts of the critical eigenvalues, and is a random variable vector. The covariance matrix $\mathbf{C}$ may be singular when the parameter variation random processes are perfectly correlated, which is not possible in practice. As such, the calculation of probabilities will not be affected when using the *generalized Tetrachoric series*, and this series will be used for calculating probability of power system small signal stability properties.

We normalize the single variant PDFs using the transformation given above. As such, we obtain $\mathbf{Z}_\alpha = \begin{bmatrix} \alpha_{z1} & \alpha_{z2} & \cdots & \alpha_{zn} \end{bmatrix}^T$, and hence $E(\mathbf{Z}_\alpha) = \mathbf{M} = \mathbf{0}$ which is a reasonable assumption, as the underlying random process of the power system parameter variations is assumed to be of zero mean, ergodic, and stationary. Further, if we normalize real parts of the critical eigenvalue covariance matrix $\mathbf{C}$ by dividing each row i with its leading diagonal of each row $\sqrt{\mathbf{C}_{ii}}$ and column j by $\sqrt{\mathbf{C}_{jj}}$, we will get the *normalized covariance matrix* (or correlation coefficient matrix) $\mathbf{Q}$ with unity diagonal elements and correlation coefficients q_{ij} for its off-diagonal terms. We will be able to simplify the CF in the above equation to give the PDF

$$f(\mathbf{Z}_\alpha) = \frac{1}{\sqrt{(2\pi)^n}} e^{-\frac{\mathbf{Z}_\alpha \mathbf{Q}^{-1} \mathbf{Z}_\alpha^T}{2}}. \tag{5.149}$$

The multivariate normal PDF is actually a multi-dimensional extension of the scalar normal distribution PDF. By taking the inverse Fourier transform of the above equation using the *Fourier Inversion Theorem* [216], we are able to get the multivariate normal PDF as

$$f(\mathbf{Z}_\alpha) = \frac{1}{\sqrt{(2\pi)^n}} \int_{-\infty}^{\infty} e^{-\frac{\mathbf{Z}_\alpha \mathbf{Q}^{-1} \mathbf{Z}_\alpha^T}{2}} e^{-j\mathbf{Z}\mathbf{\Omega}} d\mathbf{\Omega}. \tag{5.150}$$

It is worth noting that the normalized covariance matrix $\mathbf{Q}$ is symmetric. By writing the exponential in (5.150) as the exponential product of n^2, we can derive the following

$$f(\mathbf{Z}_\alpha) = \frac{1}{\sqrt{(2\pi)^n}} \int_{-\infty}^{\infty} e^{-j\mathbf{Z}\boldsymbol{\Omega}} \prod_{i=1}^{n} \prod_{k=1}^{n} e^{-q_{ik}\omega_i\omega_k} \, d\boldsymbol{\Omega}. \tag{5.151}$$

The above equation is the multivariate normal PDF extension of the individual PDFs. In order to calculate the joint probability of dynamic small signal stability, we are interested in the following *joint* probability

$$P(\alpha_{z1}, \alpha_{z2}, \cdots, \alpha_{zn} < u_1, u_2, \cdots, u_i, \cdots, u_n), \tag{5.152}$$

with the variables u_i in the form

$$u_i = -\frac{\bar{\alpha}_i}{\sigma_{\alpha_i}}. \tag{5.153}$$

From single variant normal distribution and its extension to a multivariate case, the mathematical models for the distribution of the real parts of the critical eigenvalues are derived. This model will allow us to calculate the small signal stability region. The probability of small signal stability will be evaluated with the generalized Tetrachoric series in the following sections.

5.5.5 PROBABILITY CALCULATIONS

With the multivariate normal PDF of the real parts of the eigenvalues available, the probability of the system under small signal stability conditions with stochastic load flow injections can then be derived. In this section, the small signal stability region will be calculated using the generalized Tetrachoric series based on the techniques introduced in previous sections.

It has to be made clear that the following assumptions are made for probability calculation techniques:

1. The system parameter variations are the source of probabilistic analysis. These include generation and its control systems parameters, load forecast variations, and transmission network parameters variations, etc.; and
2. It is assumed that the parameter variation is of linear relationship with the eigenvalue variations, i.e., $\Delta\boldsymbol{\Lambda} = \mathbf{S}\Delta\boldsymbol{\Gamma}$, where $\mathbf{S}$ is a linear sensitivity matrix.

In the following sections we will first introduce Hermite polynomials as the basis for probability calculation for multivariable cases, and then describe the generalized Tetrachoric series as a way to find the *joint* probability of the real parts of the critical eigenvalues lying in the open left half of the complex plane. Probability calculation is presented to give a clear picture of using the Tetrachoric series. Finally, a case study is performed on the New England 39-Bus Test System.

Prior to formulating the probability calculation, an important class of polynomials known as *Hermite polynomials* will be introduced. In this section, the definitions and some basic properties of the Hermite polynomials will be covered.

The Hermite polynomials are defined as [215]

$$\left(\frac{d}{dx}\right)^n e^{-\frac{x^2}{2}} = (-1)^n H_n(x) e^{-\frac{x^2}{2}}, \tag{5.154}$$

for $n \in \{0, 1, 2, \cdots\}$ and $H_n(x)$ as a polynomial in x with degree n. Differentiating the above equation and comparing coefficients, we are able to obtain

$$\begin{aligned}
H_0(x) &= 1,\\
H_1(x) &= x,\\
H_2(x) &= x^2 + 1,\\
H_3(x) &= x^3 - 3x,\\
H_4(x) &= x^4 - 6x^2 + 3,\\
H_5(x) &= x^5 - 10x^3 + 15x,\\
H_6(x) &= x^6 - 15x^4 + 45x^2 - 15,\\
&\vdots
\end{aligned} \tag{5.155}$$

and the Hermite polynomials can also be obtained via the following recursive formula for all $n \geq 1$ using

$$H_n(x) = xH_{n-1}(x) - (x-1)H_{n-2}(x). \tag{5.156}$$

This recursive formula is used to generate the required powers of the Hermite polynomial. It is worth noting that the Hermite polynomials $H_n(x)$ are sequences of *orthogonal polynomials* associated with the normal distribution.

Next, the probabilistic small signal region is formulated using *Generalized Tetrachoric Series* [217]. This is made possible with the derived joint multivariate PDF as derived in (5.153). To calculate the joint probability

$$P(\alpha_{z1}, \alpha_{z2}, \cdots, \alpha_{zn} \geq u_1, u_2, \cdots, u_n), \tag{5.157}$$

we integrate (5.152) with respect to $\mathbf{X}_\alpha$ to obtain [177]

$$P(\mathbf{Z}_\alpha \geq \mathbf{U}) = \frac{1}{\sqrt{(2\pi)^n}} \int_{-\infty}^{\infty} \int_{u}^{\infty} e^{-j\mathbf{Z}\boldsymbol{\Omega}} d\mathbf{Z}_\alpha \prod_{i=1}^{n} \prod_{k=1}^{n} e^{-q_{ik}\omega_i\omega_k} d\boldsymbol{\Omega}, \tag{5.158}$$

where $\mathbf{U} = \begin{bmatrix} u_1 & u_2 & \cdots & u_n \end{bmatrix}^T$. The inner multi-dimensional integral can be evaluated directly. The outer integral is computed by rewriting $e^{-q_{ik}\omega_i\omega_k}$ as a Taylor series expansion using Hermite polynomials. This gives the following Tetrachoric series on the right-hand side

$$\begin{aligned}
P(\mathbf{Z}_\alpha \geq \mathbf{U}) &= \prod_{k=1}^{n} \frac{1}{\sqrt{(2\pi)}} e^{-\frac{u_k^2}{2}} \sum_{a_1=0}^{\infty} \sum_{a_2=0}^{\infty} \cdots \sum_{a_m=0}^{\infty} \frac{q_{12}^{a_1} q_{13}^{a_2} q_{n-1,n}^{a_m}}{a_1! a_2! \cdots a_m!} \cdots \\
&\quad \times H_{r1}(u_1) H_{r2}(u_2) \cdots H_{rn}(u_n),
\end{aligned} \tag{5.159}$$

where $m = n(n-1)/2$, and q_{12}, q_{13}, $\cdots$, $q_{n-1,n}$ are the elements in the normalized correlation coefficient matrix $\mathbf{Q}$ with diagonal entries equal to one. $H_r(u_k)$ is a Hermite polynomial with order r and argument u_k. Rewriting (5.159) with the following substitutions

$$H_r(\mathbf{U}) = \prod_{r=1}^{n} H_{r1}(u_1)H_{r2}(u_2)\cdots H_{rn}(u_n), \tag{5.160}$$

$$\Phi_k(\mathbf{U}) = \prod_{k=1}^{n} \frac{1}{\sqrt{(2\pi)}} e^{-\frac{u_k^2}{2}}, \tag{5.161}$$

we get

$$P(\mathbf{Z}_\alpha \geq \mathbf{U}) = \Phi_k(\mathbf{U}) \sum_{a_1=0}^{\infty} \sum_{a_2=0}^{\infty} \cdots \sum_{a_m=0}^{\infty} \frac{q_{12}^{a_1} q_{13}^{a_2} q_{n-1,n}^{a_m}}{a_1! a_2! \cdots a_m!} H_r(\mathbf{U}). \tag{5.162}$$

(5.162) is the probability of the situation that the real part of system critical eigenvalues is greater than a specified level as defined by vector $\mathbf{U}$. It will be used to reveal the probabilistic small signal stability range of a power system subject to system parameter uncertainties.

Using the generalized Tetrachoric series, the probabilistic small signal stability region can be calculated. This probability is determined when the real parts of the critical eigenvalues λ_i are all on the left half complex plane. Before we can determine the required probability $P(\mathbf{X}_\alpha < \mathbf{0})$ using $P(\mathbf{Z}_\alpha \geq \mathbf{U})$, the Hermite polynomial's order r_i will first have to be calculated. A table listing all the off-diagonal entries of the normalized correlation coefficient matrix $\mathbf{Q}$ is first constructed, along with a list of summation indices as can be seen from Table 5.18 [177]. To determine the order of the Hermite polynomials, form the sum of the indices corresponding to the entries which have a subscript containing the number 1.

The order r_1 is obtained by subtracting 1 from

$$r_1 = a_1 + a_2 + a_3 + \ldots + a_{n-1} - 1. \tag{5.163}$$

Similarly, r_2 can be obtained as

$$r_2 = a_1 + a_n + a_{n+1} + \ldots + a_{2n-3} - 1. \tag{5.164}$$

As such, other numbers of order r_i can be obtained the same way by counting the number i subscript and subtracting 1. Hence, with the complementary property of probability, we can evaluate the following

$$\begin{aligned} P(\mathbf{X}_\alpha < \mathbf{0}) &= 1 - P(\mathbf{X}_\alpha \geq \mathbf{0}) \\ &= 1 - P(\mathbf{Z}_\alpha \geq \mathbf{U}). \end{aligned} \tag{5.165}$$

The probability calculation in (5.159) is convergent for $|q_{ij}| < 1$ where $|q_{ij}|$ is the off-diagonal of the normalized covariance matrix $\mathbf{C}_{\Delta\Lambda}$.

TABLE 5.18
Hermite Polynomial and Its Indices

Entry	Index
q_{12}	a_1
q_{13}	a_2
q_{14}	a_3
$\vdots$	$\vdots$
q_{1n}	a_{n-1}
q_{23}	a_n
q_{24}	a_{n+1}
$\vdots$	$\vdots$
$q_{n-1,n}$	a_m

5.5.6 SMALL SIGNAL STABILITY REGION

The New England system as shown Figure 5.3 is studied as an example to illustrate the techniques presented. The New England system is composed of ten synchronous machines, thirty-nine buses, thirty-four transmission lines, nine IEEE Type 1 Automatic Voltage Regulator (AVR) and exciters, nine speed governors and turbines, as well as one Power System Stabilizer (PSS). Bus 39 is a slack bus in this case study. Under certain loading conditions, there are a total of nine critical eigenvalues identified as shown in Table 5.19. These critical eigenvalues will be used for probabilistic small signal stability analysis in the sequel.

TABLE 5.19
Critical Eigenvalues and Their Locations at 830MW Level

No	Bus Location	Critical Eigenvalue	Frequency	Damping Ratio
1	11	$-0.3676+8.7533j$	1.39	0.0420
2	13	$-0.4021+8.6764j$	1.38	0.0464
3	15	$-0.3131+8.4747j$	1.35	0.0370
4	21	$-0.2740+7.4602j$	1.19	0.0368
5	23	$-0.0001+6.9946j$	1.11	−0.0002
6	25	$-0.2475+6.9946j$	1.11	0.0356
7	27	$-0.2501+6.3504j$	1.01	0.0394
8	29	$-0.2609+5.9670j$	0.95	0.0433
9	33	$-0.2796+3.7787j$	0.61	0.0725

To calculate the small signal stability probability, we want the joint probability such that all the real parts of the critical eigenvalues are in the open left half complex plane. This probability is given by the joint probability $P(\alpha_1 < 0, \alpha_2 < 0, \cdots, \alpha_9 < 0)$, or $P(\mathbf{X}_\alpha < \mathbf{0})$ and when normalized can be rewritten as $P(\alpha_{z1} < u_1, \alpha_{z2} < u_2, \cdots, \alpha_{z9} < u_9)$, or $P(\mathbf{Z}_\alpha < \mathbf{U})$ where $\mathbf{U} = \begin{bmatrix} u_1 & u_2 & \ldots & u_9 \end{bmatrix}^T$.

The probability $P(\mathbf{Z}_\alpha \geq \mathbf{U})$ can be obtained from the generalized Tetrachoric series

$$\begin{aligned} P(\mathbf{Z}_\alpha \geq \mathbf{U}) &= \prod_{k=1}^{12} \frac{1}{\sqrt{(2\pi)}} e^{-\frac{u_k^2}{2}} \sum_{a_1=0}^{\infty} \sum_{a_2=0}^{\infty} \cdots \sum_{a_{36}=0}^{\infty} \frac{q_{12}^{a_1} q_{13}^{a_2} \cdots q_{8,9}^{a_{36}}}{a_1! a_2! \cdots a_{36}!} \cdots \\ &\quad \times H_{r1}(u_1) H_{r2}(u_2) \cdots H_{r9}(u_9), \end{aligned} \tag{5.166}$$

where q_{12}, q_{13}, $\cdots$, $q_{8,9}$ are the elements in correlation coefficient matrix $\mathbf{Q}$ with diagonal entries equal to one. $H_r(u_k)$ is a Hermite polynomial with order r and argument u_k. Rewriting the above equation using the following substitutions

$$\begin{aligned} H_r(\mathbf{U}) &= \prod_{r=1}^{9} H_{r1}(u_1) H_{r2}(u_2) \cdots H_{r9}(u_9), \\ \Phi_k(\mathbf{U}) &= \prod_{k=1}^{n} \frac{1}{\sqrt{2\pi}} e^{-\frac{u_k^2}{2}}, \end{aligned} \tag{5.167}$$

we get

$$P(\mathbf{Z}_\alpha \geq \mathbf{U}) = \Phi_k(\mathbf{U}) \sum_{a_1=0}^{\infty} \sum_{a_2=0}^{\infty} \cdots \sum_{a_m=0}^{\infty} \frac{q_{12}^{a_1} q_{13}^{a_2} \cdots q_{8,9}^{a_{36}}}{a_1! a_2! \cdots a_{36}!} H_r(\mathbf{U}). \tag{5.168}$$

To determine the order r_i of the Hermite polynomial needed, we use the information derived according to Table 5.18 and as shown in Table 5.20. The entries of Table 5.20

TABLE 5.20
Hermite Polynomial and Its Indices of the New England System

Entry	Index
q_{12}	a_1
q_{13}	a_2
q_{14}	a_3
$\vdots$	$\vdots$
$q_{1,9}$	a_8
q_{23}	a_9
q_{24}	a_{10}
$\vdots$	$\vdots$
$q_{8,9}$	a_{36}

are of the off-diagonal entries of $\mathbf{Q}$. To determine the order of the Hermite polynomials, we form the sum of the indices corresponding to the entries which have a subscript containing the number i. The order r_i is then obtained by subtracting 1 from the sum.

Based on such information and (5.163)–(5.164), the orders of the Tetrachoric series of the New England system are

$$\begin{aligned} r_1 &= a_1+a_2+a_3+a_4+a_5+a_6+a_7+a_8-1, \\ r_2 &= a_1+a_9+a_{10}+a_{11}+a_{12}+a_{13}+a_{14}+a_{15}-1, \\ r_3 &= a_2+a_9+a_{16}+a_{17}+a_{18}+a_{19}+a_{20}+a_{21}-1, \\ r_4 &= a_3+a_{10}+a_{16}+a_{22}+a_{23}+a_{24}+a_{25}+a_{26}-1, \\ r_5 &= a_4+a_{11}+a_{17}+a_{22}+a_{27}+a_{28}+a_{29}+a_{30}-1, \\ r_6 &= a_5+a_{12}+a_{18}+a_{23}+a_{27}+a_{31}+a_{32}+a_{33}-1, \\ r_7 &= a_6+a_{13}+a_{19}+a_{24}+a_{28}+a_{31}+a_{34}+a_{35}-1, \\ r_8 &= a_7+a_{14}+a_{20}+a_{25}+a_{29}+a_{32}+a_{34}+a_{36}-1, \\ r_9 &= a_8+a_{15}+a_{21}+a_{26}+a_{30}+a_{33}+a_{35}+a_{36}-1, \end{aligned} \tag{5.169}$$

and the probability $P(\mathbf{Z}_\alpha \geq \mathbf{U})$ can be evaluated.

If the eigenvalues are highly decoupled or uncorrelated with $q_{12}, q_{13}, \cdots, q_{8,9}$ being close to 0, the summation to infinity for probability can be reduced to

$$P(\mathbf{Z}_\alpha \geq \mathbf{U}) = \Phi_k(\mathbf{U}) \sum_{a_1=0}^{t} \sum_{a_2=0}^{t} \cdots \sum_{a_{36}=0}^{t} \frac{q_{12}^{a_1} q_{13}^{a_2} q_{8,9}^{a_{36}}}{a_1! a_2! \cdots a_{36}!} H_r(\mathbf{U}), \tag{5.170}$$

and $5 < t < 10$ is chosen for fast convergence. The required joint probability $P(\mathbf{X}_\alpha < \mathbf{0})$ can hence be obtained from

$$\begin{aligned} P(\mathbf{X}_\alpha < \mathbf{0}) &= 1 - P(\mathbf{X}_\alpha \geq \mathbf{0}) \\ &= 1 - P(\mathbf{Z}_\alpha \geq \mathbf{U}). \end{aligned} \tag{5.171}$$

To illustrate the actual computational results, three different cases are used. Because of the iterative nature of the Tetrachoric series, the computational cost grows with the increasing number of critical eigenvalues considered. In the following sections, two, three, and all nine critical eigenvalues are considered to compute the probabilistic stability regions of the New England 39-Bus Test System.

First, we perform the probability computation with respect to two critical eigenvalues. The first two critical eigenvalues are selected to compute the probability of the real parts of the eigenvalues less than 0, with $\mathbf{U} = \begin{bmatrix} u_1 & u_2 \end{bmatrix}^T = \begin{bmatrix} 0.01 & 0.05 \end{bmatrix}^T$ using the eigenvalue sensitivity computed in Table 5.21. The probability is computed as

$$\begin{aligned} P(\mathbf{X}_\alpha < \mathbf{0}) &= 1 - P(\mathbf{Z}_\alpha \geq \mathbf{U}) \\ &= 0.8511. \end{aligned} \tag{5.172}$$

Next, we form the probability computation with respect to three critical eigenvalues. The first three critical eigenvalues are selected to compute the probability of their

TABLE 5.21
Sensitivity Computation with Respect to K_A at Buses 31 and 32

No.	Oscillation Mode	Sensitivity Factor w.r.t K_A at Bus 31	Sensitivity Factor w.r.t K_A at Bus 32
1	$-0.3678 \pm j8.7547$	-7.5393×10^{-7}	-1.5675×10^{-7}
2	$-0.4031 \pm j8.6747$	8.753×10^{-7}	-4.2894×10^{-6}
3	$-0.3139 \pm j8.4773$	7.8492×10^{-7}	1.5992×10^{-6}
4	$-0.2745 \pm j7.4595$	0.00010969	8.8151×10^{-5}
5	$0.0013 \pm j6.9647$	3.0866×10^{-5}	1.3184×10^{-5}
6	$-0.2493 \pm j6.9965$	-3.8684×10^{-5}	4.8562×10^{-6}
7	$-0.2507 \pm j6.3571$	0.00012536	6.177×10^{-5}
8	$-0.2600 \pm j5.9958$	1.2609×10^{-5}	-8.4461×10^{-6}
9	$-0.2798 \pm j3.8493$	0.00044289	0.00022885

real part distribution with $\mathbf{U} = \begin{bmatrix} u_1 & u_2 & u_3 \end{bmatrix}^T = \begin{bmatrix} 0.0053 & 0.0072 & 0.0678 \end{bmatrix}^T$ using the eigenvalue sensitivity computed in Table 5.21.

The probability is computed as

$$\begin{aligned} P(\mathbf{X}_\alpha < \mathbf{0}) &= 1 - P(\mathbf{Z}_\alpha \geq \mathbf{U}) \\ &= 0.83896. \end{aligned} \tag{5.173}$$

Finally, we perform the probability computation with respect to *all* nine critical eigenvalues. Now, all nine critical eigenvalues are selected to compute the probability of their real part distribution with $\mathbf{U} = \begin{bmatrix} 0.0731 & 0.0578 & 0.0040 & 0.0677 & \cdots \\ 0.0569 & 0.0256 & 0.0377 & 0.0296 & 0.1475 \end{bmatrix}^T$ using the eigenvalue sensitivity computed in Table 5.21. Similarly, the probability is computed as

$$\begin{aligned} P(\mathbf{X}_\alpha < \mathbf{0}) &= 1 - P(\mathbf{Z}_\alpha \geq \mathbf{U}) \\ &= 0.81362. \end{aligned} \tag{5.174}$$

As such, we can observe from above that we are able to perform the probabilistic small signal stability assessment of the systems's critical eigenvalues with respect to its non-deterministic parametric variations. It should be noted that this is done assuming *static* loads. In the following section, we will extend the concepts to include small signal stability assessments using *dynamic* loads and study the corresponding effects of load changes on small signal stability.

5.6 IMPACT OF INDUCTION MOTOR LOAD

Loads are essentially *random* by nature. In order to model them with confidence and to describe them at any given season and time has been recognized to be very difficult. Undrill and Laskowski [218] used a dynamic Induction Motor (IM) model to represent the loads. Recent papers by Kao [219] and Pereira et al. [191] conclude that the composite static and dynamic load model gives the best simulation results in comparison with the field results. The static model of the IM is presented using a curve-fitting technique in [220]. Recent advances in identification of IM loads from power system composite load models are presented in [221][222]. A comprehensive discussion on various issues of load modeling can be found in [223], which focuses on the development of physically based load models based on field tests and natural disturbances data. The fitted model is simulated from the actual dynamic response of the motor and is realistic. Here, a detailed model for the IM is used to investigate the effect of the composite load on the system stability.

Uncertainty with respect to operating conditions, e.g., nodal injection, PV-voltage by probabilistic methods using the first- and second-order eigenvalue sensitivity analyses, are considered in [178][181][183]. The probabilistic collocation method relating uncertain parameters to the quantities of interest by polynomial relationships addresses parametric uncertainties [224]. This technique is computationally very efficient if the total number of uncertain parameters is small. The concept of asymptotic stability robustness for model (e.g., change in operating conditions) and parameter (e.g., change in network parameters) uncertainties using eigenvalue sensitivity analysis is presented in [225], where two indices derived from sensitivity matrices are used to measure asymptotic small signal stability robustness. However, it becomes computationally quite expensive to calculate sensitivity with respect to parameters for the large system. The comparison of analytical and numerical methods to calculate eigenvalue sensitivity has been reported in [41][205]. It is established that a numerical approach can compute all the sensitivity factors in a one-time run, whereas the analytical method has to compute this by each parameter; hence the former method is computationally efficient for the large number of parameters in a big system. The sensitivity analysis is used for ranking loads in a power system.

The load that is highly sensitive with respect to its parameters should be modeled carefully in the system [226]. Load sensitivity provides a useful insight into proper system operation and control, to prevent instability. Proper load sensitivity analysis provides computational efficiency in measuring selected important loads for sensitivity and stability analysis. By far, most of the load sensitivity works are mainly concerned with the generic load model, which is essentially an exponential recovery model. However, modeling all the loads as static or dynamic loads may not truly reflect the load ranking. Moreover, it is widely recognized that a power system can be modeled with static and dynamic components. This is usually described as a ZIP (constant impedance, constant current and constant power) load and IM loads [227]. Different load structures may have different impacts on system dynamic characteristics. It is necessary to evaluate the combination of load, which is most sensitive to system stability.

With this motivation in mind, we aim to present a comprehensive load ranking analysis framework in this section to handle both static and dynamic loads) when subjected to different load structures. In particular, we

1. study the effect of the location of a composite load on the overall sensitivity

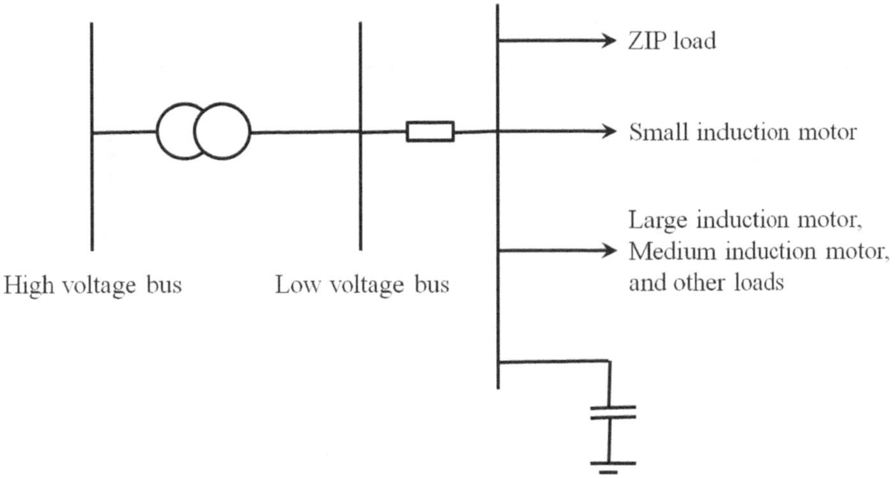

FIGURE 5.5 Composite load model consisting of a ZIP load and an IM load.

and the eigenvalue movement of the system;

2. find the most and least sensitive parameters of the critical eigenvalues of the system;
3. study the effect of the parametric variation on the mobility of the critical modes of the system;
4. study the effect of the increase in the number of IMs in the system on the overall sensitivity of the load; and
5. study the effect of the variation of the composition of load on overall sensitivity.

5.6.1 COMPOSITE LOAD MODEL FOR SENSITIVITY ANALYSIS

The load model considered is the composition of a constant impedance Z, constant current I, and constant power P (ZIP) load and an IM load as shown in Figure 5.5.

The load model is expressed as

$$S_{tot} = (1-\alpha)S_{ZIP} + \alpha S_{ind}, \tag{5.175}$$

where α is the percentage of IM load in the composite load model. S_{ZIP} and S_{ind} are the complex powers of the static ZIP and IM loads, respectively,

$$\begin{aligned} S_{ZIP} &= P_{ZIP} + jQ_{ZIP} \\ &= \left[A_P\left(\frac{V}{V_o}\right)^2 + B_P\left(\frac{V}{V_o}\right) + C_P\right]P^o_{ZIP} \\ &\quad + \left[A_Q\left(\frac{V}{V_o}\right)^2 + B_Q\left(\frac{V}{V_o}\right) + C_Q\right]Q^o_{ZIP}, \end{aligned} \tag{5.176}$$

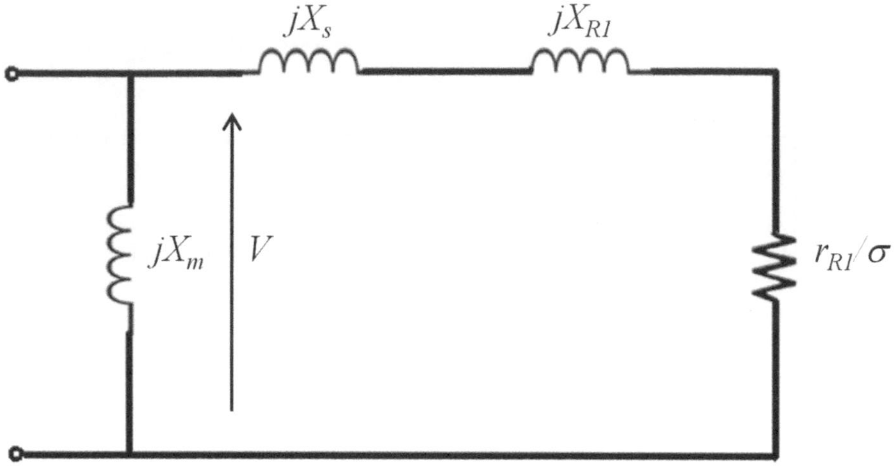

FIGURE 5.6 IM transient-state equivalent circuit.

where A_P, B_P, C_P, A_Q, B_Q, and C_Q are coefficients that represent the constant impedance, constant current, and constant power portion of the real and reactive power of the ZIP load, respectively. For the purpose of our study, only the constant power portion of the ZIP load is considered.

The IM model used is in compliance with the IEEE Task Force [190]. The IM part of the load can be "small" or "large" motors depending on the specific load characteristics to be modeled. As a general rule, the "small" motor runs at a much higher slip than most "large" motors above about 30 kW [228][229]. A third-order model is usually adequate for aggregated motors in the bulk system dynamic simulation.

Stator transients are neglected and electrical transients in one rotor per axis are represented. The transient-state equivalent circuit consisting of a voltage behind transient impedance is shown in Figure 5.6.

The model dynamics can be described by the following three differential equations

$$\dot{E}'_{mx} = \frac{\omega_o \sigma E'_{my} - (E'_{mx} + (X_o - X')I_{my})}{T'_o}, \tag{5.177}$$

$$\dot{E}'_{my} = \frac{-\omega_o \sigma E'_{mx} - (E'_{my} + (X_o - X')I_{mx})}{T'_o}, \tag{5.178}$$

$$\dot{\sigma}_m = \frac{T_m - T_e}{2H}, \tag{5.179}$$

where E'_{mx} and E'_{my} are the real and imaginary axis voltage behind the stator resistance, and I_{mx} and I_{my} are the real and imaginary axis current to be defined in a while. σ_m is the slip of the IM, ω_o is the synchronous frequency in rad/s, and X_o, X', and T'_o are rotor open-circuit reactance, blocked short-circuit reactance, and transient open-circuit time constant, respectively, and are expressed below. T_m and T_e are the mechanical and electrical torque represented respectively by,

$$X_o = X_s + X_m, \tag{5.180}$$

$$X' = X_s + \frac{X_t X_m}{X_t + X_m}, \tag{5.181}$$

$$T_o' = \frac{X_f + X_m}{\omega_o R_f}, \tag{5.182}$$

$$T_m = (a_m + b_m\sigma_m + c_m\sigma_m^2)T_m^o, \tag{5.183}$$

$$T_e = \frac{E_{mx}' I_{mx} + E_{my}' I_{my}}{\omega_o}, \tag{5.184}$$

$$I_{my} = \frac{-R_s(E_{mx}' - V_{mx}) + X'(E_{my}' - V_{my})}{R_s^2 + X_m^2}, \tag{5.185}$$

$$I_{my} = \frac{X'(E_{mx}' - V_{mx}) - R_s(E_{my}' - V_{my})}{R_s^2 + X_m^2}, \tag{5.186}$$

where X_s, X_m, X_r, R_r, R_s, and H are stator impedance, magnetizing reactance, rotor impedance, rotor resistance, stator resistance, and inertia of the IM, respectively [228]. V_{mx} and V_{my} are the real and imaginary axis terminal voltages. a_m, b_m, and c_m are the torque coefficients of the torque slip characteristics.

5.6.2 MOTOR LOAD PARAMETER SENSITIVITY ANALYSIS

Small signal stability is the ability of the power system to maintain synchronism under small disturbances. To investigate the small signal stability of a power system, we need to model the dynamic components (e.g., generators) and their control systems (such as excitation control systems and speed governor systems) in detail. The accuracy of power system stability analysis depends on the accuracy of the models used. More reliable results about system transfer capability and associate economic benefits can be calculated with better models. Modeling of system components such as generators and transmission systems are well developed; however, there are still considerable uncertainties in load modeling. More research is needed to build better load models that can contribute to better overall system stability assessment results. In this section, we look at the impacts of load modeling structures on power system small signal stability characteristics. It is more demanding for more reliable system models, especially load models, when operating under an open access environment under current deregulation.

Similarly, the generators are modeled as fourth order synchronous machines, as explained by the set of equations below

$$\begin{aligned}\dot{\delta} &= \omega_o(\omega-1), &(5.187)\\ \dot{\omega} &= \frac{P_m - P_e - D(\omega-1)}{M}, &(5.188)\\ \dot{E}'_q &= \frac{E_{fd} - E'_q - (X_d - X'_d)I_d}{T'_{do}}, &(5.189)\\ \dot{E}'_d &= \frac{(X_q - X'_q)I_q - E'_d}{T'_{qo}}, &(5.190)\end{aligned}$$

where the symbols are of the typical synchronous generators and have the usual meaning as in current literature [201].

The IEEE Type-1 Exciter [201][230] is used in this section. The first step for sensitivity and stability analysis is to model the power system in the form of differential algebraic equations of the form

$$\begin{aligned}\dot{\mathbf{X}} &= \mathbf{F}(\mathbf{X},\mathbf{Y},\mathbf{u}),\\ 0 &= \mathbf{G}(\mathbf{X},\mathbf{Y},\mathbf{u}), \end{aligned} \qquad (5.191)$$

where $\mathbf{X}$ is a vector of state variables, $\mathbf{Y}$ is a vector of algebraic variables and system parameters, and $\mathbf{u}$ is the control input. These equations can be linearized and rearranged as

$$\Delta\dot{\mathbf{X}} = \mathbf{A}\Delta\mathbf{X}, \qquad (5.192)$$

where $\mathbf{A}$ is a system matrix and is calculated as explained in the following section.

Although the expression of the system matrix is well established in several references [201][230], it is important to investigate the effect of the load model on the overall system matrix. The effect of composite load modeling on the system matrix should be emphasized, and the system matrix with IM loads can be built by

$$\begin{bmatrix}\Delta\dot{X}_g\\ \Delta\dot{X}_m\end{bmatrix} = \begin{bmatrix}A_g & 0\\ 0 & A_m\end{bmatrix}\begin{bmatrix}\Delta X_g\\ \Delta X_m\end{bmatrix} + \begin{bmatrix}C_g & 0\\ 0 & C_m\end{bmatrix}\begin{bmatrix}\Delta V_g\\ \Delta V_m\end{bmatrix}, \qquad (5.193)$$

where V_g and V_m are the terminal voltage vectors of generators and IM load, respectively. X_g and X_m are state vectors for the generator (including exciter and PSS) and motor, respectively, and are represented by

$$\begin{aligned}X_g &= \begin{bmatrix}\delta & \omega & E_q & E_d & V_{ex1} & V_{ex2} & V_{ex3} & E_{fd}\end{matrix}\\ &\qquad \begin{matrix}V_{ps1} & V_{ps2} & V_{ps3}\end{bmatrix}^T, &(5.194)\\ X_m &= \begin{bmatrix}E'_{mx} & E'_{my} & \sigma_m\end{bmatrix}^T, &(5.195)\end{aligned}$$

where V_{ex1}, V_{ex2}, V_{ex3}, E_{fd} and V_{ps1}, V_{ps2}, V_{ps3} are exciter and PSS variables, respectively. The matrices A_g, C_g, A_m, and C_m are block diagonal, and are numerically dependent on the system operating point and the machine parameters. The impact of an IM on the overall system small-signal stability is implicitly reflected by submatrices A_m and A_g, which represent the dynamics of the other dynamic components of the system, i.e., mainly the generators and associated excitation systems.

The linear representation of the coupling characteristics between the system dynamic devices and the network can be expressed as

$$\begin{bmatrix} \Delta I_g \\ \Delta I_l \end{bmatrix} = \begin{bmatrix} W_g & 0 \\ 0 & W_m \end{bmatrix} \begin{bmatrix} \Delta X_g \\ \Delta X_m \end{bmatrix} + \begin{bmatrix} N_g & 0 \\ 0 & N_l \end{bmatrix} \begin{bmatrix} \Delta V_g \\ \Delta V_m \end{bmatrix}, \tag{5.196}$$

where I_g and I_l are the injection currents from generators and composite loads (consisting of static load and the dynamic IM load), respectively. N_g represents the generator matrix. The matrix N_l represents the overall effect of the composite load (i.e., the static load matrix N_s and the IM matrix N_m). Matrices W_g and W_m represent the dependence of the generator and the composite load injection currents with the corresponding state vectors, respectively.

A complete transmission linear representation of the transmission network is given as

$$\begin{bmatrix} \Delta I_g \\ \Delta I_l \end{bmatrix} = \begin{bmatrix} Y_{gg} & Y_{gl} \\ Y_{lg} & Y_{ll} \end{bmatrix} \begin{bmatrix} \Delta V_g \\ \Delta V_l \end{bmatrix}, \tag{5.197}$$

where Y_{gg}, Y_{gl}, Y_{lg}, and Y_{ll} are the matrices derived from the admittance matrix of the system.

Using the above equations and eliminating the algebraic variables, the **A** matrix can be expressed as

$$\mathbf{A} = \begin{bmatrix} A_g & 0 \\ 0 & A_m \end{bmatrix} + \begin{bmatrix} C_g & 0 \\ 0 & C_m \end{bmatrix} \begin{bmatrix} Y_{gg} - N_g & Y_{gl} \\ Y_{lg} & Y_{ll} - N_l \end{bmatrix}^{-1} \begin{bmatrix} W_g & 0 \\ 0 & W_m \end{bmatrix}. \tag{5.198}$$

Now, recall from (2.81) and (5.103) that

$$\frac{\partial \lambda_i}{\partial \mathbf{K}_j} = \frac{\mathbf{w}_i^T \frac{\partial \mathbf{A}}{\partial \mathbf{K}_j} \mathbf{v}_i}{\mathbf{w}_i^T \mathbf{v}_i}, \tag{5.199}$$

with the main setback being obtaining the term $\frac{\partial \mathbf{A}}{\partial \mathbf{K}_j}$. For a motor load parameter K_j, differentiating the **A** matrix above would lead to a sparse matrix as the differentiation of all other matrices except the IM matrices (e.g., A_m, C_m, N_l, and W_m) would lead to zeros

$$\begin{aligned} \frac{\partial \mathbf{A}}{\partial K_j} &= \begin{bmatrix} 0 & 0 \\ 0 & \frac{\partial A_m}{\partial K_j} \end{bmatrix} + \begin{bmatrix} 0 & 0 \\ 0 & \frac{\partial C_m}{\partial K_j} \end{bmatrix} \begin{bmatrix} Y_{gg} - N_g & Y_{gl} \\ Y_{lg} & Y_{ll} - N_l \end{bmatrix}^{-1} \begin{bmatrix} W_g & 0 \\ 0 & W_m \end{bmatrix} \\ &\quad - \begin{bmatrix} C_g & 0 \\ 0 & C_m \end{bmatrix} \begin{bmatrix} Y_{gg} - N_g & Y_{gl} \\ Y_{lg} & Y_{ll} - N_l \end{bmatrix}^{-1} \begin{bmatrix} 0 & 0 \\ 0 & -\frac{\partial N_l}{\partial K_j} \end{bmatrix} \cdots \\ &\quad \times \begin{bmatrix} Y_{gg} - N_g & Y_{gl} \\ Y_{lg} & Y_{ll} - N_l \end{bmatrix}^{-1} \begin{bmatrix} W_g & 0 \\ 0 & W_m \end{bmatrix} \\ &\quad + \begin{bmatrix} C_g & 0 \\ 0 & C_m \end{bmatrix} \begin{bmatrix} Y_{gg} - N_g & Y_{gl} \\ Y_{lg} & Y_{ll} - N_l \end{bmatrix}^{-1} \begin{bmatrix} 0 & 0 \\ 0 & \frac{\partial W_m}{\partial K_j} \end{bmatrix}. \end{aligned} \tag{5.200}$$

Eigenvalue sensitivity with respect to different motor load parameters can be obtained with (5.199) and (5.200). The *Overall Sensitivity* (OS) OS_p of load p with respect to *all* the modes can be defined as

$$OS_p = \sum_{\lambda_i} \sum_{K_j} \left| \frac{\partial \lambda_i}{\partial K_j} \right|. \tag{5.201}$$

For a particular operating point, critical eigenvalues can be identified based on their modes of oscillation. Basically, only the critical eigenvalues are used for stability analysis. In this chapter, they are also used to obtain overall sensitivity as well. It should be noted that in order to assess the system small signal stability over a wider range of operating points, the computation needs to be repeated at different operation points to obtain an overall picture of the system small signal stability property [231][232].

Sensitivity analysis is an important step for power system stability analysis, such as probabilistic small-signal stability analysis [41]. Moreover, another important implementation of sensitivity analysis is the study of the impact of different locations of IM loads on the system (critical) eigenvalues. Different methods have been developed to reduce the computational cost for analyzing the power system with complex IM loads [233]. In this chapter, not all buses need to be modeled by IM loads. Therefore selecting the best locations that can best model the dynamics brought to the system by IM loads is an important step for overall sensitivity analysis. IM loads modeled at different buses may have different impacts on the system (critical) eigenvalues. Generally, buses located far away from major generators tend to be more sensitive in the way they are being modeled. In this section, we show that damping of local modes associated with generators is less if the composite load including the IM is located far away from the generator.

Similarly, the proposed method is tested on the standard New England (ten machines and thirty-nine buses) system as shown in Figure 5.4. All the generators are fourth-order models with the IEEE Type-1 exciter. Generator 10 has one PSS to improve the damping of one of the local modes, which was poorly damped. The system is modeled as described in the previous section and at a given operating point; it is observed that the system has eight local modes and one inter-area mode of oscillation. Here, the operating point is kept constant. Keeping the same operating point refers to the condition where the generation and the loading level remain unaltered, but the composition varies. Base case refers to the condition where there is no IM load in the system, i.e., all loads are modeled by a constant power load. Composite load refers to the combination of a certain percentage of IM load and the rest is ZIP load. Various studies are performed on the system and they are discussed below.

We now desire to compute the impact of location of composite loads. At a given operating point, the location impact of the composite load on the system small-signal stability is investigated. This is done by first investigating the impact of composition of the load on OS and then observing the eigenvalue movement occurring because of the composite load at different locations. The load at the bus is kept constant; however, its composition is varied. There are seventeen load buses in the system and every time only one bus is modeled as a composite load having 30% IM load and 70% ZIP

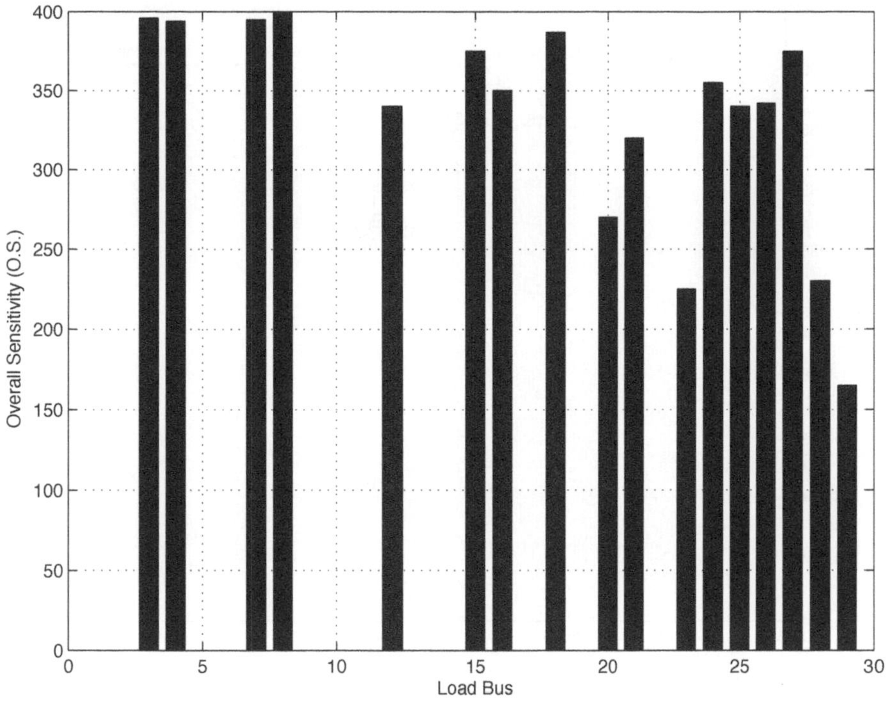

FIGURE 5.7 Effect on the OS of the load modeled as a composite load with a certain percentage of IM and ZIP loads.

load, whereas others are kept as 100% ZIP load. Hence the simulations are performed seventeen times. Sensitivity is observed with respect to load parameters. It is worth noting that the magnitude of the OS is a relative value and only signifies the relative sensitivity of the loads.

It appears from Figure 5.7 that the location of a load at Bus 29 would affect the system the least, whereas the locations of Buses 3, 4, 6, 7, and 18 are highly sensitive. The rest of the load buses can be classified as the second rated sensitive buses with respect to load parameters. Hence it can be concluded that the composite load when connected to Bus 29 has the least effect on the system small signal stability because of parametric variations.

Next, we investigate the effect of motor load locations on eigenvalue movement. The movement of critical eigenmodes because of the different composition of the load is investigated in this section. The eigenmode at the base case and the composite load at each bus are represented by "+" and "*", respectively. It is clear from Figures 5.8 and 5.9 that the effect on the movement of the least damped local mode and the inter-area mode of oscillation is noticeable when the load at Bus 29 is modeled as a composite load. There is eigenmode shift in both of the modes.

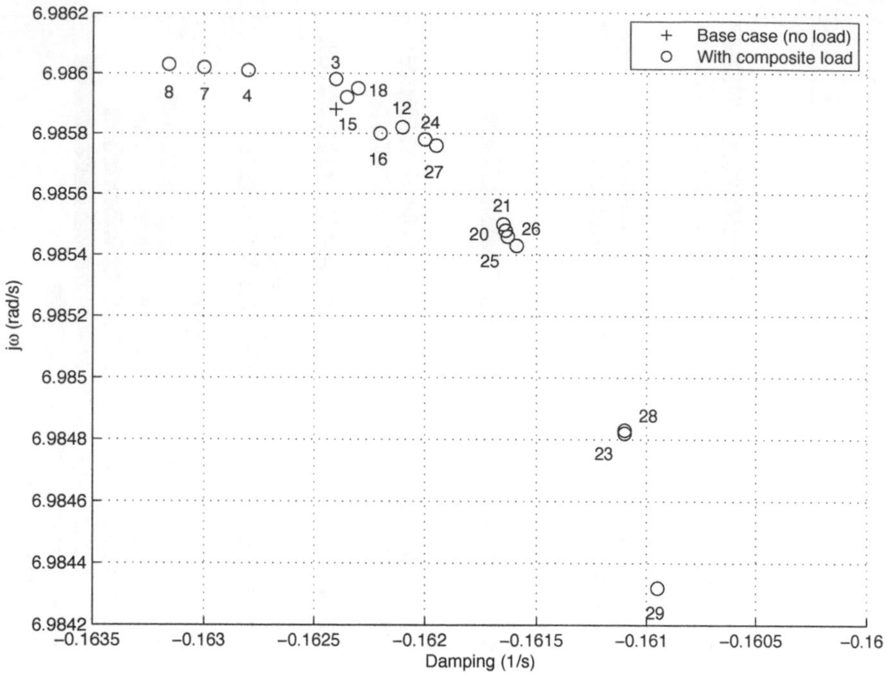

FIGURE 5.8 Least damped local mode movement with a change in the location of composite load from Bus 3 to Bus 29.

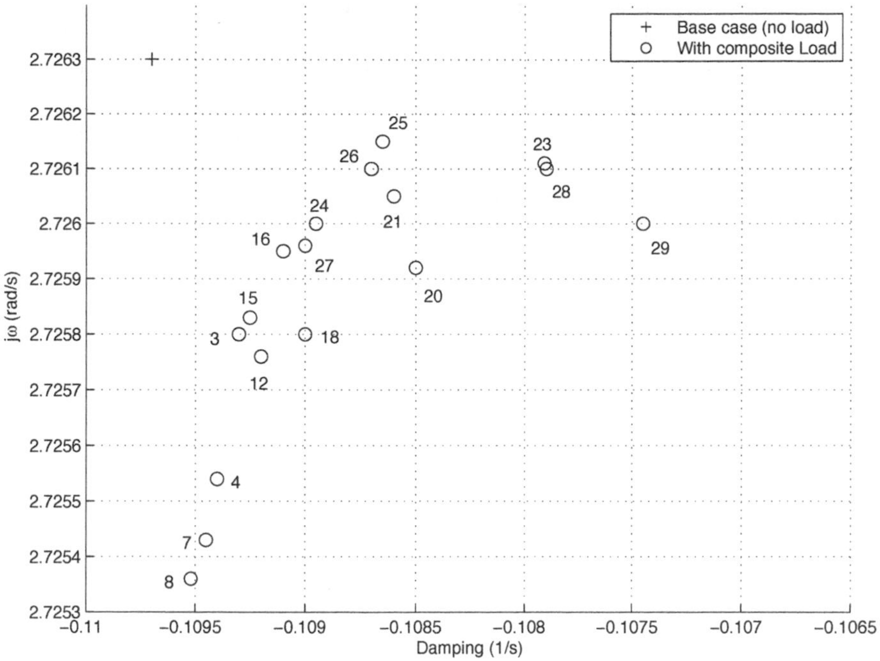

FIGURE 5.9 Least damped inter-area mode movement with a change in the location of composite load from Bus 3 to Bus 29.

This essentially means that when the load at Bus 29 is modeled as a composite load, the eigenvalue is shifted towards the right, as compared to the case when the load at Bus 3 is modeled in the same way. It depicts the position of the IM in the system where the particular system eigenmode is close to the imaginary axis and hence prone to small signal instability. It appears from the topology of the network that the damping of the local mode associated with the generator is reduced when the composite load is located far away from that generator.

Similar movements are observed for other local modes as well. For the purpose of simplicity, only the local mode that is least damped and the inter-area mode are shown in Figures 5.8 and 5.9.

The effect of parametric variations in the composite loads on the critical modes of the system is observed in this section. In this case, the load at Bus 3 is modeled as a composite load (50% IM+50% ZIP), keeping all other loads as 100% ZIP loads. The sensitivity of each critical mode with respect to each of the IM load parameters is calculated as shown in Table 5.22.

TABLE 5.22

Parametric Sensitivity of the Critical Eigenvalues for Composite Load at Bus 3 (50% IM)

No.	Critical Eigenvalue	Frequency (Hz)	Damping	$\frac{\partial\lambda_i}{\partial R_s}(10^{-4})$	$\frac{\partial\lambda_i}{\partial X_s}(10^{-4})$
1	$-0.1624 \pm j6.9859$	1.11	**0.023**	776	815
2	$-0.2206 \pm j7.0176$	1.11	0.031	225	268
3	$-0.2405 \pm j6.8933$	1.09	0.034	184	188
4	$-0.3154 + \pm 8.4846$	1.35	0.037	32	28
5	$-0.2897 + \pm 7.4850$	1.19	0.038	0.102	0.083
6	$-0.1097 + \pm 2.7263$	**0.43**	0.040	658	617
7	$-0.3678 + \pm 8.7524$	1.39	0.041	0.067	0.023
8	$-0.2712 + \pm 5.9810$	0.95	0.045	4	12
9	$-0.4205 + \pm 8.6990$	1.38	0.048	107	114

No.	$\frac{\partial\lambda_i}{\partial R_r}(10^{-4})$	$\frac{\partial\lambda_i}{\partial X_r}(10^{-4})$	$\frac{\partial\lambda_i}{\partial X_m}(10^{-4})$	$\frac{\partial\lambda_i}{\partial H}(10^{-4})$	$\sum_{K_j}\left\|\frac{\partial\lambda_i}{\partial K_j}\right\|(10^{-4})$
1	**4423**	765	40	0	**6819**
2	**1507**	250	13	0	2295
3	**1050**	176	9	0	1608
4	**131**	26	1	0	219
5	**0.281**	0.076	0.004	0	0.546
6	**3704**	582	34	0	**5595**
7	**0.718**	0.019	0.007	0	0.834
8	**291**	12	3	0	321
9	**891**	108	8	0	1227

The sensitivity of the inter-area mode with respect to load parameters for the composite load at Bus 3 is 0.55. The sensitivity of the least damped mode is very important as this mode is prone to move to the right half of the s-plane, eventually leading to unstable behavior of the system under any small changes. (One of the local modes that is least damped is related to the first generator.) Sensitivity with respect to this mode is 0.68. A similar pattern is observed for the composite loads at other buses as well. For instance, when the load at Bus 4 is modeled as a composite load (70% IM+30% ZIP) while maintaining the rest of the loads as 100% ZIP load, a similar result is observed as depicted in Table 5.23. Again, the inter-area mode and the less-damped mode are most sensitive to the variations in the IM load parametric variations.

TABLE 5.23
Parametric Sensitivity of the Critical Eigenvalues for Composite Load at Bus 4 (70% IM)

No.	Critical Eigenvalue	Frequency (Hz)	Damping	$\frac{\partial\lambda_i}{\partial R_s}(10^{-4})$	$\frac{\partial\lambda_i}{\partial X_s}(10^{-4})$
1	$-0.1587 \pm j6.9999$	1.11	**0.022**	4096	4276
2	$-0.2222 \pm j7.0131$	1.11	0.031	1420	1472
3	$-0.2426 \pm j6.8956$	1.09	0.035	854	86
4	-0.3154 ± 8.4851	1.35	0.037	168	146
5	-0.2897 ± 7.4850	1.19	0.038	0.558	0.448
6	-0.1105 ± 2.7362	**0.43**	0.040	323	2988
7	-0.3678 ± 8.7524	1.39	0.042	4	2
8	-0.2785 ± 5.9814	0.95	0.045	18	58
9	-0.4202 ± 8.7011	1.38	0.048	556	59

No.	$\frac{\partial\lambda_i}{\partial R_r}(10^{-4})$	$\frac{\partial\lambda_i}{\partial X_r}(10^{-4})$	$\frac{\partial\lambda_i}{\partial X_m}(10^{-4})$	$\frac{\partial\lambda_i}{\partial H}(10^{-4})$	$\sum_{K_j}\left\vert\frac{\partial\lambda_i}{\partial K_j}\right\vert(10^{-4})$
1	**2302**	4010	208	0	**35610**
2	**8256**	1382	74	0	12604
3	**4792**	808	44	0	7358
4	**662**	136	6	0	1118
5	**1.467**	0.416	0.014	0	2.905
6	**17704**	2818	164	4	**26904**
7	**42**	2	0	0	50
8	**1618**	6	14	0	1768
9	**4736**	562	42	0	6488

The inter-area mode of oscillation is not well damped and the sensitivity of this mode with respect to the IM parameters is very high. The less damped local mode also has very high sensitivity. It appears that the sensitivity is high if damping of that mode is low.

It is important to identify the parameters of the IM load that are most and least sensitive to system stability. This study would help in identifying the parameters that should be dealt carefully while doing the field testing and hence would help in proper power system planning.

Again, the load at Bus 3 is modeled as a composite load (50% IM+50% ZIP), keeping all the other loads as 100% ZIP loads as in the above section. The change in the particular eigenmode with a small change in IM load parameter is calculated using (5.200) and (5.201) for each parameter. The result is presented in Table 5.22. It is found that R_r is the most sensitive parameter as $\frac{\partial \lambda_i}{\partial R_r}$ is maximum for all the critical modes, whereas H is the least. A similar pattern is observed when the loads at other buses are modeled as composite loads.

5.6.3 PARAMETRIC CHANGES AND CRITICAL MODES MOBILITY

The effect of the increase in a particular parameter of the IM load is observed in the mobility of the eigenmode. This is achieved by modeling the load at Bus 3 as a composite load (10% IM+90% ZIP) and then the parameters (X_s, X_m, X_r, R_r, R_s and H) of the IM are changed individually. Every time, only one parameter is increased while keeping the rest constant one at a time. Each load parameter is increased by 2% for up to three steps and the eigenmodes are plotted. The effect of increasing each IM load parameter on the mobility of the eigenmode is depicted in Figure 5.10.

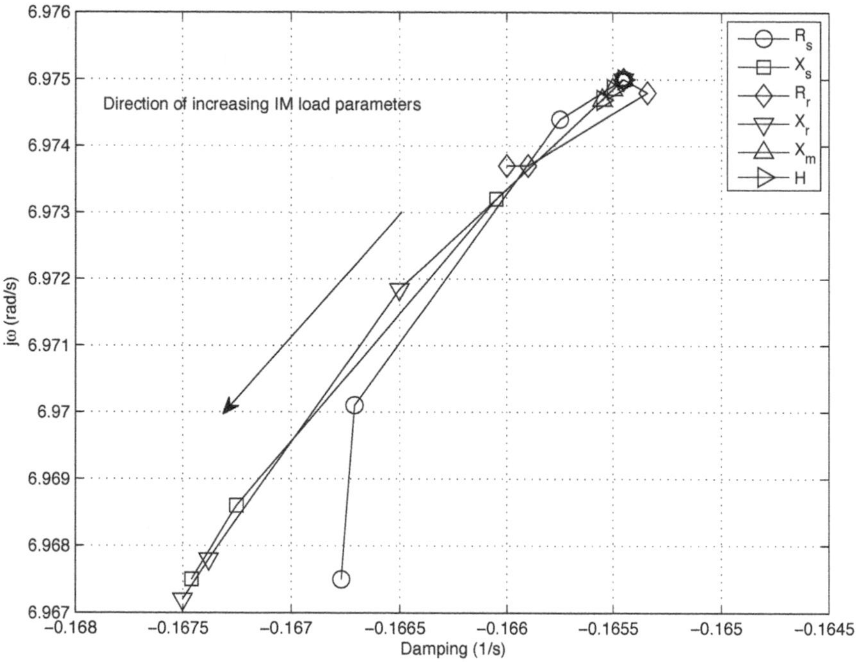

FIGURE 5.10 Root loci of the local mode for the given operating condition for the different values of the IM load parameters.

Increasing the parameters changes the damping of the least damped local mode. For example, if parameter X_r is changed to $1.06 \times X_r$, the eigenvalue would change from $-0.1655+j6.9755$ (top right corner) to $-0.1675+j6.967$ (bottom left corner) as shown in Figure 5.10. A similar effect is observed in the inter-area mode of oscillation, which is shown in Figure 5.11.

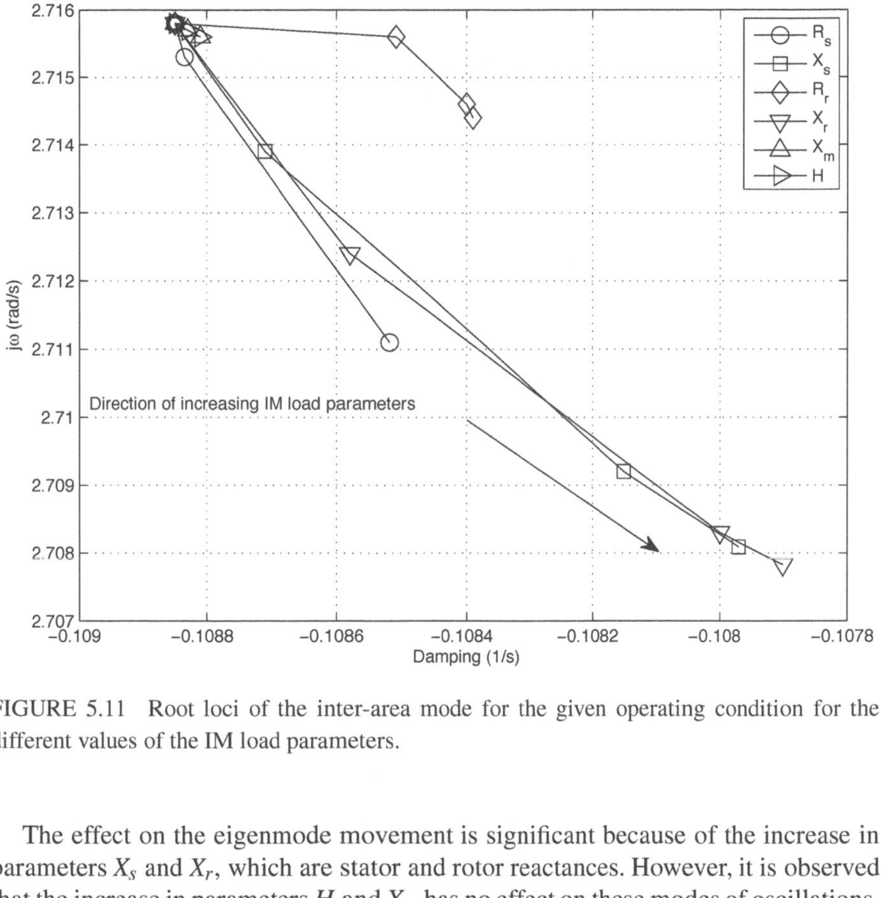

FIGURE 5.11 Root loci of the inter-area mode for the given operating condition for the different values of the IM load parameters.

The effect on the eigenmode movement is significant because of the increase in parameters X_s and X_r, which are stator and rotor reactances. However, it is observed that the increase in parameters H and X_m has no effect on these modes of oscillations.

5.6.4 EFFECT OF THE NUMBER OF IMS ON OVERALL SENSITIVITY (WITH 30% IM LOAD)

When all the motors are modeled identically (as a large IM), the effect of the increasing number of IM loads in the system is observed by modeling the load of one bus as a composite load and rest of the loads as a constant power load, and then sequentially increasing the number of buses having composite loads. The results are summarized in Table 5.24.

The Standard Deviation (SD) of the OS of the load bus is calculated with the variable number of IMs in the system, and it is observed that the OS of Bus 12 and Bus 26 do not show much variations. This is due to the fact that there are not many IMs

TABLE 5.24
OS of the Composite Loads with the Variable Number of IMs Modeled in the System

Load Bus	Number of IMs in the System 1	2	3	4	5	6	7	8	9
3	0.76	0.11	0.61	0.35	1.52	0.27	1.05	0.58	2.02
4	0	1.06	0.42	4.30	2.88	2.05	2.08	6.33	2.28
7	0	0	0.05	0.14	0.99	0.22	1.08	0.19	0.68
8	0	0	0	3.99	3.95	2.67	2.49	6.07	2.33
12	**0**	**0**	**0**	**0**	**0.00**	**0.00**	**0.00**	**0.00**	**0.00**
15	0	0	0	0	0	0.39	0.98	2.33	1.25
16	0	0	0	0	0	0	1.07	0.56	1.53
18	0	0	0	0	0	0	0	0.57	0.45
20	**0**	**0**	**0**	**0**	**0**	**0**	**0**	**0**	**17.33**
21	0	0	0	0	0	0	0	0	0
23	0	0	0	0	0	0	0	0	0
24	0	0	0	0	0	0	0	0	0
25	0	0	0	0	0	0	0	0	0
26	0	0	0	0	0	0	0	0	0
27	0	0	0	0	0	0	0	0	0
28	0	0	0	0	0	0	0	0	0
29	0	0	0	0	0	0	0	0	0

Load Bus	Number of IMs in the System 10	11	12	13	14	15	16	17	SD
3	0.30	0.72	0.32	1.75	0.17	0.92	1.25	1.95	0.631
4	3.55	1.16	3.93	2.66	1.97	3.14	9.32	4.54	2.271
7	0.19	0.26	0.18	0.80	0.68	0.49	0.51	0.73	0.348
8	3.61	2.61	3.66	10.32	7.11	2.76	9.66	6.28	3.059
12	**0.00**	**0.00**	**0.00**	**0.00**	**0.00**	**0.00**	**0.00**	**0.00**	**0.001**
15	1.76	1.72	3.06	2.57	1.99	1.81	4.10	1.70	1.228
16	0.96	1.72	0.74	2.20	0.94	0.95	0.82	1.34	0.692
18	0.56	0.13	0.58	0.76	0.56	0.27	0.74	0.31	0.293
20	**1.32**	**2.47**	**6.00**	**15.68**	**3.07**	**3.74**	**3.62**	**6.96**	**5.373**
21	0.71	0.56	2.13	1.55	1.29	1.06	2.52	1.36	0.844
23	0	0.26	0.21	1.62	0.21	0.76	0.10	0.87	0.446
24	0	0	2.03	1.00	1.38	0.38	1.71	1.30	0.716
25	0	0	0	0.60	0.09	0.32	0.10	0.78	0.234
26	0	0	0	0	0.12	0.08	0.38	0.56	0.159
27	0	0	0	0	0	0.50	0.06	1.93	0.476
28	0	0	0	0	0	0	0.25	1.08	0.265
29	0	0	0	0	0	0	0.00	1.57	—

in the surroundings of these loads, and if there are some, there is not much interaction among them. This is clear from Figure 5.12 as well. Variations in the OS of the load at Bus 8 and Bus 20 are also remarkably high, which shows that motors in the vicinity of these load buses do interact with each other and hence affect the overall parametric sensitivity. Furthermore, it is necessary to observe the patches of loads that do interact with each other leading to increased OS.

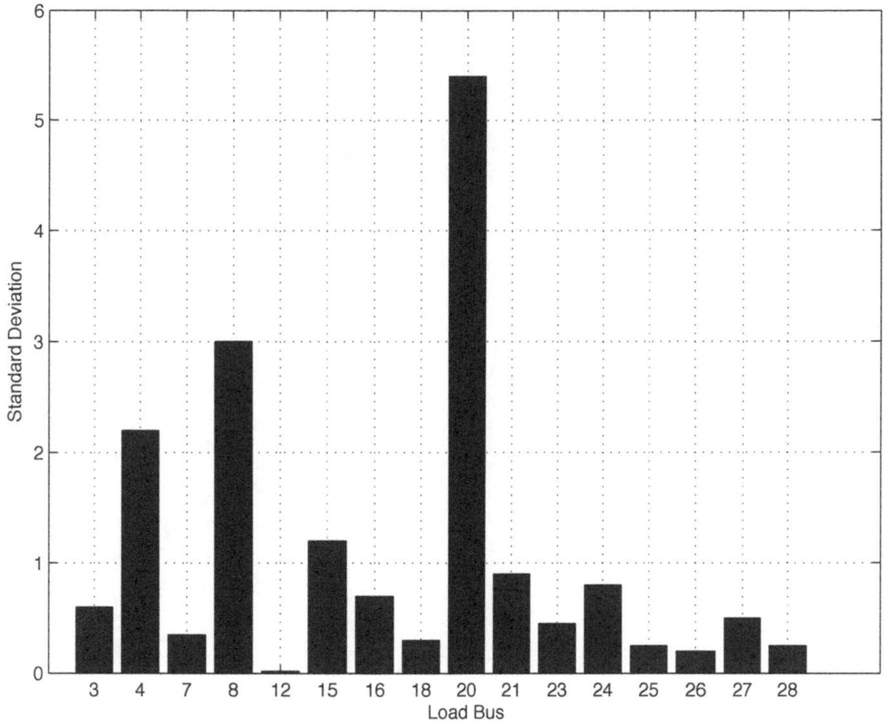

FIGURE 5.12 S.D. of the motor loads with a variable number of IMs in the system.

Figures 5.13 and 5.14 give a closer view of the interaction of certain groups of IM loads in the system. Overall sensitivity with all the buses is modeled as composite loads and hence leading to a total of seventeen motors in the system is depicted in Figure 5.13. It is evident that the OS of the load at Bus 27 and at Bus 28 increases with the introduction of dynamic motor load (30% IM) in Bus 29. It is also evident from the topology of the New England system that Load 28 and Load 27 are in the vicinity of Load 29 and are connected by G9. Hence, it appears that the increase in the number of IMs in the vicinity would affect the overall sensitivity of each of the motors in that vicinity and these motors tend to be more sensitive to the variations in the load parameters.

A similar observation can be made in other patches of motors near Buses 15, 20, and 21 as depicted in Figure 5.14. As such, these results show that there is an increased cumulative impact of the total number of nearby IM loads on the overall sensitivity of the load.

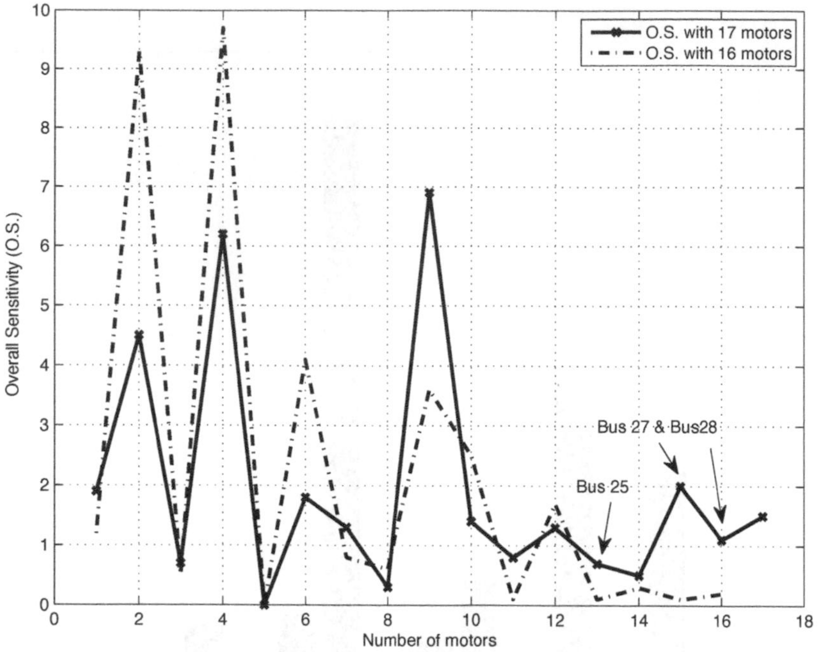

FIGURE 5.13 OS of the load with 16 and 17 motors in the system.

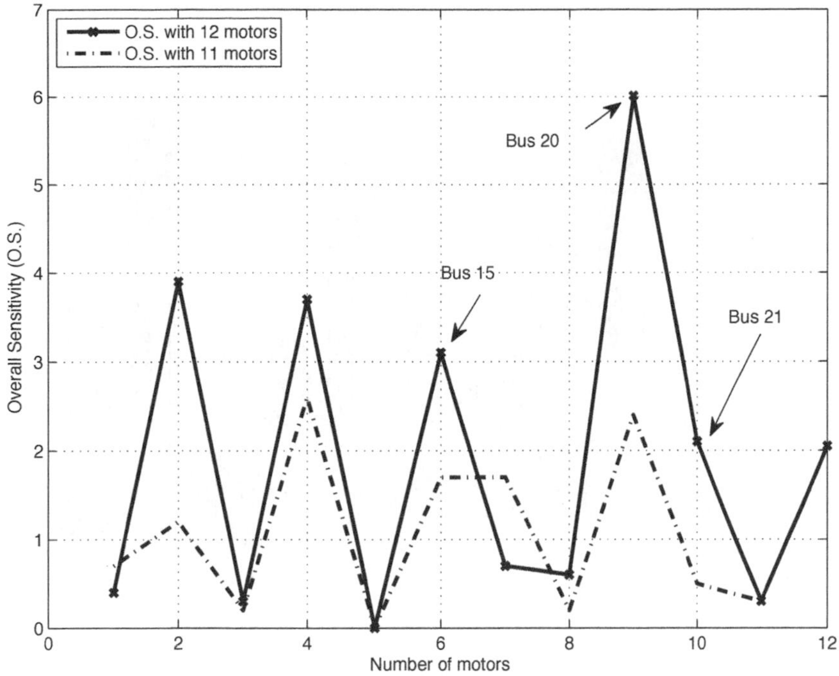

FIGURE 5.14 OS of the load with 11 and 12 motors in the system at Bus 23 and Bus 24.

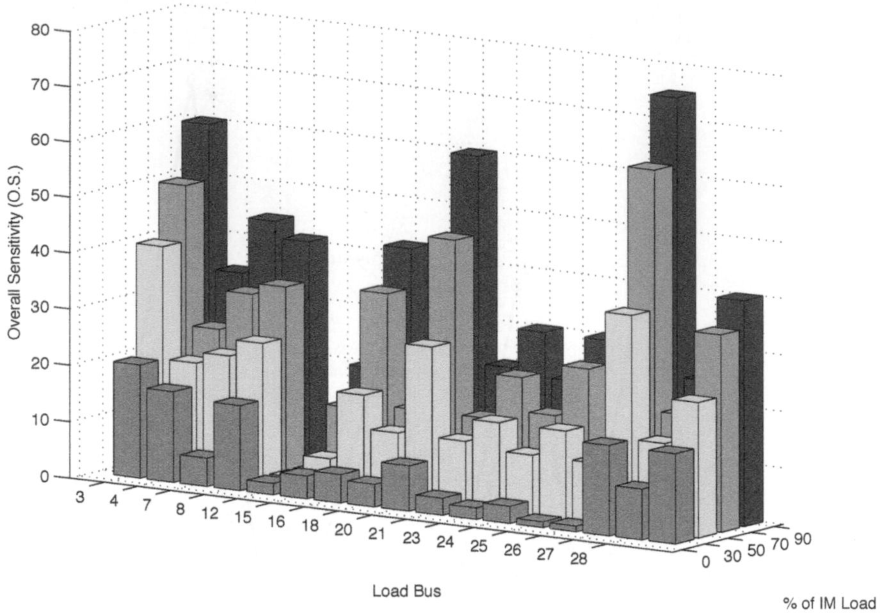

FIGURE 5.15 Overall sensitivity of the load with a different percentage of dynamic load component.

5.6.5 EFFECT ON OVERALL SENSITIVITY WITH DIFFERENT PERCENTAGES OF IM LOAD IN THE COMPOSITE LOAD

The effect of different compositions of the load on the OS is investigated in this section. All the load buses are modeled with 30% IM load and the rest are ZIP loads. The components of IM are increased to 50%, 70%, and 90%, and the OS is observed. With the increase in the dynamic component of the composite load, OS is increased as shown in Figure 5.15. It can be seen that loads having more dynamic components are found to be more sensitive to the load parameters in comparison to loads having less dynamic components represented by the IM load. It can also be seen from Figure 5.15 that most of the relative rankings of load sensitivity with a particular composition of dynamic and static loads remain basically the same. For example, the load at Bus 27 remains the most sensitive across 30–90% IM load. However, the load at Bus 8 has relatively less effect on the change in the OS across the same range. Moreover, Bus 2 has negligible effect on the OS.

Findings for this case study can be summarized as follows:

1. The effect of the location of the composite load on the system on the eigenvalue sensitivity is observed, and the locations that are most and least sensi-

tive are identified. These depend on system topology, and are useful for the planning and operation of the power system. However, it appears that the farther the composite loads are from the generators contributing the oscillatory modes, the smaller is the damping of those modes. From the observations, it will be easier to find locations where proper detailed load modeling is necessary for system stability studies;

2. The relative magnitude of the sensitivity with respect to the load parameters is calculated, to find the contribution of different oscillatory modes to the overall sensitivity of the load. It is observed that the load is most sensitive to the least damped mode. This further emphasizes the importance of load modeling in system stability analysis; and
3. The effect of the vicinity of the motor on the overall sensitivity of the loads is studied as well. This essentially means that the motors in a certain patch in the system affect each other and contribute to the increase in the sensitivity of each of the motors in that vicinity, including it's own sensitivity.

5.7 DISCUSSION

It can be seen clearly that probabilistic small signal stability assessment is a systematic approach which includes:

1. system modeling and system state matrix computation;
2. eigenvalue computation and identification of critical modes;
3. eigenvalue sensitivity computation with respect to system parameters; and
4. advanced probabilistic techniques such as the Tetrachoric series and Hermite polynomials.

These four stages are critical for computing the small signal stability probabilities for large scale interconnected systems.

In particular, the computation of the probability that the system is stable is made possible with the probability density function of the real parts of the eigenvalues (with known mean and covariances) based on multivariate random variables and the generalized Tetrachoric series. Generalization of the Tetrachoric series requires a normalized covariance matrix of a set of multivariate random variables such as power system parameters. Hermite polynomials are used in the Tetrachoric series. The order of the Hermite polynomials is computed from the entries of the normalized covariance matrix and their corresponding indices in Taylor Series expansion.

With the techniques described so far, we now have a rather complete tool to compute the probabilistic power system small signal stability subject to small system parameter variations and IM loads. The probabilistic small signal stability is more important for system planning when considering the system parameter uncertainties without extensive repeated computations on individual equilibrium points and various contingency cases. It also provides comprehensive probabilistic stability information which is more realistic in practice than deterministic information, considering the variety of uncertainties in a power system, especially in a deregulated environment.

This forms the theoretical foundation and identifies future research in order to achieve a comprehensive probabilistic small-signal stability assessment methodology. The results are used to help Regional Transmission Organizations (RTOs) and Independent System Operators (ISOs) perform planning studies under the open access environment. Also, the results can be used as a valuable reference for utility power system small-signal stability assessment probabilistically and reliably for any realistic interconnected networked engineering systems.

5.8 CONCLUSION

Power systems have been undergoing dramatic changes recently following the deregulation of the industry. The formally vertically integrated power industry has been deregulated into different entities including generation companies, transmission companies, and distribution companies, etc. Electricity markets have been formed in many countries such as the United States, United Kingdom, and Australia. Under the deregulated environment, power systems are experiencing more and more uncertainties in both system operation and planning areas. Such uncertainties are the result of increased demand, intense competition, and deregulation, etc. Traditionally, power system small signal stability is performed in a deterministic way where different system equilibrium and a list of contingencies have to be evaluated in order to achieve an overall picture of the system stability margin. However, with the increasing uncertainties, such a traditional approach is facing more and more difficulties, and therefore requires the use of probabilistic approaches. These motivate us to develop a systematic approach for the probabilistic small signal stability assessment, which includes the following four major steps:

1. forming the state matrix for overall system;
2. performing eigenvalue analysis and identifying critical oscillation modes;
3. computing sensitivity factors of critical eigenvalues with respect to system non-deterministic parameters; and
4. calculating the overall system small signal stability index using Tetrachoric series.

This approach is applied to a New England 39 Bus test system, and the following are achieved:

1. building the mathematical models of overall system for small-signal stability assessment purposes (dynamic characteristics of the generators, excitation systems, governor systems, etc.);
2. performing eigenvalue analysis of the state matrix and identifying the critical eigenvalues;
3. developing the algorithms to calculate the sensitivity matrix of the system eigenvalues to the non-deterministic random variables;
4. computing the probability density function of the critical eigenvalues given that the probabilistic description of random variables is known;
5. proposing the overall system probabilistic small signal stability index; and

6. applying Tetrachoric series to the calculation of the overall system probabilistic small signal stability index with constant and composite loads.

We restated the importance of composite load modeling, in particular, dynamic load models with IMs in power system stability assessment as well. Misrepresentation of the actual load could lead to quite different stability assessment results. Based on detailed system dynamic analysis, a comprehensive study is performed on load sensitivity considering IM loads in a composite load model. Various scenarios are observed to understand the effect of composite load modeling on the systems small signal stability characteristics. The main findings of this chapter and their implications in power system small signal stability analysis are also given following the detailed comprehensive case studies. Using the presented approach, the most sensitive composite load location can be found for further stability analysis. Significant parameters of the IM load are identified based on sensitivity analysis. It is also observed that the vicinity of motors has an increasing effect on the overall sensitivity, and the increasing dynamic component of the load has a significant effect on certain loads in a power system. These findings provide useful information for further system stability analysis, as evidenced through the system oscillation analysis in this chapter.

We have used the data in *real* matrices for intelligent diagnosis and prognosis of realistic engineering applications including Dominant Feature Identification (DFI) for advanced feature extraction to predict tool wear, Modal Parametric Identification (MPI) of critical resonant modes to improve complex mechatronics R&D, and probabilistic small signal stability assessment of large-scale interconnected power systems. In the next chapter, we will use *binary* or *Boolean* matrices whose entries consist of 1s and 0s only for discrete event command and control for distributed networked teams with multiple missions.

6 Discrete Event Command and Control

The perusal of real matrices for intelligent diagnosis and prognosis of some realistic industrial networked engineering systems is discussed in previous chapters. In this chapter, we will introduce the use of *binary* or *Booolean* matrices for *event-triggered* discrete event systems, and an example of Discrete Event Control (DEC) is also given for military missions.

During mission execution in military applications, the US Army Training and Doctrine Command (TRADOC) Pamphlet 525-66 Battle Command and Battle Space Awareness capabilities prescribe expectations that networked teams will perform in a reliable manner under changing mission requirements, varying resource availability and reliability, and resource faults, etc. In this chapter, a Command and Control (C2) structure is presented that allows for computer-aided execution of the networked team decision-making process, control of force resources, shared resource dispatching, and adaptability to change based on battlefield conditions. A mathematically justified networked computing environment is provided called the Discrete Event Control (DEC) framework. DEC has the ability to provide the logical connectivity among all team participants including mission planners, field commanders, war-fighters, and robotic platforms. The proposed data management tools are developed and demonstrated on a simulation study and an implementation on a distributed Wireless Sensor Network (WSN). We also present a simulation study of a realistic military ambush attack using Future Combat Systems (FCSs). The results show that the tasks of multiple missions are correctly sequenced in real-time, and that shared resources are suitably assigned to competing tasks under dynamically changing conditions without conflicts and bottlenecks.

6.1 INTRODUCTION

The US Army TRADOC Pamphlet 525-66 identifies Force Operating Capabilities required for the Army to fulfill its mission for a networked war-fighter concept. Two such capabilities are Battle Command and Battle-Space Awareness, for which there are expectations that networked teams will perform in a reliable manner under changing mission requirements, varying resource reliability, and resource faults. Battlefield or disaster area teams may be heterogeneous networks consisting of interacting humans, ground sensors, and Unmanned Aerial Vehicles (UAVs) or Unmanned Ground Vehicles (UGVs). Such scenarios should provide intelligent task sequencing for multiple missions, synchronization of efforts for multiple missions, and shared services of resources to augment the capabilities of the remote-site mission commander and on-site war-fighter. This requires a scalable, deployable, and mobile networking capability that supports mission tailoring, force responsiveness and agility, ability to change missions without exchanging forces, and general adaptability to changing battlefield conditions.

In this chapter, we present a computer programmable Command and Control (C2) structure that allows for execution of the decision-making process, control of force resources, and adaptability to change. We describe a rigorous mathematically justified networked computing environment that has the potential to provide the logical connectivity among all team participants including mission planners, field commanders, war-fighters, and robotic platforms. Included are data management tools to ensure that the tasks of multiple missions are correctly sequenced in real-time and that shared resources are suitably assigned to competing tasks under dynamically changing conditions.

A rule-based Discrete Event Controller (DEC) has been developed [234] and applied to various engineering and manufacturing applications ranging from schedule planning [235][236][237] to reducing the product life-cycle in prototype designs, as well as Wireless Sensor Networks (WSNs) [59]. The DEC matrix-based formulation [58] is portable and easily implemented on any platform. It facilitates industries adapting quickly to fast-evolving market conditions for transition between inventory and products with minimal human intervention [238].

Based on this DEC, we develop here a C2 structure. DEC is easily programmed on a laptop digital computer. DEC is based on *matrices* that contain two types of information. Mission task requirements are prescribed by mission commanders in terms of a Task Sequencing Matrix (TSM). This allows commanders to convey purpose without providing detailed direction on how to perform the task or mission. Resource assignments to the tasks are prescribed by field commanders or the war-fighter in terms of a Resource Assignment Matrix (RAM). As missions change or are added, the task sequencing matrices are easily reconfigured. Multiple missions can be programmed by multiple mission commanders into the same networked team. As resources fail or are added, the resource assignment matrices are easily reconfigured in real-time time.

DEC allows for synchronizing forces and war-fighting functions in time, space, and purpose to accomplish multiple simultaneous missions that may change dynamically. The matrix formulation of DEC allows for rigorous mathematical analysis of the performance of the networked team, and reveals problems such as bottlenecked resources and shared-resource blocking phenomena. DEC guarantees proper sequencing of the competing tasks of multiple missions, assigning appropriate resources immediately as they become available and resolving conflict situations. DEC can be programmed into networked microprocessors using a novel "or/and" matrix Boolean algebra that allows programming of rule-based decisions in streamlined software algorithms. Computer or Personal Digital Assistant (PDA) user interfaces can allow automatic generation of the TSMs given the requirements of mission commanders, and of the RAMs by field commanders.

The rest of the chapter is organized as follows. Section 6.2 describes the proposed Command and Control (C2) structure for distributed networked teams with multiple missions. Section 6.3 details the conjunctive rule-based DEC with state and output equations, and bottlenecks, blocking, and deadlocks are avoided. Section 6.4 introduces the disjunctive rule-based DEC, and the properness of the disjunctive DEC is proven rigorously. Section 6.5 simulates the proposed conjunctive DEC2 structure, and the effectiveness is verified with implementation results on a WSN, as well as simulations on a military ambush attack using FCSs. Our conclusion and future work directions are summarized in Section 6.6.

6.2 DISCRETE EVENT C2 STRUCTURE FOR DISTRIBUTED TEAMS

In this section, we describe a Command and Control (C2) structure for programming multiple missions into heterogeneous teams of distributed agents, and controlling the performance of these missions in real-time. This is a decision-making DEC that contains rules to sequence the tasks in each mission, and to assign resources to those tasks. DEC has a message-passing architecture that is in conformance with Joint Architecture for Unmanned Ground Systems (JAUGS) [239], and is an efficient means to realize the high-level Observe, Orient, Decide, and Act (OODA) loops of 4D/RCS [240]. DEC is able to coordinate the sequencing of operations between multiple soldiers and robots efficiently and without conflict, thus contributing to the concept of Safeops.

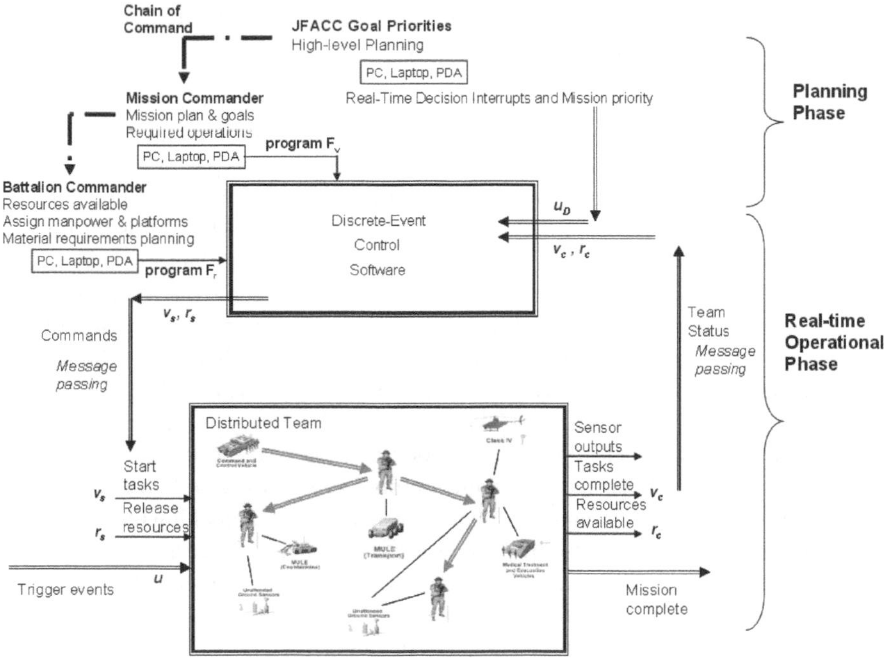

FIGURE 6.1 C2 rule–based discrete event control for distributed networked teams.

DEC allows fast mission programming of distributed teams, and facilitates rapid deployment of man/machine teams, wireless sensor networks, and other event-based systems. DEC provides a seamless C2 architecture that facilitates quickly turning any deployed team into a tactical unit [241]. The DEC runs on a computer and functions as a feedback controller in real-time. See Figure 6.1. As a feedback controller, DEC obtains information from each networked agent about which tasks that agent has just completed, and which of its resources are currently available. This information about team status can be transmitted via a message-passing protocol over a wireless sensor network (WSN), or over the internet [59]. Then, given such information from all active nodes, DEC computes which mission tasks *could* be performed next. Then, based on priority measures or war-fighter decision input, DEC decides which tasks the team *should* perform next. Based on this, it sends message-based commands to each agent to perform certain tasks or release certain resources. All this is accomplished efficiently using a computer software DEC tool to be described.

The commands sent by DEC to the team agents could be command inputs into semi-autonomous machine nodes, and could be in the form of messages for decision assistance over a PDA for human agents.

The DEC can be programmed on a digital computer and requires very small code for implementation. The keys to the ease of use and implementation of DEC [234] are formal mathematical computations based on *matrices* that contain two types of information (TSM and RAM below), and the use of a non-standard *matrix or/and* algebra. The functionality of DEC has two phases:

1. *Planning/Programming Phase.* Mission task requirements are prescribed by mission commanders in terms of a TSM. This allows commanders to convey purpose without providing detailed direction on how to perform the task or mission. Next, resource assignments to the missions tasks are prescribed by field commanders or the war-fighter in terms of RAM. All this information could be entered via Graphical User Interfaces on laptops, handheld PDA, etc. As missions change or are added, these matrices are easily reconfigured in real time. Multiple missions can be programmed by multiple mission commanders into the same networked team that shares the same resources. This is effectively the world modeling phase of 4D/RCS [240].
2. *Operational Phase.* The DEC will automatically poll active agents for their status at each event update and properly sequence the tasks of all programmed missions, and assign the required resources. Conflicting requests for resources are automatically handled so as to avoid blocking phenomena. During operation, as resources fail or are added, the resource assignment matrix is easily reconfigured in real-time time to allow uninterrupted mission performance in spite of resource failures. At any time, additional missions may be programmed into the team or deleted.

The Programming Phase is discussed in this section, and the Operational Phase in the next.

A mission should achieve desired goal states that are triggered by external events occurring within the frame of discernment of the distributed team. Mission task requirements are specified by mission commanders in terms of desired responses to special trigger events, and accomplishment of required goals. A mission is defined as a sequence of script behaviors [242], here called tasks, that is triggered by prescribed events, and results in prescribed goal states. Mission commanders should be able to convey purpose without providing detailed direction on how to perform the tasks of the mission as dictated in the US Army TRADOC Pamphlet 525-66. A sample mission scenario, taken from [240], is shown in Figure 6.2.

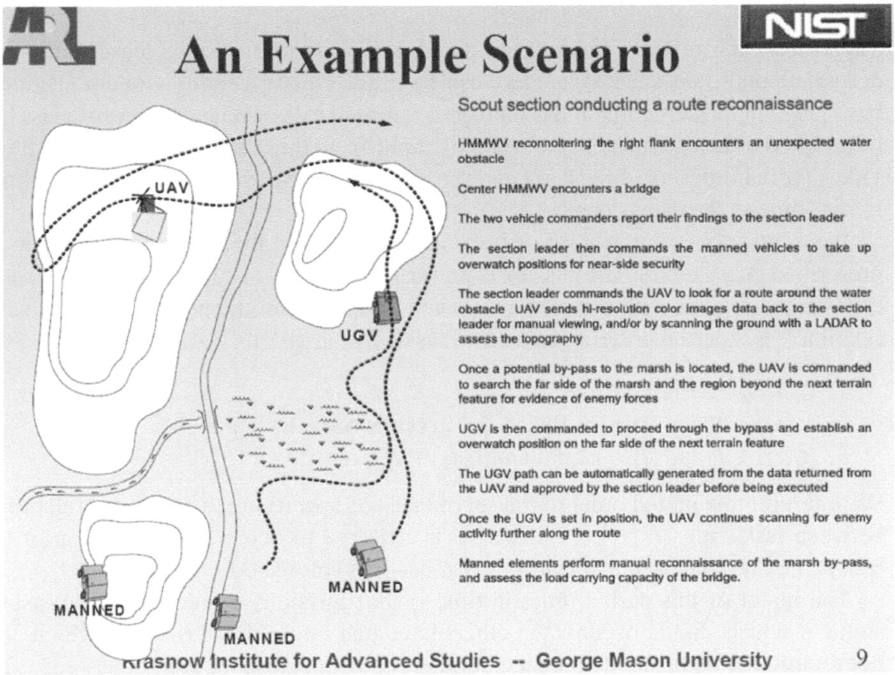

FIGURE 6.2 Sample mission scenario from [240].

6.2.1 TASK SEQUENCING MATRIX (TSM)

Given the basic mission requirements specified by mission commanders, a sequence of scripts or tasks that perform the mission is constructed. A grammar-based method for doing this is given [242], where the sequence of tasks for a mission is determined using a planning function such as A-Star search. In 4D/RCS [240], a task analysis is used to create a task decomposition tree. In fact, given basic elemental tasks and the required high-level goals, there are many software planners that can fill in the detailed sequence of steps needed to attain the goals. Notably effective are the so-called Hierarchical Task Network (HTN) planners [235][236][237] which decompose goals into sequences of primitive actions and compound actions, e.g., tasks, that are required to attain those goals. On the other hand, detailed task sequences could also be constructed by aides to the mission commander.

Each mission task has a well-defined initial state and start event, and ends with a well-defined exit [242]. The tasks are fired by rules of the form

Rule i: IF (the tasks required as immediate precursors to task *i* are compete)
AND (the resources required for task *i* are available)
THEN perform task *i*.

Therefore, each mission can be considered as a sequence of rules prescribing under what conditions each of its tasks can be fired. This is a semi-autonomous rule-base [243] in the sense that it has hard-programmed rules which can nevertheless be interrupted or resequenced in real-time by input from the war-fighter. The rules provide a partial ordering of the tasks and the actual task sequence is determined online in real-time as the events occur.

We define a *mission* as a set of tasks that is triggered by events and results in a prescribed goal state(s). The mission is prescribed in terms of a *strict partial ordering* of tasks that begins with detected events and ends up with the goal states. A binary relation P is a partial ordering on a set of tasks $\{t_i\}$ in $\{t_i\}$ if

1. $(t_i, t_i) \notin P$ (irreflexive);
2. $(t_i, t_j) \in P$ and $(t_j, t_i) \in P$ then $t_i = t_j$ (antisymmetry); and
3. $(t_i, t_j) \in P$ and $(t_j, t_k) \in P$ then $(t_i, t_k) \in P$ (transitive).

We interpret this partial order of the set of tasks as specifying the temporal relations between tasks, e.g., $(t_i, t_j) \in P$ if task t_i is required to occur immediately prior to task t_j, i.e., task t_j can only occur if task t_i has just completed.

The intent of this partial order in time is that missions should consist of tasks, some of which should occur when others have just finished, but many of which are not required to be in any definite temporal order with respect to each other (e.g., see Figure 6.2). This gives the mission commander great freedom to prescribe only those causal relations between tasks which are tactically important. This allows commanders to convey purpose without providing detailed direction on how to perform the task or mission. Since the missions consist only of partial orderings of tasks, then it will be the responsibility of the DEC to decide the actual ordering of tasks in real time as the events unfold and the resources needed for the tasks become available. The mechanisms for performing this during the Operational Phase are described in the next section.

To capture the mission and its tasks in a convenient and computable form, define the Task Sequencing Matrix (TSM), which has element (i, j) equal to "1" if task t_j is a required immediate precursor for task t_i. Note that multiple 1's in a single row i indicate that multiple tasks are required as immediate precursors for task t_i. TSM was used by Steward [244], Warfield [56], and Wolter et al. [245], and others to sequence the tasks required in manufacturing assembly and part processing.

TSM is a mapping from tasks to tasks. We would like to construct a mathematically formal rule-based DEC that runs as software code on a C2 computer and is capable of sequencing the tasks and assigning resources dynamically in a networked team. In the next section it is shown that this is possible if each task is fired by a rule. Therefore, to introduce the rule base, decompose the TSM as

$$TSM = S_v \cdot F_v, \tag{6.1}$$

where the *input TSM* F_v (loosely called simply the TSM) is a mapping from the tasks to the set of rules, and the *output TSM* S_v is a mapping from the rules to the task space.

This matrix multiplication denoted by the small dot is not the standard matrix multiply, but occurs in the *or/and* algebra, where multiply means *.and.* and addition means *.or.* It is easy to write a programming function to multiply matrices in or/and. This function is part of the simple computational machinery of DEC and is shown in the next section (see Table 6.3).

The input TSM F_v maps the tasks $\{t_j\}$ to a rule base consisting of a set of rules $\{x_i\}$. It has entry (i, j) of 1 if rule i requires task j as a immediate precursor to its firing. Multiple ones in a row lead to rules of the form

IF (task j_1 has finished AND task j_2 has finished AND task j_3 has finished)
THEN fire rule i.

The output TSM S_v maps the rule base $\{x_j\}$ to the tasks $\{t_i\}$. It has entry (i, j) of 1 if task i is to be started when rule j fires. Often, there is one rule to fire each task, so that S_v is essentially the identity matrix.

To illustrate, a sample mission is shown in Table 6.1. This mission could be programmed by any user of the networked sensor team. This scenario has a Wireless Sensor Network (WSN) consisting of unattended ground sensors (UGS) and mobile robots (R), possibly UGV or UAV. When the trigger event u^1, (e.g., here, a chemical attack) is detected by a UGS (specifically, sensor *UGS1* here), a prescribed sequence of tasks is carried out that includes taking further sensor readings and dispatching mobile robots to gather additional information. The mission ends when sensor *S2* takes a measurement, either verifying a threat or declaring a false alarm. Each task has a label, displayed in the second column.

The input TSM F_v corresponding to this Mission 1 is F_v^1 as

$$\begin{array}{c} \\ x_1^1 \\ x_2^1 \\ x_3^1 \\ x_4^1 \\ x_5^1 \\ x_6^1 \\ x_7^1 \\ x_8^1 \\ x_9^1 \end{array}
\begin{array}{c}
\begin{array}{ccccccccccc} t_1 & t_2 & t_3 & t_4 & t_5 & t_6 & t_7 & t_8 & t_9 & t_{10} & t_{11} \end{array} \\
\left[\begin{array}{ccccccccccc}
0 & 0 & 0 & 0 & 0 & 0 & 0 & 0 & 0 & 0 & 0 \\
1 & 1 & 0 & 0 & 0 & 0 & 0 & 0 & 0 & 0 & 0 \\
0 & 0 & 1 & 1 & 0 & 0 & 0 & 0 & 0 & 0 & 0 \\
0 & 0 & 0 & 0 & 1 & 0 & 0 & 0 & 0 & 0 & 0 \\
0 & 0 & 0 & 0 & 0 & 1 & 0 & 0 & 0 & 0 & 0 \\
0 & 0 & 0 & 0 & 0 & 0 & 1 & 1 & 0 & 0 & 0 \\
0 & 0 & 0 & 0 & 0 & 0 & 0 & 0 & 1 & 0 & 0 \\
0 & 0 & 0 & 0 & 0 & 0 & 0 & 0 & 0 & 1 & 0 \\
0 & 0 & 0 & 0 & 0 & 0 & 0 & 0 & 0 & 0 & 1
\end{array}\right]
\end{array}$$

TABLE 6.1
Mission 1–Task Sequence for Deployed WSN

Mission 1	Task Label	Task Description
Input 1	*EVENT* u^1	*UGS1* launches chemical alert
Task 1	*S4m*^{1}	*UGS4* takes measurement
Task 2	*S5m*^{1}	*UGS5* takes measurement
Task 3	*R1gS2*^{1}	*R1* goes to *UGS2*
Task 4	*R2gA*^{1}	*R2* goes to location A
Task 5	*R1rS2*^{1}	*R1* retrieves *UGS2*
Task 6	*R1lis*^{1}	*R1* listens for interrupts
Task 7	*R1gS1*^{1}	*R1* goes to *UGS1*
Task 8	*R2m*^{1}	*R2* takes measurement
Task 9	*R1dS2*^{1}	*R1* deploys *UGS2*
Task 10	*R1m*^{1}	*R1* takes measurement
Task 11	*S2m*^{1}	*S2* takes measurement
output	y^1	Mission 1 completed

and

$$F_u^1 = \begin{array}{c} \\ x_1^1 \\ x_2^1 \\ x_3^1 \\ x_4^1 \\ x_5^1 \\ x_6^1 \\ x_7^1 \\ x_8^1 \\ x_9^1 \end{array} \begin{array}{c} u1 \\ \begin{bmatrix} 1 \\ 0 \\ 0 \\ 0 \\ 0 \\ 0 \\ 0 \\ 0 \\ 0 \end{bmatrix} \end{array}, \tag{6.2}$$

where the eleven columns are labeled in order corresponding to the tasks. Each row of input TSM corresponds to a rule, e.g., the second row says

IF (task 1 AND task 2 have just been completed) THEN (fire rule 2).

Note that the event that caused the mission to initiate firing of its tasks is considered as an external input and has its own input matrix F_u. This allows the trigger events to be considered as external inputs to the DEC in the next section.

The output TSM matrix S_v for this example is detailed in [59], and tells which tasks to perform when each rule fires. It is close to an identity matrix, since essentially task *i* is fired by rule *i*.

Though TSM is the matrix considered in [244][56], we have decomposed it into two portions, namely $TSM = S_v \cdot F_v$, with F_v the (input) TSM and S_v the output

TSM. The input TSM F_v and output TSM S_v capture all the temporal precedence relations between the tasks in a mission. Moreover, they map to a *rule base* that corresponds to the rows of the former and the columns of the latter. As we shall see in the next section, this mapping to a rule base is a key idea responsible for our DEC formulation, which allows formal computations for efficient on-line real-time task sequencing and dynamic resource assignment in team networks. Moreover, it is shown in [235][236][237] that the outputs of HTN planners can in fact be directly placed into the format of matrices F_v and S_v.

6.2.2 RESOURCE ASSIGNMENT MATRIX (RAM)

As just shown, mission commanders indicate intent by prescribing certain task precedences needed to perform a mission with desired goals. As such, missions consist of partially ordered tasks, whose orderings capture the tactically important aspects of a mission. The details of actual task sequencing are left to the DEC to perform during the Operational Phase in real time, as events unfold and resources become available. That is, during the Operational Phase the DEC converts the partial ordering provided by mission commanders into a total ordering that directs the actual sequencing of the tasks performed by the networked team in real time. This Operational Phase mechanism will be described in the next section.

Meanwhile, to complete the Programming Phase, resources must be assigned to the tasks. The resources capable of performing the tasks are assigned by field commanders or the war-fighter, who are familiar with the onsite situation. Task capabilities of robotic resources are also available in ARL CIP Agent Registry, a database holding information about the services that registered devices offer in the network community [246]. In this section, we detail how the resources capable of performing the tasks are prescribed.

To capture the assignments of the available resources to the mission tasks in a convenient and computable form, define the Resource Assignment Matrix (RAM), which has element (i, j) equal to "1" if resource r_j may be used to accomplish task t_i. Note that multiple 1's in a single row i indicate that multiple resources are required for task t_i. RAM has been used by Kusiak [247] and others in manufacturing and elsewhere.

Multiple 1's in column j of RAM indicate that resource r_j is needed for multiple tasks. Such *shared resources* are important in the team, as they may be competent or versatile resources that are in high demand. However, shared resources can lead to bottlenecks or catastrophic failures of the team if they are not properly assigned in real time to tasks, or *dispatched.* The dispatching of shared resources has been considered under many topics including bottlenecks, deadlocks, and other blocking phenomena [248]. DEC presented in this chapter can accommodate the dispatching of shared resources to avoid conflicts and blocking, as shown in the next section.

RAM can be assigned by field commanders who know which resources can perform which tasks. RAM could also be constructed by software tools that use pricing strategies or payoff matrix ideas, which result in the optimal assignment of resources to tasks given certain prescribed cost functions [249].

As situations change and resources fail or additional resources become available, the resources capable of performing the tasks may change. The DEC framework presented in the next section can accommodate time-varying RAM. Thus, RAM can be modified as resources fail or are added to the team.

To construct a mathematical formulation for a DEC that has guaranteed performance and can work in real time, we must introduce a rule base. Therefore, factor the RAM matrix into two portions according to

$$RAM = S_v \cdot F_r, \tag{6.3}$$

with S_v the output TSM matrix defined above and dot denotes *or/and* matrix multiply. The input RAM F_r maps the resources $\{r_j\}$ to a rule base $\{x_i\}$ as defined in the previous subsection. Thus, F_r has entry (i,j) equal to 1 if resource j is required to fire rule i. Output matrix S_v maps the rule base $\{x_j\}$ to the tasks $\{t_i\}$. It has entry (i,j) of 1 if task i is to be started when rule j fires.

Though others have used RAM in analysis [247], we have further decomposed it to map from resources to tasks through a rule base. As seen in the next section, this is one of the keys that makes our DEC useful for real-time control of networked teams.

As such, the input RAM for the Mission 1 in Table 6.1 is

$$F_r^1 = \begin{array}{c} \\ x_1^1 \\ x_2^1 \\ x_3^1 \\ x_4^1 \\ x_5^1 \\ x_6^1 \\ x_7^1 \\ x_8^1 \\ x_9^1 \end{array} \begin{array}{c} \begin{array}{ccccccc} R1 & R2 & S1 & S2 & S3 & S4 & S5 \end{array} \\ \left[\begin{array}{ccccccc} 0 & 0 & 0 & 0 & 0 & 1 & 1 \\ 1 & 1 & 0 & 0 & 0 & 0 & 0 \\ 0 & 0 & 0 & 0 & 0 & 0 & 0 \\ 0 & 0 & 0 & 0 & 0 & 0 & 0 \\ 1 & 1 & 0 & 0 & 0 & 0 & 0 \\ 0 & 0 & 0 & 0 & 0 & 0 & 0 \\ 0 & 0 & 0 & 0 & 0 & 0 & 0 \\ 0 & 0 & 0 & 1 & 0 & 0 & 0 \\ 0 & 0 & 0 & 0 & 0 & 0 & 0 \end{array}\right] \end{array}, \tag{6.4}$$

where the columns correspond to the available resources and the rows to the rules. Thus, to fire rule 1, one requires sensors *S4* and *S5* to be currently available, *etc.*, *i.e.*, rule 1 fires when the event u^1 occurs (as detected by sensor *S1*, see F_u matrix) and if *S4* and *S5* are available. Then, tasks *S4m* and *S5m* are initiated, according to the output TSM S_v.

Define likewise an output RAM S_r that maps from the rule base to the resources. This matrix has entry (i,j) of 1 if resource i is to be released when rule j fires. Then the matrix that maps from resources to resources is given by

$$G_r = S_r \cdot F_r, \tag{6.5}$$

where G_r is the Resource Dependency Matrix (RDM). G_r defines the so-called resource graph, whose nodes are the resources and whose edges (j,i) correspond to entries of 1 in entry (i,j) of G_r. This graph is indispensable in studying conflict and deadlock avoidance in systems with shared resources [248][250].

6.2.3 PROGRAMMING MULTIPLE MISSIONS

Multiple missions can be programmed into the same networked team of agents and resources. Using DEC, the various missions in a team can be programmed by different mission commanders. Each one does not need to know about other missions running in the network, or about the resources required by the other missions. All missions use the same common pool of networked team resources.

At any time, additional missions can be programmed by other mission commanders, without having to know which missions are already programmed to the resource network. At any time, the resources assigned to the tasks can be changed as resources fail or are added to the network.

As seen in the next section, during the Operational Phase DEC effectively and fairly sequences the tasks of all programmed missions and assigns the required resources on-line in real time as events occur and as resources become available. DEC programs multiple missions into a heterogeneous team of multiple networked resources.

Suppose several missions are prescribed, with Mission i having its task ordering given by input TSM F_v^i and its required resources for the tasks given by the input RAM F_r^i. Then the overall TSM and RAM are given by the block matrix compositions

$$F_v = \begin{bmatrix} F_v^1 & 0 & 0 \\ 0 & F_v^2 & 0 \\ & & \\ 0 & 0 & \ddots \end{bmatrix},$$

$$F_r = \begin{bmatrix} F_r^1 \\ F_r^2 \\ \vdots \end{bmatrix}, \tag{6.6}$$

and similarly for S_v and S_r. Note that the mission task sequences are independent, each using its own tasks, so that F_v is block diagonal. However, all the missions use the same resources available in the networked team, and so have commensurate columns of their RAMs.

DEC facilitates *mission transferability* between teams by capturing mission information in the TSM, which can easily be moved and programmed into another network.

In illustration, consider the same WSN of UGS and mobile robots used in the example above for Mission 1. Suppose the network maintenance technician programs into the same network a Mission 2 that is involved with charging the batteries of the nodes. Such a Mission 2 appears in Table 6.2. Trigger event u^2 is a low battery event.

TABLE 6.2
Mission 2–Task Sequence for Deployed WSN

Mission 2	**Task Label**	**Task Description**
input	*EVENT* u^2	*UGS3* batteries are low
Task 1	$S1m^2$	*UGS1* takes measurement
Task 2	$R1gS3^2$	*R1* goes to *UGS3*
Task 3	$R1cS3^2$	*R1* charges *UGS3*
Task 4	$S3m^2$	*UGS3* takes measurement
Task 5	$R1dC^2$	*R1* docks the charger
output	y^2	Mission 2 completed

The input TSM and input RAM for this mission are

$$
F_v^2 = \begin{array}{c} \\ x_1^2 \\ x_2^2 \\ x_3^2 \\ x_4^2 \\ x_5^2 \\ x_6^2 \end{array}
\begin{array}{c} \begin{array}{ccccc} t_1 & t_2 & t_3 & t_4 & t_5 \end{array} \\
\left[\begin{array}{ccccc}
0 & 0 & 0 & 0 & 0 \\
1 & 0 & 0 & 0 & 0 \\
0 & 1 & 0 & 0 & 0 \\
0 & 0 & 1 & 0 & 0 \\
0 & 0 & 0 & 1 & 0 \\
0 & 0 & 0 & 0 & 1
\end{array}\right] \end{array},
$$

$$
F_r^2 = \begin{array}{c} \\ x_1^2 \\ x_2^2 \\ x_3^2 \\ x_4^2 \\ x_5^2 \\ x_6^2 \end{array}
\begin{array}{c} \begin{array}{ccccccc} R1 & R2 & S1 & S2 & S3 & S4 & S5 \end{array} \\
\left[\begin{array}{ccccccc}
0 & 0 & 1 & 0 & 0 & 0 & 0 \\
1 & 0 & 0 & 0 & 0 & 0 & 0 \\
0 & 0 & 0 & 0 & 0 & 0 & 0 \\
0 & 0 & 0 & 0 & 1 & 0 & 0 \\
1 & 0 & 0 & 0 & 0 & 0 & 0 \\
0 & 0 & 0 & 0 & 0 & 0 & 0
\end{array}\right] \end{array}. \tag{6.7}
$$

The overall TSM and RAM for both the missions now in the network are

$$
F_v = \begin{bmatrix} F_v^1 & 0 \\ 0 & F_v^2 \end{bmatrix},
$$

$$
F_r = \begin{bmatrix} F_r^1 \\ F_r^2 \end{bmatrix}. \tag{6.8}
$$

The missions in a team can be programmed by different mission commanders. Each one does not need to know about other missions running in the network, or about the resources required by the other missions. All missions use the same common pool of networked resources. At any time, additional missions can be programmed by other mission commanders, without having to know which missions are already programmed to the resource network. At any time, the resources assigned to the tasks can be changed as resources fail or are added to the network.

In a similar fashion, additional missions are easily programmed into this WSN. In the next section, we show how to develop a DEC that effectively and fairly sequences the tasks of all programmed missions and assigns the required resources on-line in real time as events occur and as resources become available.

6.3 CONJUNCTIVE RULE–BASED DISCRETE EVENT CONTROLLER (DEC)

This section describes the DEC software controller that runs in the Operational Phase to sequence the tasks and assign the team resources in real time. This is a decision-making DEC that contains rules to sequence the tasks in each mission, and to assign resources to those tasks. See Figure 6.1. DEC has a message-passing architecture that is in conformance with Joint Architecture for Unmanned Ground systems (JAUGS) [239], and is an efficient means to realize the Observe, Orient, Decide, Act (OODA) loops of 4D/RCS [240]. DEC is able to coordinate the sequencing of operations between soldiers and robots efficiently and without conflict, thus contributing to the concept of Safeops.

In the previous section we saw that missions are programmed into the network by specifying TSMs F_v, S_v and RAMs F_r, S_r. The TSMs give a partial order for the tasks in each mission, however, the tasks are coupled through the shared resources as captured in RAM. Based on those constructions, it is now desired to construct a rule-based discrete event controller that can be programmed in software and which effectively reacts to external events sensed, sequences the tasks of a networked team, and assigns their available resources in such a way that the missions are accomplished without interference or blocking phenomena. Tasks of priority missions should have priority assignment of requisite resources. These can be elegantly resolved with the introduction of the DEC state equation using conjunctive reasoning.

6.3.1 DEC STATE EQUATION

In terms of the constructions just given we are now in a position to define such a DEC. Define the task vector v, resource vector r, and rule state vector x as

$$\begin{aligned} v_c &= \begin{bmatrix} t_1 & t_2 & \cdots & t_{N_T} \end{bmatrix}^T, \\ r_c &= \begin{bmatrix} r_1 & r_2 & \cdots & r_{N_R} \end{bmatrix}^T, \\ x &= \begin{bmatrix} x_1 & x_2 & \cdots & x_{N_X}, \end{bmatrix}^T \end{aligned} \tag{6.9}$$

respectively, where the set of tasks is $\{t_i\ ;\ i=1,N_T\}$, the set of resources is $\{r_i\ ;\ i=1,N_R\}$, and the set of rules is $\{x_i;\ i=1,N_X\}$. The rule state vector x has 1's in positions i corresponding to the rules that are currently enabled to fire. Define v_c as the *task completion vector* which contains 1's in positions i corresponding to the tasks t_i that have just been accomplished, and r_c as the *resource available vector* containing 1's in positions i corresponding to the resources r_i that are currently available. These are the outputs of the networked team passed through messages to the DEC. See Figure 6.1.

For instance, $v_c = \begin{bmatrix} 1 & 0 & 0 & 1 & 0 & \cdots \end{bmatrix}^T$ signifies that tasks 1 and 4 have just been performed, while $r_c = \begin{bmatrix} 0 & 0 & 1 & 1 & 0 & \cdots \end{bmatrix}^T$ signifies that resources 4 and 5 are currently available. Define the *external event vector*

$$u = \begin{bmatrix} u_1 & u_2 & \cdots & u_{nu} \end{bmatrix}^T \tag{6.10}$$

to contain 1's in positions corresponding to events u_i that have just occurred.

In terms of the TSM and RAM matrices defined above, define the DEC rule base state equation as

$$\bar{x} = F_v \cdot \bar{v}_c \oplus F_r \cdot \bar{r}_c \oplus F_u \cdot \bar{u} \oplus F_D \cdot \bar{u}_D, \tag{6.11}$$

where F_u is an input matrix that specifies which external events are to be used to launch each mission. The input u_D is a *conflict resolution control input* that decides which task to perform in the event that multiple tasks are enabled at a given time. It allows real-time interrupts and priority dispatching for urgent missions, and assigns shared resources in such a way as to avoid blocking phenomena including deadlocks and bottlenecks. See [249].

In this equation, all matrices and vectors are binary, i.e., having entries of either 0 or 1. Dot denotes matrix multiply, and $\oplus$ denotes matrix addition, with all operations carried out in the or/and algebra, where multiplication is replaced by *.and.*, and addition by *.or.*. The overbar denotes negation of all entries of a vector. Operations in the or/and algebra are easily programmed, and a routine that carries out a matrix multiply in the or/and algebra is given in Table 6.3.

TABLE 6.3
Matrix Multiply in the OR/AND Algebra

```
Matrix Multiply
C = A . B

for i= 1,I
  for j= 1,J
    c(i,j)=0
    for k= 1,K
      c(i,j)= c(i,j) .OR. ( a(i,k) .AND. b(k,j) )
    end
  end
end
```

The DEC equation contains the required mission task partial orderings and the resources required for each task, and essentially captures the world model in 4D/RCS [240].

6.3.2 DEC OUTPUT EQUATIONS

Based on the rule state vector, the task start equation

$$v_s = S_v \cdot x \tag{6.12}$$

computes the *task start vector* v_s, which has 1's in positions i corresponding to those tasks that can now be started. The resource release equation

$$r_s = S_r \cdot x \tag{6.13}$$

computes the *resource release vector* r_s, which has 1's in positions i corresponding to those resources that can now be released as their tasks have been completed. These output equations are also computed in the or/and algebra.

6.3.3 DEC AS A FEEDBACK CONTROLLER

DEC functions as a *feedback controller*, as shown in Figure 6.1. It runs as a software tool on a laptop or other computer. It is very easy, for instance, to program DEC in software. Basic code is given in [58]. The operation of DEC is as follows. At each event iteration, all active agents in the team send updates of their tasks just completed and resources currently available over the internet or via WSN. The current task completion vector v_c and the current resource available vector r_c are constructed by taking this information from all active agents in the team. They are considered as the *outputs* of the networked team to the DEC.

The DEC then uses state equation (6.11) to compute the rule state vector x. The resulting entries of 1 in the rule state vector show which rules are enabled to fire, as having all their requisite precursor tasks done and all their required resources available. Thus, the tasks in (6.12) corresponding to these active rules could now be performed. Now among all the tasks that could fire, DEC selects the tasks to actually fire by consulting the mission priorities, or by querying local field commanders via PDA. Finally, command inputs are sent by the DEC telling agents which tasks to start (v_s) and which resources to release and make available (r_s). For autonomous machines, this information is sent as commands to their internal controllers. For human nodes, the information can be sent as decision assistance via a handheld PDA—e.g., "go to point A," "contact node B and provide certain information," etc.

Note that agent nodes need only communicate to DEC when they have a *change* in tasks' completion statuses, or a *change* in resource availability; i.e., when an event occurs. On the other hand, DEC only communicates to those nodes which should next fire tasks or release resources.

6.3.4 FUNCTIONALITY OF THE DEC

The intent of the DEC is that it should provide a mathematically rigorous software tool for implementing a rule-based supervisory controller that sequences the tasks and assigns the resources of a networked team all in real time as events unfold, given at each event iteration the measured network information about which tasks have just completed and which resources are available. This is all dependent on the TSM F_v and the RAM F_r, which have been programmed into the DEC software respectively by the mission commanders and the field commanders.

The first result shows that DEC state equation (6.11) actually does compute which tasks to start based on rules of the form

IF (all tasks required as immediate precursors to rule i have just been completed) AND (all resources required by rule i are available) THEN fire rule i.

Define the tasks as $\{t_j\ ;\ j=1,N_t\}$, i.e., the elements of task vector v, and the resources as $\{r_j\ ;\ j=1,N_r\}$, the elements of resource vector r. Define T_i as the set of tasks that are immediate precursors to rule i, and R_i as the set of resources required to fire rule i. The next result verifies the proper functioning of the DEC equation (6.11), while also showing the need for the negation overbars on the vectors in (6.11).

Theorem 6.1

The ith rule, i.e., i^{th} row, of (6.11) is equivalent to

$$x_i = \bigcap_{j\in T_i} t_j \cap \bigcap_{j\in R_i} r_j, \tag{6.14}$$

■

where $\cap$ denotes logical *and*. That is, rule state x_i is true (equal to 1) if all task vector elements t_j required for rule i are true and all resource vector elements r_j required for rule i are true.

Proof 6.1

Let $\cap$ denote *and* and $\cup$ denote *or*. Overbar denotes negation. Define the elements of matrix F_v by f_{ij}^v and of F_r by f_{ij}^r. By the definition of matrix operations in the or/and algebra, one has

$$\bar{x}_i = \left(\bigcup_{j=1}^{N_T} f_{ij}^v \cap \bar{t}_j\right) \cup \left(\bigcup_{j=1}^{N_R} f_{ij}^r \cap \bar{r}_j\right). \tag{6.15}$$

Now successive applications of De Morgan's Laws yields

$$\begin{aligned} x_i &= \overline{\left(\bigcup_{j=1}^{N_T} f_{ij}^{v} \cap \bar{t}_j\right) \cup \left(\bigcup_{j=1}^{N_R} f_{ij}^{r} \cap \bar{r}_j\right)} \\ &= \overline{\left(\bigcup_{j=1}^{N_T} f_{ij}^{v} \cap \bar{t}_j\right)} \cap \overline{\left(\bigcup_{j=1}^{N_R} f_{ij}^{r} \cap \bar{r}_j\right)} \\ &= \left(\bigcap_{j=1}^{N_T} \overline{f_{ij}^{v} \cap \bar{t}_j}\right) \cap \left(\bigcap_{j=1}^{N_R} \overline{f_{ij}^{r} \cap \bar{r}_j}\right) \\ &= \left(\bigcap_{j=1}^{N_T} \bar{f}_{ij}^{v} \cup t_j\right) \cap \left(\bigcap_{j=1}^{N_R} \bar{f}_{ij}^{r} \cup r_j\right) \\ &= \left(\bigcap_{j=1}^{N_T} 1 \cup t_j\right) \cap \left(\bigcap_{j=1}^{N_R} 1 \cup r_j\right). \end{aligned} \tag{6.16}$$

But elements f_{ij}^{v} and f_{ij}^{r} are equal to zero if task t_j, respective resource r_j, is *not* needed to fire rule i. Then, $\bar{f}_{ij}^{v}$ and $\bar{f}_{ij}^{r}$ are equal to 1, so that for those elements one has $\bar{f}_{ij}^{v} \cup t_j = 1$ and $\bar{f}_{ij}^{r} \cup r_j = 1$ whether the corresponding task or resource element is true or not. Conversely, elements f_{ij}^{v} and f_{ij}^{r} are equal to 1 if task v_j, respective resource r_j, is needed to fire rule i. Then, $\bar{f}_{ij}^{v}$ and $\bar{f}_{ij}^{r}$ are equal to 0, so that for those elements one has $\bar{f}_{ij}^{v} \cup t_j = 1$ and $\bar{f}_{ij}^{r} \cup r_j = 1$ only if the corresponding task or resource element is true. Therefore the last equation is equivalent to

$$\begin{aligned} \therefore x_i &= \left(\bigcap_{j=1}^{N_T} t_j\right) \cap \left(\bigcap_{j=1}^{N_R} r_j\right) \\ &= \bigcap_{j \in T_i} t_j \cap \bigcap_{j \in R_i} r_j. \end{aligned} \tag{6.17}$$

■

6.3.5 PROPERNESS AND FAIRNESS OF THE DEC RULE BASE

Consistency of a rule base is compromised by circularity and conflicts. In this section, the properties of the proposed DE-C2 networked computing environment which has the potential to communicate among all team participants including mission planners, field commanders, war-fighters, and robotic platforms is provided. Our claims are mathematically rigorous, justified with lemmas and theorems.

The next result shows that the rule base constructed by programming the mission tasks into the TSM and the task resources into RAM produces a rule base of good structure if and only if *each mission* is properly defined. It is shown that improper definition of a single mission assigned to the team can cause improper execution of all missions assigned to the team. Then, the DEC cannot fairly assign the team resources and ensure sequential execution of the tasks in the missions. In extreme cases, improper definition of one mission can tie up resources so that other missions are blocked.

A rule has the form

IF antecedent THEN consequent,

where the antecedent consists of task clauses and resource clauses such as

(all tasks required as precursors to rule i are done)
AND (all resources required by rule i are available),

and the consequent consists of a single clause such as

fire rule i.

Problems with rule bases fall into three categories of the 3Cs, namely Consistency, Completeness, and Conciseness. We follow [251].

Lemma 6.1

DEC rule base has no circularity or conflicts. ■

Proof 6.2

The rule base is based on clauses such as task completion, and resource availability, and does not contain any of their negations. Therefore it cannot have conflicts. Matrix F_v is block diagonal. Each block defines a strict partial ordering and so is lower block triangular with zero diagonal. Therefore there are no circular chains of rules. ■

Conciseness is compromised by the presence of rules that logically serve no purpose. These include redundant rules, subsumed rules, and unnecessary IF conditions.

Lemma 6.2

The overall rule base defined by all the missions is concise if and only if task sequencing matrix of each mission corresponds to a concise rule base for the tasks. ■

Proof 6.3

Matrix F_v is block diagonal, each block of which defines a rule base. Therefore, two rules in different blocks cannot be redundant, as they cannot have the same antecedent.

No two rules in a single block are redundant by hypothesis. Therefore there are no redundant rules in DEC. The antecedent of a rule in one block cannot contain the antecedent of a rule in another block. By hypothesis, the antecedent of a rule in one block does not contain the antecedent of a rule in the same block. Therefore, there are no subsumed rules. There are no unnecessary IF conditions since, by construction, no clause is the negation of any other. ■

Completeness refers to knowledge gaps in the rule base. Such gaps make it impossible to achieve the prescribed goals given the triggering events that have occurred. Knowledge gaps include unreachable conclusions, dead end goals, dead end IF conditions, and missing rules. By using the block diagonal structure of F_v it is straightforward to prove the following.

Lemma 6.3

The overall rule base defined by all the missions is complete if and only if task sequencing matrix of each mission corresponds to a complete rule base for the tasks. ■

Proof 6.4

These results make it clear that the mission planner has a great responsibility in properly defining his task prior to programming it into the team. Improper definition of one mission among many assigned to a team can cause blocking, thereby tying up resources and making it impossible to complete other missions in the team. ■

It has just been shown that if each mission individually is properly defined, the DEC in (6.11) guarantees proper sequencing of tasks in each mission and proper assignment of the required resources. However, the resources of the team are shared by all the missions. Therefore, there may exist bottlenecks or blocking phenomena if the resources are not properly assigned in real time. This is called the shared resource *dispatching* problem. Particularly detrimental is the occurrence of deadlocks, where some tasks cannot gain access to their required resources because those resources are indefinitely held up by other tasks. This indefinitely halts the involved missions and will not allow their completion.

It is shown in [248] that DEC allows easy analysis of potential deadlocks. Then the dispatching or conflict resolution input u_D in (6.11) (see Figure 6.1) may be selected to avoid deadlock situations. In this fashion, the DEC can guarantee proper performance and completion of all missions, as long as each mission is properly defined in the sense shown in Lemmas 6.1–6.3 above.

6.4 DISJUNCTIVE RULE–BASED DISCRETE EVENT CONTROLLER (DEC)

By far, the DEC controller stipulated in (6.11)–(6.13) is a rule-based controller that has conjunctive rules, i.e., the *and* operator in the antecedent. This is the standard form assumed in many applications including Petri Nets and much analysis of rulebases [251]. However, in operations research applications such as the resource assignment problem, one uses disjunctive reasoning, where the antecedents contain the *or* operator. In the resource assignment problem one has available q resources or equipments, any one of which can be assigned to one of p tasks. Then one has the ith rule firing if any one of the listed resources is available.

Consider the set of resources r_j. To capture the possible assignments of the available resources to the mission tasks in a convenient and computable form, define the disjunctive-input Resource Assignment Matrix (dRAM) F_{rd} which has entry (i, j) equal to 1 if resource j can accomplish rule i. F_{rd} maps the resources r_j to the rule base x_i. Input dRAM captures information about which resources can possibly be used for each task. Here, it is assumed that only one of the possible resources listed in row i of F_{rd} need be available to accomplish task i.

Define the state vector for disjunctive resource assignment x_{rd} by

$$x_{rd} = F_{rd} r_c, \tag{6.18}$$

where multiplication is carried out in and/or algebra. Note that there is no negation overbar. The next result is obvious.

Lemma 6.4

Define R_{di} as the set of resources, any one of which can accomplish rule i. Then the ith rule, i.e. the ith row of x_{rd} is equivalent to

$$(x_r)_i = \bigcup_{r_j \in R_{di}} r_j, \tag{6.19}$$

where $\cup$ denotes logical *or*. Rule state $(x_r)_i$ is true (equal to 1) if any one of the resources $r_j \in R_{di}$ are true. This is known as *disjunctive resource assignment.* ■

Define now the overall rule base vector by

$$\bar{x} = \bar{x}_v + \bar{x}_r + \bar{x}_{rd}, \tag{6.20}$$

that is

$$\bar{x} = F_v \bar{v}_c + F_r \bar{r}_c + \overline{F_{rd} r_c}, \tag{6.21}$$

where $+$ denotes logical *or*, and overbar denotes negation. Then one has the following obvious result.

Theorem 6.2

Define T_i as the set of tasks that are required as immediate precursors to rule i, R_i as the conjunctive set of resources that are all required to fire rule i, and R_{di} as the disjunctive set of additional resources, any one of which can accomplish rule i in addition to all the required resources R_i. The ith row of x is=

$$x_i = \bigcap_{t_j \in T_i} t_j \cap \bigcap_{r_j \in R_i} r_j \cap \left(\bigcup_{r_j \in R_{di}} r_j \right). \tag{6.22}$$

This is known as the *conjunctive/disjunctive rulebase for task sequencing and resource assignment.* ■

Now, rule state x_i is true (equal to 1) if all task vector elements t_j required for rule i are true and all resource vector elements $r_j \in R_i$ required for rule i are available, while any one of the resources in R_{di} is available. It is to be noted that the DEC described in (6.22) is more general than a Petri Net (PN). In fact, the first two terms are *equivalent* to a PN, while the third term allows additional flexibility using *or* statements in the rulebase.

It is interesting to note that setting $F_{rd} = 0$ one obtains the standard rulebase DEC used for manufacturing sequencing and resource assignment, i.e., a PN. In this case, one is concerned about blocking phenomena and deadlocks due to multiple 1's in the columns of F_r. On the other hand, setting $F_v = 0$, $F_r = 0$, and F_{rd} equal to the matrix of 1's, one achieves the usual resource assignment problem commonly confronted in operations research. In this case, one optimally assigns a single resource to each task using techniques such as linear programming or bid/auction methodology.

The formulation here provides the framework to confront both situations at once. It is worth noting that it is possible to extend resource assignment methods to the case where F_{rd} has some zero elements. This corresponds to the case where certain resources can accomplish only a subset of the tasks, i.e., *heterogeneous resource assignment*, where the resources can have different capabilities.

6.5 DEC SIMULATION AND IMPLEMENTATION

For simplicity but without loss of generality, the conjunctive DEC detailed earlier in Section 6.3 is chosen here for verification with simulation and implementation studies.

DEC is easy to implement using computer simulation software. The basic code necessary to implement the DEC shown in Figure 6.1 is based on the or/and multiply routine in Table 6.3, and is given in [58]. Messages are passed (using Wifi, internet, WSN, or PDA, etc.) from team nodes to the C2 computer whenever an event occurs, i.e., any node finishes a task or has a new resource made available. New task events are placed into task completion vector v_c, while new resource events are placed into resource available vector r_c. Then the DEC state equation (6.11) is evaluated using the software, and the tasks to be started are computed using (6.12), while resources to be released are computed using (6.13). If several tasks are enabled, user-specified priority decisions or deadlock considerations are used to select which task to actually fire. Messages are passed back to the team nodes detailing which tasks to perform next vs and which resources to release r_s, as commands into machine nodes, or as decision aids via PDA to human agents.

6.5.1 SIMULATION OF NETWORKED TEAM EXAMPLE

The DEC was run on the networked team example whose TSM and RAM were shown above. This is a simulation on a digital computer. The resulting event traces are seen in Figure 6.3. Chemical attack event $u1$ occurred at time 8 min, and the low battery event at time 3 min. The progress of the two missions through the team as the resources are assigned and the tasks are performed is clearly seen. In the task traces, an "up" means a task is being performed, while in the resource traces, a "down" means the resource is being used.

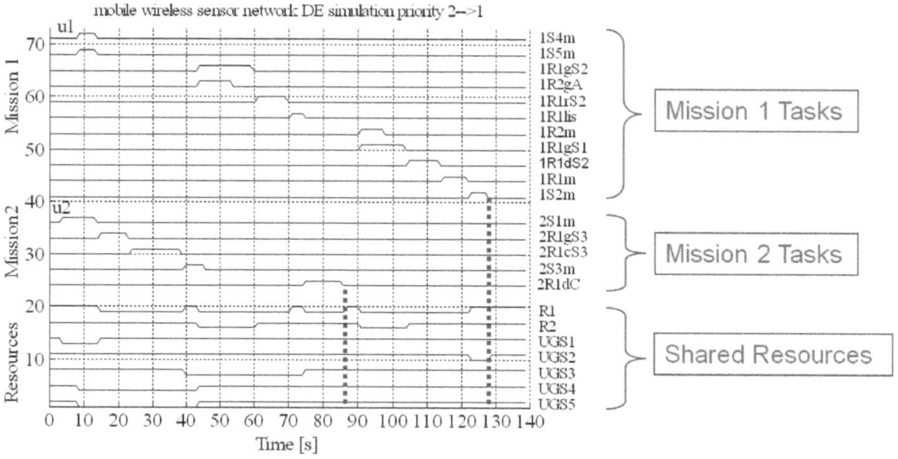

FIGURE 6.3 Simulation results of DEC sequencing mission tasks in the networked team example.

In the figure, Mission 1 terminates at 128 min, while Mission 2 terminates at 87 min. In Figure 6.4, the priority of the missions in changed, so that Mission 1 is given a higher priority. Thus, when there is a request for the same resource by two tasks, one from each mission, DEC will now assign the Mission 1 task first. This is accomplished in DEC by proper choice of the conflict resolution input u_D in (6.11) (See also Figure 6.1). Details are given in [59]. Now, Mission 1 takes less time and terminates at 62 min.

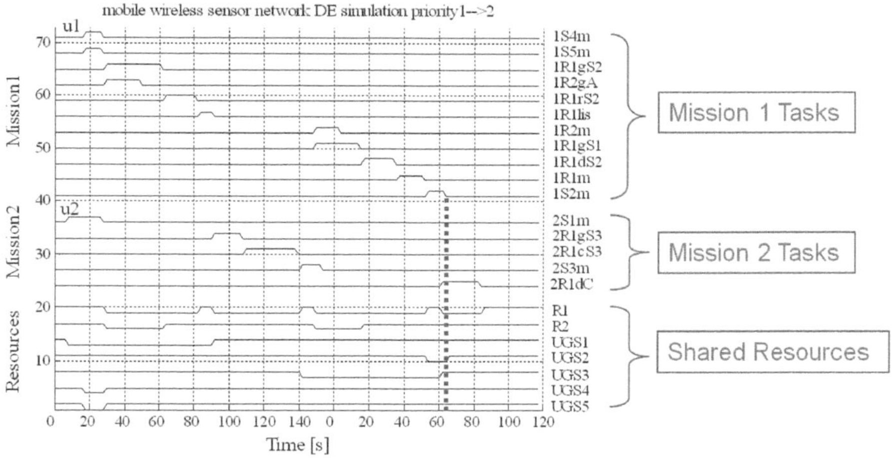

FIGURE 6.4 Simulation results of DEC sequencing mission tasks with increased Mission 1 priority.

6.5.2 IMPLEMENTATION OF NETWORKED TEAM EXAMPLE ON ACTUAL WSN

It is very easy to implement DEC on an actual networked team. In fact, the same code is used for both simulation and implementation. The DEC was implemented on a WSN team of mobile robots and UGS at the University of Texas at Arlington's Automation & Robotics Institute. Details of the hardware are given in [59]. A 3D user interface depicted the motions of the robots during the mission execution. The panoramic view during the mission execution is shown in Figure 6.5.

The actual event traces observed during the experimental implementation are shown in Figure 6.6. They bear a close resemblance to the simulated event traces above.

6.5.3 SIMULATION OF MULTIPLE MILITARY MISSIONS USING FCS

In this example, the army Future Combat Systems (FCS) [252][253] is used to design the simulated networked military team. FCS was the United States Army's principal modernization program from 2003 to early 2009, and is envisioned to create new brigades equipped with new manned and unmanned vehicles. The vehicular agents are also linked by an unprecedented fast and flexible battlefield network, and are aided by various pieces of other gear. Typical resources in FCS include:

1. *Unattended Ground Sensor (UGS)*: Perimeter defense, surveillance, target acquisition, chemical, bio, radiological nuclear early warning;
2. *XM51 Non-Line of Sight Launch System (XM51 NLOS-LS)*: Self contained tactical firing with 15 loitering attack or precision attack missiles;

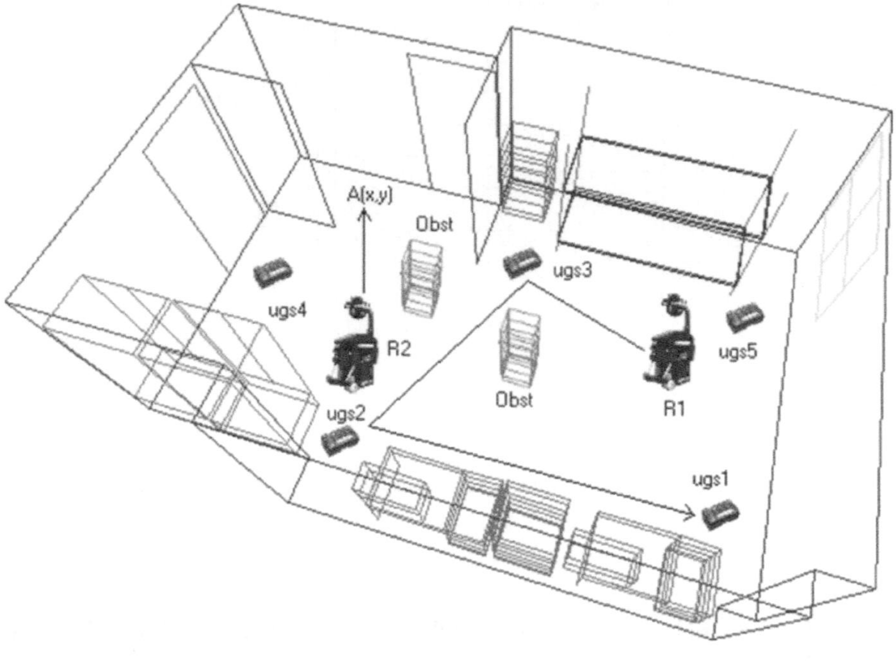

FIGURE 6.5 DEC virtual reality interface panoramic view of the configuration of the mobile WSN during real-world experiments.

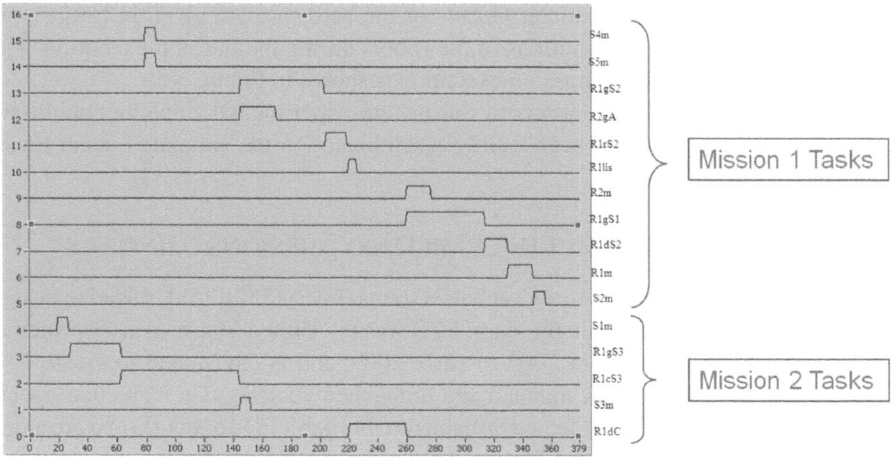

FIGURE 6.6 Experimental results showing the task event trace of the WSN.

3. *XM156 Class I Unmanned Aerial Vehicle (XM156 UAV)*: Controlled by dismounted soldiers, provides recon, surveillance, target acquisition (weighs less than 15lbs);
4. *Multifunctional Utility/Logistics and Equipment vehicle (MULE)*: Multifunctional Utility/Logistics Equipment vehicle, autonomous. Can be either: assault, transport, mine detector vehicle; and
5. Joint Tactical Radio System (JTRS) or Transformational Satellite System (TSAT): Secured and advanced communication systems.

To illustrate, a sample mission is shown in Table 6.4. We consider an army FCS

TABLE 6.4
Suppressing Enemy Troop 1

Mission I	Task Label	Resource	Task Description
input	u^1		Tr1 arrives at B
Task 1	$UBrB1XM^1$	*UB*	*UB* reports to B1 and XM51 about Tr1 arrival at B
Task 2	$B1block^1$	*B1*	20% of *B1* goes to A (rear blocking) and another 20% of *B1* goes to C (front blocking)
Task 3	$XMfire\_1^1$	*XM51*	*XM51* fires a number of missiles to destroy Tr2
Task 4	$UBrXM^1$	*UB*	*UB* takes measurement and reports to *XM51* about percentage of desired damage achieved after the first fire
Task 5	$XMfire\_2^1$	*XM51*	*XM51* fires a number of missiles to further destroy Tr1
Task 6	$B1attack^1$	*B1*	*B1* at A (20%), C (20%), and station (60%) face-to-face attacks Tr1 from three directions, occupy whole target area
Output	y^1		Mission 1 completed

team which consists of two UGSs (UB and UF), two brigades (B1 and B2) including armed soldiers and tanks, one XM51 NLOS-LS (XM51), and one T-SAT (TSAT). The proposed FCS team is to perform two realistic offensive missions against two enemy troops using Ambush tactic, which is a long-established military tactic where the aggressors (the ambushing force) use concealment to attack a passing enemy. The simulated battlefield is in the middle of tropical forest, where two enemy troops (Tr1 and Tr2) are moving toward meeting point D from two different roads (A→D and G→D). It is known that Tr1 (represented by big square) is equipped with heavier defensive systems than Tr2 (represented by small square). The proposed FCS team is deployed inside the forest and equipped with anti-radar tools so that Tr1 and Tr2 are not capable of detecting them. For simplicity but without loss of generality, we assume that both up-link and down-link channels of T-SAT operate at very high bandwidth and are always available. This is a realistic assumption. The XM51 NLOS-LS is placed far away from the battlefield and communicates with the team through T-SAT. It is equipped with very high precision lethal attack missiles which are able to destroy enemy targets from a distance. The proposed battlefield is shown in Figure 6.7.

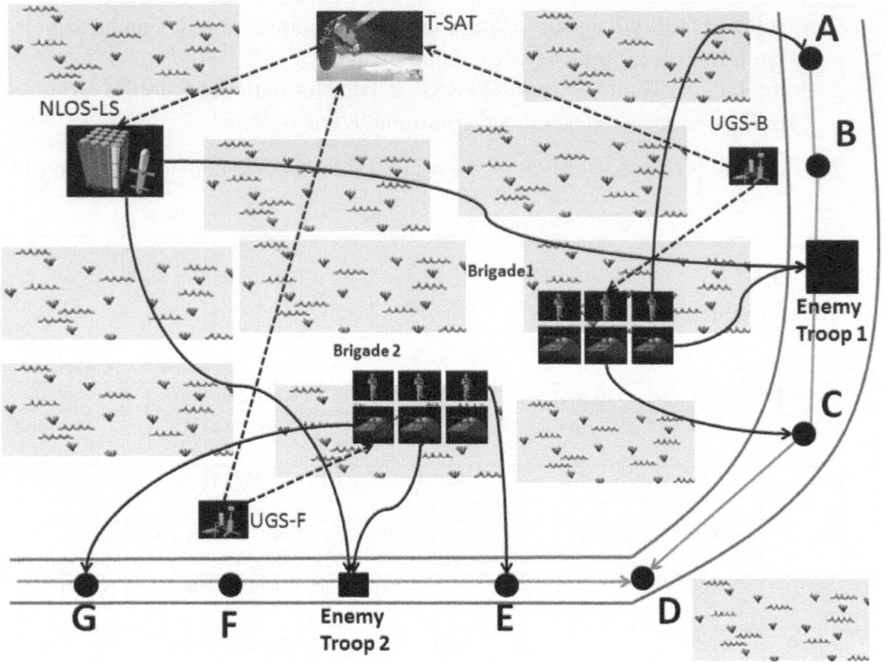

FIGURE 6.7 Simulated battlefield with networked military team using ambush attack tactics.

The input TSM F_v corresponding to this Mission I is

$$F_v^1 = \begin{array}{c} \\ x_1^1 \\ x_2^1 \\ x_3^1 \\ x_4^1 \\ x_5^1 \\ x_6^1 \\ x_7^1 \end{array} \begin{array}{c} \begin{array}{cccccc} t_1 & t_2 & t_3 & t_4 & t_5 & t_6 \end{array} \\ \begin{bmatrix} 0 & 0 & 0 & 0 & 0 & 0 \\ 1 & 0 & 0 & 0 & 0 & 0 \\ 0 & 1 & 0 & 0 & 0 & 0 \\ 0 & 0 & 1 & 0 & 0 & 0 \\ 0 & 0 & 0 & 1 & 0 & 0 \\ 0 & 0 & 0 & 0 & 1 & 0 \\ 0 & 0 & 0 & 0 & 0 & 1 \end{bmatrix} \end{array}, \qquad (6.23)$$

$$F_u^1 = \begin{array}{c} \\ x_1^1 \\ x_2^1 \\ x_3^1 \\ x_4^1 \\ x_5^1 \\ x_6^1 \\ x_7^1 \end{array} \begin{array}{c} u1 \\ \left[\begin{array}{c} 1 \\ 0 \\ 0 \\ 0 \\ 0 \\ 0 \\ 0 \end{array}\right] \end{array}, \tag{6.24}$$

where the six columns are labeled in order corresponding to the tasks.

As such, the input RAM for the Mission I in Table 6.4 is

$$F_r^1 = \begin{array}{c} \\ x_1^1 \\ x_2^1 \\ x_3^1 \\ x_4^1 \\ x_5^1 \\ x_6^1 \\ x_7^1 \end{array} \begin{array}{c} \begin{array}{ccccc} UB & UF & B1 & B2 & XM \end{array} \\ \left[\begin{array}{ccccc} 1 & 0 & 0 & 0 & 0 \\ 0 & 0 & 1 & 0 & 0 \\ 0 & 0 & 0 & 0 & 1 \\ 1 & 0 & 0 & 0 & 0 \\ 0 & 0 & 0 & 0 & 1 \\ 0 & 0 & 1 & 0 & 0 \\ 0 & 0 & 0 & 0 & 0 \end{array}\right] \end{array}, \tag{6.25}$$

where the columns correspond to the available resources and the rows to the rules. Thus, to fire rule 1, one requires UGS *UB* to be currently available, i.e., rule 1 fires when the event u^1 occurs (see F_u matrix) and if *UB* is available. Then, Task $UBrB1XM^1$ is initiated according to the output TSM S_v.

In illustration, consider a realistic military Mission 2 in Table 6.5.

TABLE 6.5
Suppressing Enemy Troop 2

Mission II	Task Label	Resource	Task Description
input	u^2		Tr2 arrives at F
Task 1	$UFrB2XM^2$	*UF*	*UF* reports to B2 and XM51 about Tr2 arrival at F
Task 2	$B2block^2$	*B2*	20% of *B2* goes to G (rear blocking) and another 20% of B2 goes to E (front blocking)
Task 3	$XMfire\_1^2$	*XM51*	*XM51* fires a number of missiles to destroy Tr1
Task 4	$B2attack^2$	*B2*	*B2* G (20%), E (20%), and station (60%) face-to-face attacks Tr1 from three directions, occupy whole target area
Output	y^2		Mission 2 completed

The input TSM and input RAM for this mission are

$$F_v^2 = \begin{array}{c} x_1^2 \\ x_2^2 \\ x_3^2 \\ x_4^2 \\ x_5^2 \end{array} \overset{\begin{array}{cccc} t_1 & t_2 & t_3 & t_4 \end{array}}{\begin{bmatrix} 0 & 0 & 0 & 0 \\ 1 & 0 & 0 & 0 \\ 0 & 1 & 0 & 0 \\ 0 & 0 & 1 & 0 \\ 0 & 0 & 0 & 1 \end{bmatrix}}, \tag{6.26}$$

$$F_r^2 = \begin{array}{c} x_1^1 \\ x_2^1 \\ x_3^1 \\ x_4^1 \\ x_5^1 \end{array} \overset{\begin{array}{ccccc} UB & UF & B1 & B2 & XM \end{array}}{\begin{bmatrix} 0 & 1 & 0 & 0 & 0 \\ 0 & 0 & 0 & 1 & 0 \\ 0 & 0 & 0 & 0 & 1 \\ 0 & 0 & 0 & 1 & 0 \\ 0 & 0 & 0 & 0 & 0 \end{bmatrix}}. \tag{6.27}$$

The overall TSM and RAM for both the missions in the network are now given by

$$\begin{aligned} F_v &= \begin{bmatrix} F_v^1 & 0 \\ 0 & F_v^2 \end{bmatrix}, \\ F_r &= \begin{bmatrix} F_r^1 \\ F_r^2 \end{bmatrix}. \end{aligned} \tag{6.28}$$

The missions in a team can be programmed by different mission commanders. Each one does not need to know about other missions running in the network, or about the resources required by the other missions. All missions use the same common pool of networked resources. At any time, additional missions can be programmed by other mission commanders, without having to know which missions are already programmed to the resource network. Similarly, the resources assigned to the tasks can be changed as resources fail or are added to the network. In similar fashion, additional missions are easily programmed into this army FCS.

Two realistic missions, which are predefined by the total commander, are modelled by the proposed matrix-based DEC. Missions I and II are performed in the same tactic. The difference is that XM51 has to fire twice in order to reduce or destroy Tr1 because it is known to be armed with heavy defensive weapons *a priori*. The DEC was run on the networked team example whose TSM and RAM were shown above. The resulting event traces are seen in Figure 6.8. The higher priority is given to Mission I because enemy troop 1 is armed with heavy defensive weapons and therefore needed to be destroyed first. When there is a request for the same resource by two tasks (one from each mission), the DEC will automatically assign resources to complete the tasks in the Mission I first.

6.6 CONCLUSION

In this chapter, *binary* or *Boolean matrices* are used to construct a Discreet Event Command and Control (DEC2) structure for event-triggered systems of distributed

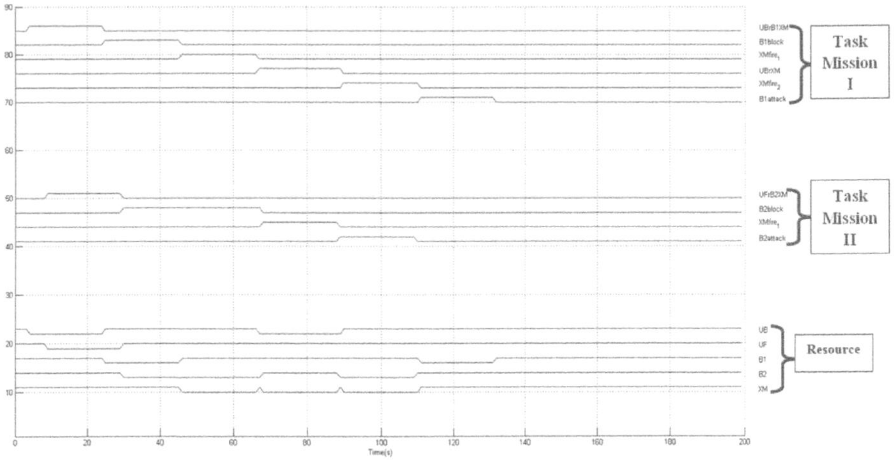

FIGURE 6.8 DEC sequencing mission tasks in the Networked Team Example. Simulation.

teams with multiple missions. Discrete Event Control operates on a rule base which requires preprogramming of the Task Sequencing Matrix (TSM) and Resource Assignment Matrix (RAM), and the tasks to be fired next can be easily computed using the DEC state/output equations which operates on boolean *and* and *or* matrix operations. It is shown that the DEC2 is a complete proper and fair framework, and is consistent, concise, and with no circularity nor conflicts. The conjunctive and disjunctive DEC are rigorously proven to have no blocking, bottleneck, and deadlock issues. Simulation and experimental results on a Wireless Sensor Network (WSN) as well as simulation results using a military ambush attack in a Future Combat System (FCS) show the effectiveness of the proposed DEC for intelligent diagnosis and prognosis of discrete event systems.

7 Future Challenges

Currently, manufacturing industries have the biggest share in energy consumption and emission of Green House Gases (GHGs). In order to improve energy efficiency on manufacturing shop floors, companies need to be equipped with advanced intelligent diagnosis and prognosis technologies with energy considerations, to track energy performance *in situ*, while providing hierarchical decision support based on accurate and real-time up-to-date information. As such, this is fast becoming the top research priority due to rising energy prices, emergence of new environmental regulation, and a growing trend in consumers towards buying "green" products and services, etc. Manufacturers will hence not only be required to compete in product quality and cost, but also in reducing energy efficiency and carbon footprint during the manufacturing processes.

7.1 ENERGY-EFFICIENT MANUFACTURING

With the current global warming situation and the desire of many companies to be good corporate citizens today, many industries have increased interest in saving energy and reducing carbon footprints. One of the biggest energy consumption in the industry sector is electric motor driven equipment. Studies carried out by the European Commission [254][255][256] state that motor-driven systems use 65–70% of all electricity consumed by industries. It is estimated that these statistics will also be representative of the UK, and further reports [257][258] show that that more than half of all electricity consumed in the UK is used to drive electric motors. It is projected that by switching to energy-efficient motor systems, EU industries would save [254]:

1. 202 billion kWh in electricity consumption (approximately 7.5% of that consumed in all sectors);
2. £3–6 billion per annum in operating costs;
3. £4 billion in environmental costs;
4. 79 million tons of CO_2 emissions (one quarter of the EU's Kyoto target);
5. 45 GW reduction in the need for new power plant capacity; and
6. 6% reduction in energy imports.

The energy costs needed to operate machinery throughout its useful life can easily exceed the original equipment cost. Although the improvement in the machine design and optimization of the control system can improve the energy efficiency, daily operations and machine maintenance play even more important roles in reducing the overall environmental impact and cost to operate these machines! The energy efficiency and machine reliability are actually closely correlated [259]. The results of tests conducted by a major process industrial user investigating the Mean Time Between Failure (MTBF) for centrifugal pumping systems is shown in Figure 7.1.

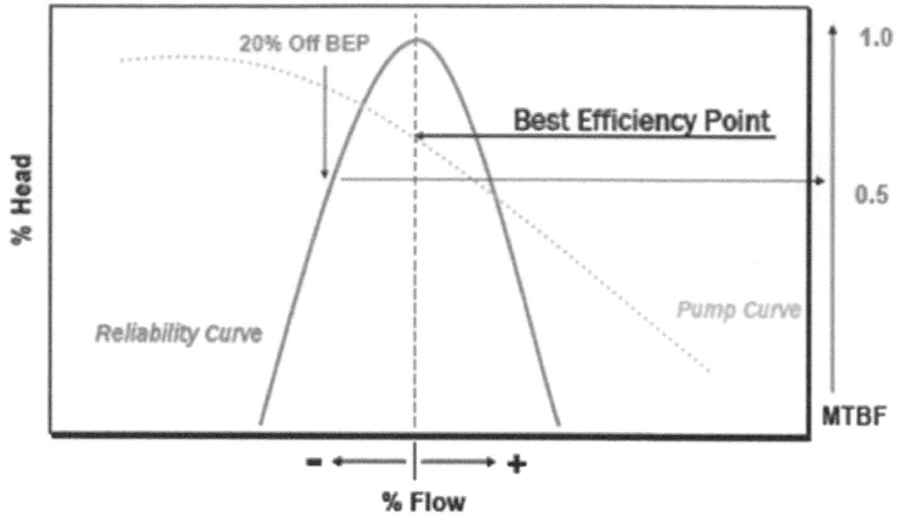

FIGURE 7.1 Relationship between Best Efficiency Point (BEP) and Mean Time Between Failure (MTBF).

Plant systems that operated off the Best Efficiency Point (BEP) by only 20% resulted in an MTBF reduction of 50%. Centrifugal pumps and other mechanical systems operating off their design point will effectively turn inefficiency into heat or vibration, leading to eventual failure. In fact, many mills spend millions of dollars on vibration detection and analysis, when they could improve both mill energy efficiency and reliability by adopting changes to assure systems operate at their designed efficiency points.

For large-scale milling machines, the utilization of the average cutting power signal is used to detect changes in the machining conditions [260]. Historic data collected in learning machine runs are collected during healthy machining runs and form the healthy buffer as shown in Figure 7.2. The boundaries of the inner detection buffer, or healthy buffer, are computed as the two extreme values in the collected samples. Beyond the healthy buffer, outer detection buffers are also introduced to represent possible transitions in power data. Since the machining process is a series of linked cutting instructions, the power consumed in neighboring sections of the NC program may influence the extremities in data measurements for the section under consideration. These outliers' area in the outer detection buffers is illustrated in Figure 7.2.

Recently, the advances in process monitoring and signal processing have brought the usefulness of multi-sensor systems closer to industrial realization. Key technologies related to machine health degradation detection and energy efficient monitoring include signal processing and feature extraction, alarm pattern identification and association, etc. Energy efficiency is not only affected by machine health on the shop

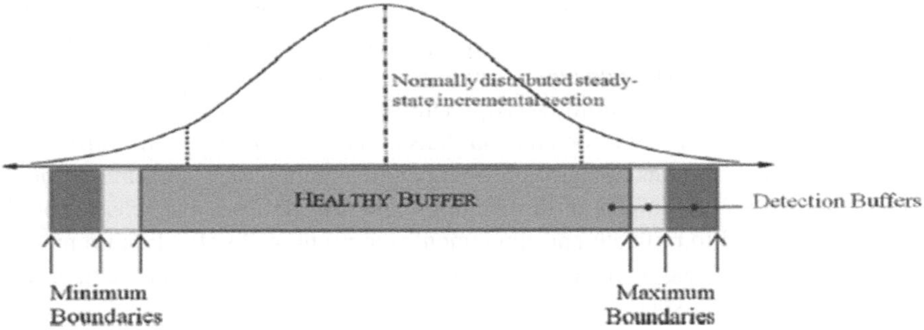

FIGURE 7.2 Power data collected in learning machine runs are used to form detection buffers.

floor, but also by the factory operation planning and maintenance. Economic and market indicators suggest that many types of US manufacturing are poised for a new period of major capacity investments as existing capacity approaches full utilization. At the same time, global trends are encouraging a shift to domestic production for domestic consumption.

With these objectives in mind, it is hence imperative to develop a framework for energy-efficient manufacturing, which includes energy consumption and optimization models, methodologies for energy efficiency monitoring, and energy key performance indicators. The framework will help manufacturing factories assess the energy usage, identify the key energy consumption machines and components, and optimize the energy utilization in the production to enhance the factories' economic competitiveness in the near future by optimizing the production life cycle and energy utilization in the manufacturing process.

However, the integration between the machine level energy management and shop floor energy efficiency optimization is still an open problem. Where we are, we are emphasizing on large-scale systems engineering by studying not only individual components, but a synergetic integration of composite systems to be considered simultaneously. As such, we aim to provide a formal decision software tool for higher level decision making and command in synergetic integration between several industrial processes and stages, for shorter time in failure and fault analysis in the entire industrial production life cycle.

7.2 LIFE CYCLE ASSESSMENT (LCA)

Life Cycle Assessment (LCA) is a well-established technique which results in both quantitative and qualitative analysis of the operating environmental merits and demerits of a process, considering all stages of the cycle. It is also commonly employed to generate novel ideas of new technologies aimed at shortening production time while reducing failure by identifying the key factors via an array of options.

The essence of LCA is the examination, identification, and evaluation of the relevant environmental implications of a material, process, product, or system across its life span, from creation to disposal or, preferably, to recreation in the same or another useful form. The Society of Environmental Toxicology and Chemistry defines the LCA process as follows [70] "The life-cycle assessment is an objective process to evaluate the environmental burdens associated with a product, process, or activity by identifying and quantifying energy and material usage and environmental releases, to assess the impact of those energy and material uses and releases on the environment, and to evaluate and implement opportunities to effect environmental improvements. The assessment includes the entire life cycle of the product, process or activity, encompassing extracting and processing of raw materials, manufacturing, transportation, and distribution, use/reuse/maintenance, recycling, and final disposal."

The term "life cycle" refers to the major activities in the course of the product's life-span from its manufacture, use, and maintenance, to its final disposal, including the raw material acquisition required to manufacture the product. Specifically, LCA is a technique to assess the environmental aspects and potential impacts associated with a product, process, or service, by:

1. Compiling an inventory of relevant energy and material inputs and environmental releases;
2. Evaluating the potential environmental impacts associated with identified inputs and releases; and
3. Interpreting the results to help decision-makers make a more informed decision.

The LCA process is a systematic, phased approach and consists of four components: goal definition and scoping, inventory analysis, impact assessment, and interpretation. The four components are detailed below [70]:

1. *Goal Definition and Scoping*. Define and describe the product, process or activity. Establish the context in which the assessment is to be made and identify the boundaries and environmental effects to be reviewed for the assessment;
2. *Inventory Analysis*. Identify and quantify energy, water and materials usage, and environmental releases (e.g., air emissions, solid waste disposal, waste water discharges, etc).
3. *Impact Assessment*. Assess the potential human and ecological effects of energy, water, and material usage and the environmental releases identified in the inventory analysis.
4. *Interpretation*. Evaluate the results of the inventory analysis and impact assessment to select the preferred product, process or service with a clear understanding of the uncertainty and the assumptions used to generate the results.

LCA is unique because it encompasses all processes and environmental releases beginning with the extraction of raw materials and the production of energy used to create the product through the use and final disposition of the product. When deciding between two or more alternatives, LCA can help decision-makers compare all major environmental impacts caused by products, processes, or services.

LCA can help decision-makers select the product or process that results in the least impact to the environment. This information can be used with other factors, such as cost and performance data to select a product or process. LCA data identifies the transfer of environmental impacts from one media to another (e.g., eliminating air emissions by creating a waste water effluent instead) and/or from one life cycle stage to another (e.g., from use and reuse of the product to the raw material acquisition phase). If an LCA were not performed, the transfer might not be recognized and properly included in the analysis because it is outside of the typical scope or focus of product selection processes.

This ability to track and document shifts in environmental impacts can help decision makers and managers fully characterize the environmental trade-offs associated with product or process alternatives. By performing an LCA, analysts can [70]:

1. Develop a systematic evaluation of the environmental consequences associated with a given product;
2. Analyze the environmental trade-offs associated with one or more specific products/processes to help gain stakeholder (state, community, etc.) acceptance for a planned action;
3. Quantify environmental releases to air, water, and land in relation to each life cycle stage and/or major contributing process;
4. Assist in identifying significant shifts in environmental impacts between life cycle stages and environmental media;
5. Assess the human and ecological effects of material consumption and environmental releases to the local community, region, and world;
6. Compare the health and ecological impacts between two or more rival products/processes or identify the impacts of a specific product or process; and
7. Identify impacts on one or more specific environmental areas of concern.

However, we should note that it is both resource and time consuming to perform an LCA. Depending upon how thorough an LCA the user wishes to conduct, gathering the data can be problematic, and the availability of data can greatly impact the accuracy of the final results. Therefore, it is important to weigh the availability of data, the time necessary to conduct the study, and the financial resources required against the projected benefits of the LCA. Also, LCA will not determine which product or process is the most cost effective or works the best. Therefore, the information developed in an LCA study should be used as one component of a more comprehensive decision process assessing the trade-offs with cost and performance.

There are a number of ways to conduct an LCA. While the methods are typically scientifically-based, the complexity of environmental systems has led to the development of alternative impact models. LCA can also be applied to products, industrial processes, and industrial facilities.

From the current framework of LCA, one can observe that there are many inputs and states. Concurrently, engineers and scientists are working toward a control model for current LCA. The suggested control model for future work will employ *modern feedback control theory* with selected states and outputs will be feedback to the controller for controlling and monitoring purposes.

To describe and predict the behavior of complex industrial systems, it is often necessary to use elaborate mathematical modeling. In the same manner, identification of the optimum operating conditions that will ensure improved process performance usually renders the use of an optimization technique essential. Historically, system optimization in chemical and process engineering applications has focused on maximizing the economic performance, subject to the certain constraints in the system. Over the past decade, optimization of environmental performance has started to be incorporated into system optimization, alongside traditional economic criteria. These approaches have mainly been focused on various waste minimization techniques. The attempts to incorporate environmental considerations into the design and optimization procedures represent the beginning of the paradigm shift in the process industry traditionally oriented towards the economic performance of the process. However, the main disadvantage of these approaches is that they concentrate on the emissions from the plant only, without considering other stages in the life cycle. Thus, it is possible for waste minimization approaches to reduce the emissions from the plant but to increase the burdens elsewhere in the life cycle, so that overall environmental impacts are increased. A general framework for the optimum LCA performance methodology comprises four steps:

1. Completion of the LCA study;
2. Formulation of the optimization problem in the context of LCA;
3. Application of operations research and control theory for optimality; and
4. Multi-criteria decision analysis and choice of the best compromise solution.

For instance, the equality constraints may be defined by energy and material balances, while the inequality constraints may describe material availabilities, heat requirements, capacities, etc. This prompts us to identify the essential *signals* for feedback and LCA.

7.3 SYSTEM OF SYSTEMS (SOS)

Recently, there has been a growing interest in a class of complex systems whose constituents are themselves complex. Performance optimization, robustness, and reliability among an emerging group of heterogeneous systems in order to realize a common goal has become the focus of various applications including military, security, aerospace, space, manufacturing, service industry, environmental systems, and disaster management [261], etc. There is an increasing interest in achieving synergy between these independent systems to achieve the desired overall system performance.

Currently, there are numerous definitions whose detailed discussion is beyond the space allotted to this chapter. Based on a recognized taxonomy of SoS, there are four types of SoS which are found in the literature today [262][263]:

1. *Virtual*. Virtual SoS lack a central management authority and a centrally

agreed upon purpose for SoS. Large-scale behavior emerges and may be desirable but this type of SoS must rely upon relatively invisible mechanisms to maintain it.

2. *Collaborative.* In collaborative SoS the component systems interact more or less voluntarily to fulfill agreed upon central purposes. The Internet is a collaborative system. The Internet Engineering Task Force works out standards but has no power to enforce them. The central players collectively decide how to provide or deny service, thereby providing some means of enforcing and maintaining standards.
3. *Acknowledged.* Acknowledged SoS have recognized objectives, a designated manager, and resources for the SoS; however, the constituent systems retain their independent ownership, objectives, funding, and development and sustainment approaches. Changes in the systems are based on collaboration between the SoS and the system.
4. *Directed.* Directed SoS are those in which the integrated system-of-systems is built and managed to fulfill specific purposes. It is centrally managed during long-term operation to continue to fulfill those purposes as well as any new ones the system owners might wish to address. The component systems maintain an ability to operate independently, but their normal operational mode is subordinated to the central managed purpose.

In the realm of open problems in SoS, there are numerous unsolved problems, and immense attention is needed by many engineers and scientists. The major issue here is that a merger between SoS and engineering needs to be made. In other words, conventional systems engineering needs to undergo a number of innovative changes to accommodate and encompass SoS. As discussed earlier, there are numerous unsolved problems in this open field of SoS which do need much attention of researchers and engineers.

References

1. G. Vachtsevanos, F. L. Lewis, M. Roemer, A. Hess, and B. Wu, *Intelligent Fault Diagnosis and Prognosis for Engineering Systems*, Wiley, 2006.
2. P. Baruah and R. B. Chinnam, "HMMs for Diagnostics and Prognostics in Machining Processes," *International Journal of Production Research*, Vol. 43, No. 6, pp. 1275-1293, March 2005.
3. H. M. Ertunc, K. A. Loparo, and H. Ocak, "Tool Wear Condition Monitoring in Drilling Operations Using Hidden Markov Models," *International Journal of Machine Tools and Manufacture*, Vol. 41, No. 9, pp. 1363-1384, 2001.
4. J. Kang, C. Feng, and H. Hu, "Tool Wear Monitoring and Pattern Recognition of Tool Failure," *Journal of Vibration: Measurement and Diagnosis*, Vol. 29, No. 1, pp. 5-9, 2009.
5. L. Wang, M. G. Mehrabi, and E. Kannatey-Asibu, Jr., "Hidden Markov Model-Based Tool Wear Monitoring in Turning," *Journal of Manufacturing Science and Engineering*, Vol. 124, No.3, pp. 651-658, August 2002.
6. K. Zhu, Y. S. Wong, and G. S. Hong, "Multi-Category Micro-Milling Tool Wear Monitoring with Continuous Hidden Markov Models," *Mechanical Systems and Signal Processing*, Vol. 23, No. 2, pp. 547-560, February 2009.
7. C. Zhang, and S. Y. Liang, "Condition-Based Maintenance of Rolling Element Bearings," *International Journal of COMADEM*, Vol. 7, pp. 3-12, 2004.
8. C. Bunks, D. McCarthy, and T. Al-Ani, "Condition-Based Maintenance of Machines Using Hidden Markov Models," *Mechanical Systems and Signal Processing*, Vol. 14, No. 4, pp. 597-612, July 2000.
9. H. Hu, M. An, G. Qin, and N. Hu, "Study on Fault Diagnosis and Prognosis Methods Based on Hidden Semi-Markov Model," *Acta Armamentarii*, Vol. 30, No. 1, pp. 69-75, 2009.
10. Z. Yang and M. Dong, "Equipment Fault Diagnosis Using Autoregressive Hidden Markov Models," *Journal of Shanghai Jiaotong University*, Vol. 42, No. 3, pp. 471-474, 2008.
11. A. Chen and G. S. Wu, "Real-Time Health Prognosis and Dynamic Preventive Maintenance Policy for Equipment Under Aging Markovian Deterioration," *International Journal of Production Research*, Vol. 45, No. 15, pp. 3351-3379, August 2007.
12. H. Guo and X. Yang, "Quantitative Reliability Assessment for Safety Related Systems Using Markov Models," *Journal of Tsinghua University (Science and Technology)*, Vol. 48, No. 1, pp. 149-155, 2008.
13. S. Song and Z. Lu, "Markov Chain-Based Line Sampling Method for Reliability Estimation of Structural Parallel System with Multiple Failure Modes," *Journal of Northwestern Polytechnical University*, Vol. 26, No. 2, pp. 234-238, 2008.
14. G. N. Fouskitakis and S. D. Fassois, "Pseudolinear Estimation of Fractionally Integrated ARMA (ARFIMA) Models with Automotive Application," *IEEE Transactions on Signal Processing*, Vol. 47, No. 12, pp. 3365-3380, December 1999.
15. S. J. Hu, and Y. G. Liu, "Process Mean Shift Detection Using Prediction Error Analysis," *Journal of Manufacturing Science and Engineering*, Vol. 120, No. 3, pp. 489-495, August 1998.
16. G. Wu, P. Liang, and X. Long, "Forecasting of Vibration Fault Series of Stream Turbine Rotor Based on ARMA," *Journal of South China University of Technology*, Vol. 33, No. 1, pp. 67-73, 2005.

17. X. Li, X. Cui, and R. Lang, "Forecasting Method for Aeroengine Performance Parameters," *Journal of Beijing University of Aeronautics and Astronautics*, Vol. 34, No. 3, pp. 253-256, 2005.
18. S. E. Oraby, A. F. Al-Modhuf, and D. R. Hayhurst, "A Diagnostic Approach for Turning Tool Based on the Dynamic Force Signals," *Journal of Manufacturing Science and Engineering*, Vol. 127, No. 3, pp. 463-475, 2005.
19. V. T. Tran, B. -S. Yang, M. -S. Oh, and A. C. C. Tan, "Machine Condition Prognosis Based on Regression Trees and One-Step-Ahead Prediction," *Mechanical Systems and Signal Processing*, Vol. 22, No. 5, pp. 1179-1193, July 2008.
20. K. Kamali and L. J. Jiang, "Using Q-learning and Genetic Algorithms to Improve the Efficiency of Weight Adjustments for Optimal Control and Design Problems," *Journal of Computing and Information Science and Engineering*, Vol. 7, No. 4, pp. 302-308, December 2007.
21. G. Levitin and S. V. Amari, "Optimal Load Distribution in Series Parallel Systems," *Reliability Engineering and System Safety*, Vol. 94, No. 2, pp. 254-260, February 2009.
22. Y. Lei, D. Djurdjanovic, J. Ni, R. Mayor, J. Lee, and G. Xiao, "System Level Optimization of Preventive Maintenance in Industrial Automation Systems," in *Proceedings of the 34th North American Manufacturing Research Conference, ASME*, Vol. 34, pp. 79-86, 2006.
23. Z. Yang, D. Djurdjanovic, and J. Ni, "Maintenance Scheduling in Manufacturing Systems Based on Predicted Machine Degradation," *Journal of Intelligent Manufacturing* Vol. 19, No. 1, pp. 87-98, February 2008.
24. V. Makis, "Multivariate Bayesian Process Control for a Finite Production Run," *European Journal of Operational Research*, Vol. 194, No. 3, pp. 795-806, May 2006.
25. Y. Pan, J. Chen, and L. Guo, "Robust Bearing Performance Degradation Assessment Method Based on Improved Wavelet Packet–Support Vector Data Description," *Mechanical Systems and Signal Processing*, Vol. 23, No. 3, pp. 669-681, April 2009.
26. Y. Ohue, A. Yoshida, and M. Seki, "Application of the Wavelet Transform to Health Monitoring and Evaluation of Dynamic Characteristics in Gear Sets," in *Proceedings of the Institution of Mechanical Engineers, Part J: Journal of Engineering Tribology*, Vol. 218, No. 1, pp. 1-11, 2004.
27. D. J. Pedregal and M. C. Carnero, "State Space Models for Condition Monitoring: A Case Study," *Reliability Engineering and System Safety*, Vol. 91, No. 2, pp. 171-180, 2006.
28. D. C. Swanson, "A General Prognostic Tracking Algorithm for Predictive Maintenance," in *Proceedings of IEEE Aerospace Conference*, Vol. 6, pp. 2971-2977, 2001.
29. M. Cao, K. W. Wang, L. DeVries, Y. Fujii, W. E. Tobler, G. M. Pietron, T. Tibbles, and J. McCallum, "Steady State Hydraulic Valve Fluid Field Estimator Based on Non-Dimensional Artificial Neural Network (NDANN)," *Journal of Computing and Information Science and Engineering*, Vol. 4, No. 3, pp. 257-270, September 2004.
30. G. Parthasarathy, S. Menon, K. Richardson, A. Jameel, D. McNamee, T. Desper, M. Gorelik, and C. Hickenbottom, "Neural Network Models for Usage Based Remaining Life Computation," *Journal of Engineering for Gas Turbines and Power*, Vol. 130, No. 1, pp. 012508.1-012508.7, 2008.
31. W. Q. Wang, M. F. Golnaraghi, and F. Ismail, "Prognosis of Machine Health Condition Using Neuro-Fuzzy Systems," *Mechanical Systems and Signal Processing*, Vol. 18, No. 4, pp. 813-831, July 2004.
32. C. K. Pang, F. L. Lewis, and T. H. Lee, "Modal Parametric Identification of Flexible Mechanical Structures in Mechatronic Systems," *Transactions of the Institute of Mea-*

surement and Control, Vol. 32, No. 2, pp. 137-154, April 2010.
33. L. G. Kelly, Chapter 5, Curve Fitting and Data Smoothing, in *Handbook of Numerical Methods and Applications*, Addison-Wesley Inc., 1967.
34. C. K. Pang, C. Y. Lim, W. E. Wong, and F. Hong, "GUI Toolkit for Learning Advanced Magnetic Recording Technologies in Ultra-High Storage Capacity Hard Disk Drives," submitted to *Journal of Educational Technology Systems*, November 2010.
35. C. K. Pang, T. S. Ng, F. L. Lewis, and T. H. Lee, "Managing Complex Mechatronics R&D: A Systems Design Approach," *IEEE Transactions on Systems, Man, and Cybernetics Part A: Systems and Humans*, in press.
36. C. L. Jiaa and D. A. Dornfeld, "A Self-Organizing Approach to the Prediction and Detection of Tool Wear," *ISA Transactions*, Vol. 37, No. 4, pp. 239-255, September 1998.
37. J. –H. Zhou, C. K. Pang, F. L. Lewis, and Z. –W. Zhong, "Intelligent Diagnosis and Prognosis of Tool Wear Using Dominant Feature Identification," *IEEE Transactions on Industrial Informatics*, Vol. 5, No. 4, pp. 454–464, November 2009.
38. J. –H. Zhou, C. K. Pang, Z. –W. Zhong, and F. L. Lewis, "Tool Wear Monitoring Using Acoustic Emissions for Dominant Feature Identification," *IEEE Transactions on Instrumentation and Measurement*, Vol. 60, No. 2, pp. 547-559, February 2011.
39. J. –H. Zhou, C. K. Pang, F. L. Lewis, and Z. –W. Zhong, "Dominant Feature Identification for Industrial Fault Detection and Isolation Applications," *Expert Systems with Applications*, Vol. 38, No. 8, pp. 10676-10684, August 2011.
40. C. K. Pang, Z. Y. Dong, P. Zhang, and X. Yin, "Probabilistic Analysis of Power System Small Signal Stability Region," in *Proceedings of the 2005 IEEE ICCA*, TA2-4.5, pp. 503–509, Budapest, Hungary, June 27-29, 2005.
41. Z. Y. Dong, C. K. Pang, and P. Zhang, "Power System Sensitivity Analysis for Probabilistic Small Signal Stability Assessment in a Deregulated Environment," *International Journal of Control Automation and Systems (Special Issue on Recent Advances in Power System Control)*, Vol. 3, No. 2 (Special Edition), pp. 355-362, June 2005.
42. Z. Y. Dong, C. K. Pang, and Z. Xu., "Investigation of Probabilistic Small Signal Stability," *Electric Power Research Institute Technical Report, Product ID #1002637*, September 2003 (funded by Electric Power Research Institute #1002637, EPRI, Palo Alto, CA, USA).
43. Y. Mishra, Z. Y. Dong, J. Ma, and D. J. Hill, "Induction Motor Load Impact on Power System Eigenvalue Sensitivity Analysis," *IET Generation, Transmission & Distribution*, Vol. 3, No. 7, pp. 690-700, July 2009.
44. F. L. Lewis, G. R. Hudas, C. K. Pang, M. B. Middleton, and C. Mcmurrough, "Discrete Event Command and Control for Networked Teams with Multiple Missions," in *Proceedings of SPIE*, Vol. 7332, 73320V, April 30, 2009.
45. C. K. Pang, G. Hudas, M. B. Middleton, C. V. Le, O. P. Gan, and F. L. Lewis, "Discrete Event Command and Control for Networked Teams with Multiple Military Missions," submitted to *The Journal of Defense Modeling and Simulation: Applications, Methodology, Technology*, November 2010.
46. J. S. Bay, *Funadamentals of Linear State Space Systems*, McGraw-Hill International Editions, 2001.
47. D. C. Meyer, *Matrix Analysis and Applied Linear Algebra*, Society for Industrial and Applied Mathematics, 2001.
48. H. Paley, *A First Course in Abstract Algebra*, International Thomson Publishing, 1966.
49. D. C. Lay, *Linear Algebra and Its Applications*, 3rd Edition, Addison Wesley, 2002.
50. C. W. de Silva, *Modeling and Control of Engineering System*, CRC Press, 2009.

51. G. E. Shilov, *Linear Algebra*, Dover Publications, 1997.
52. G. Strang, *Introduction to Linear Algebra*, 4th Edition, Wellesley-Cambridge Press, 1998.
53. P. Gonçalves, "Behaviour Modes, Pathways and Overall Trajectories: Eigenvector and Eigenvalue Analysis of Dynamic Systems," *System Dynamics Review*, Vol. 25, No. 1, pp. 35-62, 2009.
54. N. Forrester, *A Dynamic Synthesis of Basic Macroeconomic Policy: Implications for Stabilization Policy Analysis*, Ph.D. Thesis, MIT, Cambridge, MA, USA, 1982.
55. N. Forrester, "Eigenvalue Analysis of Dominant Feedback Loops," in *Proceedings of the 1983 International System Dynamics Conference, Plenary Session Papers*, System Dynamics Society, pp. 178-202, Albany, NY, USA, 1983.
56. J. N. Warfield, "Binary Matrices in System Modeling," *IEEE Transactions on Systems, Man, and Cybernetics*, Vol. 3, No. 5, pp. 441-449, September 1973.
57. B. -J. Jorgen and G. Gregory, *Digraphs: Theory, Algorithms and Application*, Springer, 2000.
58. D. A. Tacconi and F. L. Lewis, "A New Matrix Model for Discrete Event Systems: Application to Simulation," *IEEE Control Systems Magazine*, Vol. 17, No. 5, pp. 62-71, October 1997.
59. V. Giordano, P. Ballal, F. L. Lewis, B. Turchiano, and J. B. Zhang, "Supervisory Control of Mobile Sensor Networks: Math Formulation, Simulation, and Implementation," *IEEE Transactions on Systems, Man, and Cybernetics: Part B*, Vol. 36, No. 4, pp. 806-819, August 2006.
60. T. Murata, "Petri Nets: Properties, Analysis and Applications," in *Proceedings of IEEE*, Vol. 77, No. 4, pp. 541-580, April 1989.
61. T. H. Yan, X. D. Chen, and R. M. Lin, "Servo System Modeling and Reduction of Mechatronic System Through Finite Element Analysis for Control Design," *Mechatronics*, Vol. 18, No. 9, pp. 466-474, November 2008.
62. T. Iwasaki, "Integrated System Design by Separation," in *Proceedings of the 1999 IEEE International Conference on Control Applications*, pp. 97-102, Kohala Coast, HI, USA, August 22-27, 1999.
63. S. Hara, "Dynamical System Design from Control Perspective," in *Proceedings of the SICE-ICASE International Joint Conference 2006*, Bexco, Busan, Korea, October 18-21, 2006.
64. M. Grossard, C. Rotinat-Libersa, N. Chaillet, and M. Boukallel, "Mechanical and Control-Oriented Design of a Monolithic Piezoelectric Microgripper Using a New Topological Optimization Method," *IEEE/ASME Transactions on Mechatronics*, Vol. 14, No. 1, pp. 32-45, February 2009.
65. K. Chen, J. Bankston, J. H. Panchal, and D. Schaefer, "A Framework for Integrated Design of Mechatronic Systems," in *Collaborative Design and Planning for Digital Manufacturing*, (L. Wang and A. Y. C. Nee, Eds.), Springer, London, pp. 37-70, 2009.
66. N. S. Argyres and B. S. Silverman, "R&D, Organization Structure and the Development of Corporate Technological Knowledge," *Strategic Management Journal*, Vol. 25, pp. 929-958, 2004.
67. T. L. Griffith and J. E. Sawyer, "Research Team Design and Management for Centralized R&D," *IEEE Transactions on Engineering Management*, in press.
68. J. Lin and T. S. A. Ng, "A Systems Approach for Managing Concurrent Product Development Projects," in *Proceedings of the 3rd Asia-Pacific Conference on Systems Engineering (APCOSE)*, Singapore, July 20-23, 2009.
69. K. Wucherer, "On Track For The Intelligent Factory," presented at Siemens Press Con-

ference at the Hanover Fair, April 16, 2007.
70. T. E. Graedel and B. R. Allenby, *Industrial Ecology*, Prentice Hall, New Jersey, NJ, USA, 1995.
71. C. K. Pang, F. L. Lewis, S. S. Ge, G. Guo, B. M. Chen, and T. H. Lee, "Singular Perturbation Control for Vibration Rejection in HDDs Using the PZT Active Suspension as Fast Subsystem Observer," *IEEE Transactions on Industrial Electronics*, Vol. 54, No. 3, pp. 1375-1386, June 2007.
72. P. Eykhoff, *System Identification—Parameter and State Estimation*, New York: Wiley, 1981.
73. L. Ljung, *System Identification: Theory For The User*, 2nd Edition, PTR Prentice Hall, Upper Saddle River, NJ, 1999.
74. I. Kollár, "Frequency Domain System Identification Toolbox for Use with MATLAB," *MATLAB Manual*, 2001.
75. T. Mckelvey, H. Akcay, and L. Ljung, "Subspace-Based Identification of Infinite-Dimensional Multivariable Systems from Frequency-Response Data," *Automatica*, Vol. 32, No. 6, pp. 885–902, 1996.
76. B. M. Chen, T. H. Lee, K. Peng, and V. Venkataramanan, *Hard Disk Drive Servo Systems*, 2nd Edition, Springer, New York, (Advances in Industrial Control Series), 2006.
77. M. Kawafuku, K. Otsu, H. Hirai, and M. Kobayashi, "High Performance Control Design of HDD Based on Precise System Modeling using Differential Iteration Method," in *Proceedings of the 2003 American Control Conference*, pp. 4341-4346, Denver, CO, USA, June 4-6, 2003.
78. M. Kawafuku, K. Otsu, M. Iwasaki, H. Hirai, M. Kobayashi, and A. Okuyama, "High-Precision Positioning Control for Hard Disk Drives: 1st Report: High-Precision Simulator Using Differential Iteration Method," *Electrical Engineering in Japan*, in press.
79. M. H. Richardson and D. L. Formenti, "Parameter Estimation from Frequency Response Measurements Using Rational Fraction Polynomials," in *Proceedings of the 1st IMAC Conference*, pp. 1–15, Orlando, FL, USA, November 8-10, 1982.
80. M. H. Richardson and D. L. Formenti, "Global Curve Fitting of Frequency Response Measurements Using the Rational Fraction Polynomial Method," in *Proceedings of the 3rd IMAC Conference*, pp. 1–8, Orlando, FL, USA, January 28-31, 1985.
81. T. Yamaguchi, "Modelling and Control of a Disk File Head-Positioning System," in *Proceedings of the I MECH E Part I Journal of Systems & Control Engineering*, Vol. 215, No. 6, pp. 549-568, 2001.
82. C. K. Pang, S. C. Tam, G. Guo, B. M. Chen, F. L. Lewis, T. H. Lee, and C. Du, "Improved Disturbance Rejection with Online Adaptive Pole-Zero Compensation on a Φ-Shaped PZT Active Suspension," *Microsystem Technologies*, Vol. 15, Nos. 10-11, pp. 1499-1508, October 2009.
83. C. K. Pang, D. Wu, G. Guo, T. C. Chong, and Y. Wang, "Suppressing Sensitivity Hump in HDD Dual-Stage Servo Systems," *Microsystem Technologies*, Vol. 11, Nos. 8-10, pp. 653-662, August 2005.
84. G. F. Franklin, J. D. Powell, and M. L. Workman, *Digital Control of Dynamic Systems*, 3rd Edition, Addison-Wesley, 1998.
85. INCOSE, *Systems Engineering Handbook, v3.1*, INCOSE, 2007.
86. NASA, *Systems Engineering Handbook*, NASA/SP-2007-6105, NASA, 2007.
87. A. Sage, *Systems Engineering*, Wiley IEEE, 1992.
88. A. Sage and S. R. Olson, "Modeling and Simulation in Systems Engineering," *Simulation*, Vol. 76, No. 2, pp. 90-91, 2001.

89. A. Kossiakoff and W. N. Sweet, *Systems Engineering Principles and Practice*, Wiley-Interscience, 2002.
90. B. Blanchard, *System Engineering Management*, 4th Edition, Wiley, 2008.
91. C. S. Wasson, *System Analysis, Design, and Development: Concepts, Principles, and Practices*, Wiley Series in Systems Engineering and Management, Wiley-Interscience, 2005.
92. Y. Akao (Ed.), *Quality Function Deployment: Integrating Customer Requirements Into Product Design*, Productivity Press, Portland, OR, USA, 1990.
93. J. R. Hauser and D. Clausing, "The House of Quality," *Harvard Business Review*, pp. 63-73, May-June 1988.
94. V. Kotov, "Systems-of-Systems as Communicating Structures," *Hewlett Packard Computer Systems Laboratory Paper*, HPL-97-124, pp. 1–15, 1997.
95. S. J. Luskasik, "Systems, Systems-of-Systems, and the Education of Engineers," *Artificial Intelligence for Engineering Design, Analysis, and Manufacturing*, Vol. 12, No. 1, pp. 55-60, 1998.
96. Y. Rolain, R. Pintelon, K. Q. Xu, and H. Vold, "On the Use of Orthogonal Polynomials in High Order Frequency Domain System Identification and its Application to Modal Parameter Estimation," in *Proceedings of the 33rd IEEE Conference on Decision and Control*, pp. 3365-3373, Lake Buena Vista, FL, USA, December 14-16, 1994.
97. T. Atsumi, "Head-Positioning Control of Hard Disk Drives Through the Integrated Design of Mechanical and Control Systems," *International Journal of Automation Technology*, Vol.3, No.3, pp. 277-285, 2009.
98. T. Atsumi, S. Nakagawa, T. Yamaguchi, and H. Yosuke, "High-Bandwidth Servo Design for Shaping the Mechanical Resonance of the Actuator in Hard Disk Drives," *Transactions of the Japan Society of Mechanical Engineers: C*, Vol. 67, No. 664, pp. 3905-3910, 2001.
99. T. Semba, F. -Y. Huang, and M. T. White, "Integrated Servo/Mechanical Design of HDD Actuators and Estimation of the Achievable Bandwidth," *IEEE Transsactions on Magnetics*, Vol. 39, No. 5, No. 2, pp. 2588-2590, September 2003.
100. NHK SPRING, http://www.nhkspg.co.jp/eng/prod/disk.html [Online].
101. C. K. Pang, G. Guo, B. M. Chen, and T. H. Lee, "Self-Sensing Actuation for Nanopositioning and Active-Mode Damping in Dual-Stage HDDs," *IEEE/ASME Transactions on Mechatronics*, Vol. 11, No. 3, pp. 328-338, June 2006.
102. M. Kobayashi, S. Nakagawa, and S. Nakamura, "A Phase-Stabilized Servo Controller for Dual-Stage Actuators in Hard-Disk Drives," *IEEE Transactions on Magnetics*, Vol. 39, No. 2, pp. 844-850, March 2003.
103. D. T. Allen and D. R. Shonnard, *Green Engineering: Environmentally Conscious Design of Chemical Processes*, Prentice Hall, New Jersey, NJ, USA, 2002.
104. R. J. Kuo and P. H. Cohen, "Multi-Sensor Integration for On-Line Tool Wear Estimation Through Radial Basis Function Networks and Fuzzy Neural Network," *Neural Networks*, Vol. 12, No. 2, pp. 355-370, March 1999.
105. J. Dong, K. V. R. Subrahmanyam, Y. S. Wong, G. S. Hong, and A. R. Mohanty, "Bayesian-Inference-Based Neural Networks for Tool Wear Estimation," *The International Journal of Advanced Manufacturing Technology*, Vol. 30, Nos. 9-10, pp. 797-807, October 2006.
106. Y. S. Tarng, "Study of Milling Cutting Force Pulsation Applied to the Detection of Tool Breakage," *International Journal of Machine Tools and Manufacture*, Vol. 30, No. 4, pp. 651-660, 1990.
107. I. N. Tansel and C. McLaughlin, "Detection of Tool Breakage in Milling Operations - I. The Time Series Analysis Approach," *International Journal of Machine Tools and*

Manufacture, Vol. 33, No. 4, pp. 531-544, 1993.
108. C. S. Leem, D. A. Dornfeld, and S. E. Dreyfus, "Customized Neural Network for Sensor Fusion in On-Line Monitoring of Cutting Tool Wear," *Journal of Engineering for Industry*, Vol. 117, No. 2, pp. 152-159, 1995.
109. X. Li and X. Yao, "Multi-Scale Statistical Process Monitoring in Machining," *IEEE Transactions on Industrial Electronics*, Vol. 52, No. 3, pp. 924-927, June 2005.
110. J. Sun, G. S. Hong, M. Rahman, and Y. S. Wong, "Identification of Feature Set for Effective Tool Condition Monitoring by Acoustic Emission Sensing," *International Journal of Production Research*, Vol. 42, No. 5, pp. 901-918, 2004.
111. S. Orhan, A. O. Er, N. Camucu, and E. Aslan, "Tool Wear Evaluation by Vibration Analysis During End Milling of AISI D3 Cold Work Tool Steel with 35 HRC Hardness," *NDT & E International*, Vol. 40, No. 2, pp. 121-126, March 2007.
112. E. Haddadi, M. R. Shabghard, and M. M. Ettefagh, "Effect of Different Tool Edge Conditions on Wear Detection by Vibration Spectrum Analysis in Turning Operation," *Journal of Applied Science*, Vol. 8, No. 21, pp. 3879-3886, 2008.
113. X. Li, A. Djordjevich, and P. K. Venuvinod, "Current-Sensor-Based Feed Cutting Force Intelligent Estimation and Tool Wear Condition Monitoring," *IEEE Transactions on Industrial Electronics*, Vol. 47, No. 3, pp. 697-702, June 2000.
114. G. E. Dapos, "Tool Condition Monitoring and Machining Process Control," in *Proceedings of the 15th Annual Conference of IEEE IECON apos*, Vol. 3, Nos. 6-10, pp. 652-657, November 6-10, 1989.
115. T. W. Liao , Q. Zou, L. Mann, and M. E. Zodhi, "Online PCBN Tool Failure Monitoring System Based on Acoustic Emission Signatures," *IEE Proceedings: Science, Measurement and Technology*, Vol. 142, No. 5, pp. 404-410, September 1995.
116. J. -S. Kim, M. -C. Kang, B. -J. Ryu, and Y. -K. Ji, "Development of an On-Line Tool-Life Monitoring System Using Acoustic Emission Signals in Gear Shaping," *International Journal of Machine Tools & Manufacture*, Vol. 39, No. 11, pp. 1761-1777, November 1999.
117. G. Pontuale, F. A. Farrelly, A. Petri, and L. Pitolli, "A Statistical Analysis of Acoustic Emission Signals for Tool Condition Monitoring (TCM)," *Acoustical Society of America*, Vol. 4, No. 1, pp. 13-18, January 2003.
118. X. Chen and B. Li, "Acoustic Emission Method for Tool Condition Monitoring Based on Wavelet Analysis," *The International Journal of Advanced Manufacturing Technology*, Vol. 33, Nos. 9-10, pp. 968-976, July 2007.
119. P. S. Pai and P. K. R. Rao, "Acoustic Emission Analysis for Tool Wear Monitoring in Face Milling," *International Journal of Production Research*, Vol. 40, No. 5, pp. 1081-1093, 2002.
120. Y. M. Niu, Y. S. Wong, and G. S. Hong, "An Intelligent Sensor System Approach for Reliable Tool Flank Wear Recognition," *The International Journal of Advanced Manufacturing Technology* Vol. 14, No. 2, pp. 77-84, February 1998.
121. H. V. Ravindra, Y. G. Srinivasa, and R. Krishnamurthy, "Acoustic Emission For Tool Condition Monitoring in Metal Cutting," *Wear*, Vol. 212, No. 1, pp. 78-84, November 1997.
122. E. Kannatey-Asibu, Jr., and D. A. Dornfeld, "A Study of Tool Wear Using Statistical Analysis of Metal Cutting Acoustic Emission," *Wear*, Vol. 76, pp. 247-261, 1982.
123. K. P. Zhu, G. S. Hong, and Y. S. Wong, "A Comparative Study of Feature Selection for Hidden Markov Model-Based Micro-Milling Tool Wear Monitoring," *Machining Science and Technology*, Vol. 12, No. 3, pp. 348-369, July 2008.

124. S. Binsaeid, S. Asfour, S. Cho, and A. Onar, "Machine Ensemble Approach for Simultaneous Detection of Transient and Gradual Abnormalities in End Milling Using Multisensor Fusion," *Journal of Materials Processing Technology*, Vol. 209, No. 10, pp. 4728-4738, June 2009.
125. P. Bhattacharyyaa, D. Senguptaa, and S. Mukhopadhyayb, "Cutting Force-Based Real-Time Estimation of Tool Wear in Face Milling Using a Combination of Signal Processing Techniques," *Mechanical Systems and Signal Processing*, Vol. 21, No. 6, pp. 2665-2683, August 2007.
126. I. T. Jolliffe, *Principal Component Analysis*, Springer-Verlag, 1986.
127. G. P. McCabe, "Principal Variables," *Technometrics*, Vol. 26, No. 2, pp. 127-134, May 1984.
128. H. L. Tan, N. S. Chaudhari, and J. H. Zhou, "Time Series Prediction Using Principal Feature Analysis," in *Proceedings of the 3rd IEEE Conference on Industrial Electronics and Applications*, pp. 292-297, Singapore, June 3-5, 2008.
129. I. Cohen, Q. Tian, X. S. Zhou, and T. S. Huang, "Feature Selection Using Principal Feature Analysis," in *Proceedings of the 15th International Conference on Information Processing*, Rochester, NY, USA, September 22-25, 2002.
130. Y. Lu, I. Cohen, X. S. Zhou, Q. Tian, and T. S. Huang, "Feature Selection Using Principal Feature Analysis," in *Proceedings of the 15th International Conference on Multimedia*, pp. 301-304, Augsburg, Germany, September 25-29, 2007.
131. G. A. Fodor, "From the Editor-in-Chief Industrial Informatics: Predicting with Abstractions," *IEEE Transactions on Industrial Informatics*, Vol. 1, No. 1, pp. 3, February 2005.
132. F. L. Lewis, S. Jagannathan, and A. Yeşildirek, *Neural Network Control of Robot Manipulators and Nonlinear Systems*, CRC Press, 1998.
133. S. Simani, "Identification and Fault Diagnosis of a Simulated Model of an Industrial Gas Turbine," *IEEE Transactions on Industrial Informatics*, Vol. 1, No. 3, pp. 202-216, August 2005.
134. D. B. Percival and A. T. Walden, *Wavelet Methods for Time Series Analysis*, Cambridge University Press, 2000.
135. S. Gunn, "Support Vector Machines for Classification and Regression," Technical Report, *ISIS*, Department of Electronics and Computer Science, University of Southampton 1998.
136. R. Gong, S. H. Huang, and T. Chen, "Robust and Efficient Rule Extraction Through Data Summarization and Its Application in Welding Fault Diagnosis," *IEEE Transactions on Industrial Informatics*, Vol. 4, No. 3, pp. 198-206, August 2008.
137. S. P. Lloyd, "Least Squares Quantization in PCM," *IEEE Transactions on Information Theory (Special Issue on Quantization)*, Vol. 28, No. 2, pp. 129-137, March 1982.
138. T. Kanungo, D. M. Mount, N. Netanyahu, C. Piatko, R. Silverman, and A. Y. Wu, "An Efficient K-Means Clustering Algorithm: Analysis and Implementation," *IEEE Transactions on Pattern Analysis and Machine Intelligence*, Vol. 24, No. 7, pp. 881-892, July 2002.
139. J. Yan, M. Koc, and J. Lee, "A Prognostic Algorithm for Machine Performance Assessment and its Application," *Production Planning & Control*, Vol. 15, No. 8, pp. 796-801, December 2004.
140. G. Zhang, S. Lee, N. Propes, Y. Zhao, and G. Vachtsevanos, "A Novel Architecture for an Integrated Fault Diagnostic/Prognostic System," *AAAI Technical Report*, pp. 75, SS-02-03. 2002.
141. A. Muller, A. C. Marquez, and B. Lung, "On the Concept of E-Maintenance: Review and Current Research," *Reliability Engineering and System Safety*, Vol. 93, No. 8, pp.

1165-1187, August 2008.
142. T. Han and B. -S. Yang, "Development of an E-Maintenance System Integrating Advanced Techniques," *Computers in Industry*, Vol. 57, No. 6, pp. 569-580, August 2006.
143. B. Tao, H. Ding, and Y. L. Xiong, "IP Sensor and its Distributed Networking Application in E-Maintenance," *in Proceedings of the 2003 IEEE International Conference on Systems, Man, and Cybernetics*, Vol. 4, pp. 3858–3863, Washington, DC, USA, October 5-8, 2003.
144. M. Koç and J. Lee, "A System Framework for Next-Generation E-Maintenance System," *in Proceedings of Second International Symposium on Environmentally Conscious Design and Inverse Manufacturing*, Tokyo, Japan, 2001.
145. W. Zhang, W. Halang, and C. Diedrich, "An Agent-Based Platform for Service Integration in E-Maintenance," *in Proceedings of the 2003 IEEE International Conference on Industrial Technology*, Vol. 1, No. 10-12, pp. 426-433, December 2003.
146. A. C. Marquez and J. M. D. Gupta, "Contemporary Maintenance Management: Process, Framework and Supporting Pillars," *Omega*, Vol. 34, No. 3, pp. 313–326, June 2006.
147. W. Zhou, T. G. Habetler, and R. G. Harley, "Bearing Condition Monitoring Methods for Electric Machines: A General Review," *in Proceedings of the IEEE International Symposium on Diagnostics for Electric Machine*, pp. 3-6, Cracow, Poland, September 6-8, 2007.
148. P. D. McFadden and J. D. Smith, "Vibration Monitoring of Rolling Element Bearings By the High Frequency Resonance Technique--A Review," *Tribology International*, Vol. 17, pp. 3-10, 1984.
149. H. Qiu, H. Luo, and N. Eklund, "Draft: On-Board Bearing Prognostics in Aircraft Engine: Enveloping Analysis or FFT?" *in Proceedings of the IDETC/CIE*, pp. 1-7, San Diego, CA, USA, August 30–September 2, 2009.
150. Y. Lei, Z. He, and Y. Zi, "Application of an Intelligent Classification Method to Mechanical Fault Diagnosis," *Expert Systems with Applications*, Vol. 36, No. 6, pp. 9941–9948, August 2009.
151. Y. Lei, M. J. Zuo, Z. He, and Y. Zi, "A Multidimensional Hybrid Intelligent Method for Gear Fault Diagnosis," *Expert Systems with Applications*, Vol. 37, No. 2, pp. 1419–1430, March 2010.
152. J. C. Bezdek, *Pattern Recognition With Fuzzy Objective Function Algorithms*, Plenum Press, New York, 1981.
153. L. J. Heyer, S. Kruglyak, and S. Yooseph, "Exploring Expression Data: Identification and Analysis of Coexpressed Genes," *Genome Research*, Vol. 9, No. 11, pp. 1106-1115, November 1999.
154. J. -H. Wang, J. -D. Rau, and W. -J. Liu, "Two-Stage Clustering via Neural Networks," *IEEE Transactions on Neural Networks*, Vol. 14, No. 3, pp. 606-615, May 2003.
155. J. MacQueen, "Some Methods for Classification and Analysis of Observations," in *Proceedings of the Fifth Berkeley Symposium on Mathematics, Statistics, and Probability*, Vol. 1, pp. 281-297, University of California Press, 1967.
156. Y. Altintas and I. Yellowley, "In-Process Detection of Tool Failure in Milling Using Cutting Force Models," *Journal of Engineering for Industry*, Vol. 111, pp. 149-157 May 1989.
157. Y. Altintas, "In-Process Detection of Tool Breakage Using Time Series Monitoring of Cutting Forces," *International Journal of Machine Tools and Manufacture*, Vol. 28, No. 2, pp. 157-172, 1988.
158. J. H. Tarn and M. Tomizuka, "On-Line Monitoring of Tool and Cutting Conditions in

Milling," *Journal of Engineering for Industry, Transactions of the ASME*, Vol. 111, pp. 206-212, August 1989.
159. D. Y. Zhang, Y. T. Han, and D. C. Chen, "On-Line Detection of Tool Breakages Using Telemetering of Cutting Forces in Milling," *International Journal of Machine Tools and Manufacture*, Vol. 35, No. 1, pp. 19-27, 1995.
160. M. A. Elbestawi, J. Marks, and T. A. Papazafiriou, "Process Monitoring in Milling by Pattern Recognition," *Mechanical Systems and Signal Processing*, Vol. 3, No. 3, pp. 305-315, 1989.
161. Y. S. Tarng, Y. W. Hseih, and S. T. Hwang, "Sensing Tool Breakage in Face Milling with a Neural Network," *International Journal of Machine Tools and Manufacture*, Vol. 34, No. 3, pp. 341-350, 1994.
162. X. Q. Chen, H. Zeng, and W. Dietmar, "In-Process Tool Monitoring through Acoustic Emission Sensing," *SIMTech Technical Report (AT/01/014/AMP)*, 2001.
163. A. E. Diniz, J. J. Liu, and D. A. Dornfeld, "Correlating Tool Life, Tool Wear and Surface Roughness by Monitoring Acoustic Emission in Finish Turning," *Wear*, Vol. 152, No. 2, pp. 395-407, 1992.
164. K. Sunilkumar, L. Vijayaraghavan, and R. Krishnamurthy, "In Process Wear and Chip-Form Monitoring in Face Milling Operation Using Acoustic Emission," *Journal of Materials Processing Technology*, Vol. 44, Nos. 3–4, pp. 207-214, 1994.
165. P. Palanisamy, I. Rajendran, and S. Shanmugasundaram, "Prediction of Tool Wear Using Regression and ANN Models in End-Milling Operation," *The International Journal of Advanced Manufacturing Technology*, Vol. 37, Nos. 1-2, pp. 29-41, April 2008.
166. Z. Chen and X. M. Zhang, "Monitoring of Tool Wear Using Feature Vector Selection and Linear Regression," *Lecture Notes in Computer Science, Advances in Natural Computation*, Vol. 3611, pp. 1-6, 2005.
167. P. Dang, F. M. Ham, F. L. Lewis, and H. Stephanou, "A Two-Stage Neural Network for Expression Classification," *submitted to Neurocomputing*, April 2009.
168. F. M. Ham and I. Kostanic, *Principles of Neurocomputing for Science and Engineering*, McGraw-Hill, New York, USA, 2001.
169. SpectraQuest, http://www.spectraquest.com/Products/simulators_pro.shtml [Online].
170. A. Widodo, B. -S. Yang, D. -S. Gu, and B. -K. Cho, "Intelligent Fault Diagnosis System of Induction Motor Based on Transient Current Signal," *Mechatronics*, Vol. 19, No. 5, pp. 680-689, August 2009.
171. T. I. Liu, E. J. Ko, and S. L. Sha, "Intelligent Monitoring of Tapping Tools," *Journal of Materials Shaping Technology*, Vol. 8, No. 4, pp. 249-254, December 1990.
172. C. W. De Silva, *Vibration: Fundamentals and Practice*, 2nd Edition, CRC Press, 2006.
173. M. Norton and D. Karczub, *Fundamentals of Noise and Vibration Analysis for Engineers*, 2nd Edition, Cambridge, October 2003.
174. M. E. H. Benbouzid, "A Review of Induction Motors Signature Analysis as a Medium for Faults Detection," *IEEE Transactions on Industrial Electronics*, Vol. 47, No. 5, pp. 984-993, October 2000.
175. K. T. Chung and A. Geddam, "A Multi-Sensor Approach to the Monitoring of End Milling Operations," *Journal of Materials Processing Technology*, Vol. 139, Nos. 1-3, pp. 15-20, August 2003.
176. N. Tandon and A. Choudhury, "A Review of Vibration and Acoustic Measurement Methods for the Detection of Defects in Rolling Element Bearings," *Tribology International*, Vol. 32, No. 8, pp. 469-480, August 1999.
177. R. C. Burchett and G. T. Heydt, "Probabilistic Methods for Power System Dynamic

Stability Studies," *IEEE Transactions on Power Apparatus and Systems*, Vol. PAS-97, No. 3, pp. 695-702, May-June 1978.
178. K. W. Wang, C. Y. Chung, C. T. Tse, and K. M. Tsang, "Improved Probabilistic Method for Power System Dynamic Stability Studies," *IEE Proceedings Generation, Transmission & Distribution*, Vol. 147, No. 1, pp. 37-43, January 2000.
179. F. L. Pagola, I. J. Perez-Arriaga, and G. C. Verghese, "On Sensitivities, Residues and Participations: Applications to Oscillatory Stability Analysis and Control," *IEEE Transactions on Power Systems*, Vol. 4, No. 1, pp. 278-285, February 1989.
180. J. E. Van Ness and J. M. Boyle, "Sensitivities of Large Multiple-Loop Control Systems," *IEEE Transactions on Automatic Control*, Vol. AC-10, pp. 308-315, 1965.
181. K. W. Wang, C. T. Tse, X. Y. Bian, and A. K. David, "Probabilistic Eigenvalue Sensitivity Analysis and PSS Design in Multimachine Systems," *IEEE Transactions on Power Systems*, Vol. 18, No. 1, pp. 1439-1445, November 2003.
182. E. Chiado, F. Gagliardi, and D. Lauria, "Probabilistic Approach to Transient Stability Evaluation," *IEE Proceedings Generation, Transmission & Distribution*, Vol. 141, No. 5, pp. 537-544, September 1994.
183. G. J. Anders, *Probability Concepts in Electric Power Systems*, John Wiley & Sons, 1990.
184. EPRI, "Moving Toward Probabilistic Reliability Assessment Methods," *Technical Report* 1002639, 2003.
185. EPRI, "Issues and Solutions: North American Grid Planning (2000–2005)," *Technical Report* 1000058, 2006.
186. R. Billinton and R. N. Allan, *Reliability Evaluation of Power Systems*, 2nd Edition, Plenum Press, 1996.
187. R. Billinton and R. N. Allan, *Reliability Evaluation of Engineering Systems: Concepts and Techniques*, 2nd Edition, Plenum Press, 1992.
188. D. J. Hill and I. A. Hiskens, "Dynamic Analysis of Voltage Collapse in Power Systems," in *Proceedings of the* 31st *IEEE Conference on Decision and Control*, Vol. 3, pp. 2904–2909, 1992.
189. D. Karlsson and D. J. Hill, "Modeling and Identification of Nonlinear Dynamic Loads in Power System," *IEEE Transactions on Power Systems*, Vol. 9, No. 1, pp. 157–166, 1994.
190. IEEE Task Force on Load Representation for Dynamic Performance System Dynamic Performance Subcommittee and Power System Engineering Committee, "Standard Load Models for Power Flow and Dynamic Performance Simulation," *IEEE Transactions on Power Systems*, Vol. 10, No. 3, pp. 1302–1313, 1995.
191. L. Pereira, D. Kosterev, P. Mackin, D. Davies, and J. Undrill, "An Interim Dynamic Induction Motor Model for Stability Studies in the WSCC," *IEEE Transactions on Power Systems*, Vol. 17, No. 4, pp. 1108-1115, November 2002.
192. J. Ma, Z. Y. Dong, R. He, and D. J. Hill, "System Energy Analysis Incorporating Comprehensive Load Characteristics," *IEE Proceedings Generation, Transmission & Distribution*, Vol. 1, No. 6, pp. 855–863, 2007.
193. J. Arrillaga and N. R. Watson, *Computer Modeling of Electrical Power Systems* 2nd Edition, John Wiley & Sons, 2001.
194. C. Y. Chung, K. W. Wang, C. K. Cheung, C. T. Tse, and A. K. David, "Machine and Load Modeling in Large Scale Power Industries" in *Proceedings of Workshop on Dynamic Modeling Control Applications for Industries*, pp. 7-15, Vancouver, BC, Canada, April 30–May 1, 1998.
195. M. B. Durić, Z. M. Radojević, and E. D. Turković, "A Reduced Order Multimachine Power System Model Suitable for Small Signal Stability Analysis" *Electrical Power and*

Energy Systems, Vol. 20, No. 5, pp. 369-374, 1998.
196. J. D. Glover and M. Sarma, *Power System Analysis and Designs*, 2nd Edition, PWS Publishing Company, 1994.
197. S. W. Heuins and R. Herman, "A Probabilistic Model for Residentila Consumer Load" *IEEE Transactions on Power Systems*, Vol. 13, No. 3, pp. 621-625, August 2002.
198. IEEE Committee Report, "Computer Representation of Excitation Systems," *IEEE Transactions on Power Apparatus and Systems*, Vol. 87, No. 6, pp. 1460-1464, 1968.
199. IEEE Committee Report, "Dynamic Models for Steam and Hydro Turbines in Power System Studies," *IEEE Transactions on Power Apparatus and Systems*, Vol. 92, No. 6, pp. 1904-1915, 1973.
200. IEEE Committee Report, "IEEE Guide for Synchronous Generator Modeling Practices in Stability Analyses," *IEEE Standards*, pp. 1110-1991, 1991.
201. P. Kundur, *Power System Stability and Control*, McGraw-Hill Inc., 1994.
202. K. Maslo and J. Andel, "Gas Turbine Model Using in Design of Heat and Power Stations," in *Proceedings of the 2001 IEEE Porto Power Tech Conference*, Vol. 4, Porto, Portugal, September 10-13, 2001.
203. J. E. Van Ness and W. F. Goddard, "Formation of the Coefficient Matrix of a Large Dynamic System," *IEEE Transactions on Power Apparatus and Systems*, Vol. 87, No. 1, pp. 80-83, January 1968.
204. M. Weedy and B. J. Cory, *Electrical Power Systems*, 4th Edition, John Wiley & Sons Ltd., 1999.
205. Y. V. Makarov and Z. Y. Dong, *Eigenvalues and Eigenfunctions*, Volume Computational Science & Engineering, Encyclopedia of Electrical and Electronics Engineering, John Wiley & Sons, 1998.
206. H. K. Khalil, *Nonlinear Systems*, 2nd Edition, Prentice-Hall, Upper Saddle River, New Jersey, USA, 1996.
207. R. Seydel, *From Equilibrium to Chaos, Practical Bifurcation and Stability Analysis*, 2nd Edition, Springer-Verlag, New York, USA, 1994.
208. Y. V. Makarov, Z. Y. Dong, and D. J. Hill, "A General Method for Small Signal Stability Analysis," in *Proceedings of the International Conference of Power Industry Computer Applications*, pp.280-286, Columbus, OH, USA, May 11–16, 1997.
209. Y. V. Makarov, V. A. Maslennikov, and D. J. Hill, "Calculation of Oscillatory Stability Margins in the Space of Power System Controlled Parameters," in *Proceedings of the International Symposium of Electric Power Engineering*, pp. 416-422, Stockholm, Sweden, June 18–22, 1995.
210. C. -W. Tan, M. Varghese, P. Varaiya, and F. F. Wu, "Bifurcation, Chaos and Voltage Collapse in Power Systems," D. J. Hill (ed.), *Special Issue on Nonlinear Phenomena in Power Systems: Theory and Practical Implications, IEEE Proceedings*, Vol. 83, No. 11, pp. 1484-1496, November 1995.
211. H. Z. El-Din and R. T. H. Alden, "Second Order Eigenvalue Sensitivities Applied to Power System Dynamics," *IEEE Transactions on Power Systems*, Vol. 96, No. 6, pp. 1928-1936, November 1977.
212. A. Saltelli, K. Chan, and E. M. Scott, *Sensitivity Analysis*, John Wiley & Sons Ltd., 2000.
213. K. W. Wang, C. Y. Chung, C. T. Tse, and K. M. Tsang, "Probabilistic Eigenvalue Sensitivity Indices for Robust PSS Site Selection," *IEE Proceedings Generation, Transmission & Distribution*, Vol. 148, No. 6, pp. 603-609, November 2001.
214. K. W. Wang, C. T. Tse, and K. M. Tsang, "Algorithm for Power System Dynamic Stability Studies Taking Account of the Variation of Load Power," *Electric Power Systems*

Research, Vol. 46, pp. 221-227, 1998.

215. H. Cramer, *Mathematical Methods of Statistics*, Princeton University Press, 1966.
216. A. Papoulis, *Probability, Random Variables and Stochastic Processes*, McGraw-Hill Inc., 1984.
217. M. G. Kendall, "Proof of Relations Connected with the Tetrachoric Series and its Generalization," *Biometrika*, Vol. 32, No. 2, pp. 196-198, October 1941.
218. J. M. Undrill and T. F. Laskowski, "Model Selection and Data Assembly for Power System Simulation," *IEEE Transactions on Power Apparatus and Systems*, PAS-01, pp. 3333-3341, 1982.
219. W. A. Kao, "The Effect of Load Models on Unstable Low Frequency Oscillation Damping in Tai Power System Experience w/wo Power System Stabilizers," *IEEE Transactions on Power Systems*, Vol. 16, Vol. 3, pp. 463-472, August 2001.
220. K. Morison, H. Hamadani, and L. Wang, "Practical Issues of Load Modeling for Voltage Stability Studies," in *Proceedings of the 2003 IEEE PES General Meeting*, Toronto, Canada, July 2003.
221. J. Ma, D. Han, R. M. He, Z. Y. Dong, and D. J. Hill, "Research on Identifiability of Equivalent Motor in Composite Load Model," in *Proceedings of Power Tech 2007*, Lausanne, Switzerland, July 1-5, 2007.
222. J. Ma, R. He, and D. J. Hill, "Load Modeling by Finding Support Vectors of Load Data from Field Measurements," *IEEE Transactions on Power Systems*, Vol. 21, No. 2, pp. 726-735, 2006,.
223. EPRI, "Advanced Load Modelling," *Technical Report* 1007318, 2002.
224. J. Hockenberry and B. Lesieutre, "Evaluation of Uncertainty in Dynamic Simulation of Power System Models: The Probabilistic Collocation Method," *IEEE Transactions on Power Systems*, Vol. 15, No. 1, pp. 299–306, 2000.
225. E. E. S. Lima and L. F. J. Fernandes, "Assessing Eigenvalue Sensitivities," *IEEE Transactions on Power Systems*, Vol. 15, No. 1, pp. 299-306, 2000.
226. J. V. Milanovic, I. A. Hiskens, and V. A. Maslennikov, "Ranking Loads in Power Systems—Comparison of Different Approaches," *IEEE Transactions on Power Systems* Vol. 14, No. 2, pp. 614–619, 1999.
227. S. -H. Li, H. -D. Chiang, and S. Liu, "Analysis of Composite Load Models on Load Margin of Voltage Stability," in *Proceedings of the International Conference on Power System Technology*, pp. 1—7, Chongqing, China, October 22-26, 2006.
228. C. W. Taylor, *Power System Voltage Stability*, McGraw-Hill, 1994.
229. Z. Y. Dong, Y. Mishra, and P Zhang, "Power System Sensitivity Analysis Considering Induction Motor Loads," in *Proceedings of the* 8th *IET International Conference on Advances in Power System Control, Operation, and Management*, Kowloon, Hong Kong SAR, November 8-11, 2009.
230. P. W. Sauer and M. A. Pai, *Power System Dynamics and Stability*, Pearson Education, 2003.
231. Z. Y. Dong, D. J. Hill, and Y. Guo, "A Power System Control Scheme Based on Security Visualisation in Parameter Space," *International Journal of Electrical Power and Energy Systems*, Vol. 27, No. 7, pp. 488-495, 2005.
232. D. J. Hill and Z. Y. Dong, "Nonlinear Computation and Control for Small Disturbance Stability," *Invited Paper, IEEE/PES SM 2000 Panel Session on Recent Applications of Linear Analysis Techniques*, 2000.
233. J. Ma, D. Han, R. M. He, Z. Y. Dong, and D. J. Hill, "Reducing Identified Parameters of Measurement Based Composite Load Model," *IEEE Transactions on Power Systems*,

Vol. 23, No. 1, pp. 76-83, 2008.
234. F. L. Lewis, D. A. Tacconi, O. C. Pastravanu, and A. Gurel, "Method and Apparatus for Testing and Controlling a Flexible Manufacturing System," US Patent No. 6,185,469 B1, 6th February 2001.
235. B. Harris, D. J. Cook, and F. L. Lewis, "Automatically Generating Plans for Manufacturing," *Journal of Intelligent Systems*, Vol. 10, pp. 279-319, 2000.
236. B. Harris, D. J. Cook, and F. L. Lewis, "Combining Representations from Manufacturing, Machine Planning, and Manufacturing Resource Planning (MRP)," in *Proceedings of the AAAI Workshop on Representational Issues for Real-World Planning Systems*, 2000.
237. B. Harris, F. L. Lewis, and D. J. Cook, "Machine Planning for Manufacturing: Dynamic Resource Allocation and On-line Supervisory Control," *Journal of Intelligent Manufacturing*, Vol. 9, No. 5, pp. 413-430, 1998.
238. S. Bogdan, F. L. Lewis, Z. Kovacic, and J. Mireles, Jr., *Manufacturing Systems Control Design: A Matrix Based Approach (Advances in Industrial Control)*, Springer-Verlag, London, 2006.
239. S. A. Thibadoux, "Robotic Acquisition Programs–Technical and Performance Challenges," in Unmanned Ground Vehicle Technology IV, *Proceedings of SPIE 4715*, pp. 128-138, Orlando, FL, USA, 2002.
240. J. S. Albus, "Intelligent Systems Design," Plenary Talk, Army Science Conference, Orlando, FL, USA, 2008.
241. J. A. Bornstein, "Army Ground Robotics Research Program," in Unmanned Ground Vehicle Technology IV, *Proceedings of SPIE 4715*, pp. 118-127, Orlando, FL, USA, 2002.
242. H. K. Shah, V. Bahl, J. Martin, N. S. Flan, and K. L. Moore, "Intelligent Behavior Generator for Autonomous Mobile Robots Using Planning-Based AI Decision-Making and Supervisory Control Logic," in *Proceedings of SPIE 4715*, pp. 161-177, Orlando, FL, USA, 2002.
243. W. Smuda, P. Muench, G. Gerhart, and K. L. Moore, "Autonomy and Manual Operation in a Small Robotic System for Under-Vehicle Inspections at Security Checkpoints," in Unmanned Ground Vehicle Technology IV, *Proceedings of SPIE 4715*, pp. 1-12, Orlando, FL, USA, 2002.
244. D. V. Steward, "The Design Structure System: A Method for Managing the Design of Complex Systems," *IEEE Transactions on Engineering Management*, Vol. 28, No. 3, pp. 71-74, August 1981.
245. J. D. Wolter, S. Chakrabarty, and J. Tsao, "Methods of Knowledge Representation for Assembly Planning," in *Proceedings of the 18th Annual NSF Conference on Design and Manufacturing Systems Research*, pp. 463-469, Atlanta, GA, USA, January 1992.
246. S. H. Young and H. M. Nguyen, "System Design For Robot Agent Team," in Unmanned Ground Vehicle Technology IV, *Proceedings of SPIE 4715*, pp. 31–42, Orlando, FL, USA, 2002.
247. A. Kusiak and J. Ahn, "Intelligent Scheduling of Automated Machining Systems," *Computer-Integrated Manufacturing Systems*, Vol. 5, No. 1, pp. 3-14, 1992, (Reprinted in A. Kusiak (Ed.), *Intelligent Design and Manufacturing*, John Wiley, New York , NY, pp. 421-447, 1992).
248. A. Gurel, S. Bogdan, and F. L. Lewis, "Matrix Approach to Deadlock-Free Dispatching in Multi-Class Finite Buffer Flowlines," *IEEE Transactions on Automatic Control*, Vol. 45, No. 11, pp. 2086-2090, 2000.
249. H. -H. Huang, P. J. Gmytrasiewicz, and F. L. Lewis, "Combining Matrix Formulation

and Agent-Oriented Techniques For Conflict Resolution in Manufacturing Systems," in *Proceedings of ACME*, pp. 66-79, Chicago, IL, USA, 1996.
250. R. A. Wysk, N. S. Yang, and S. Joshi, "Detection of Deadlocks in Flexible Manufacturing Cells," *IEEE Transactions on Robotics and Automation*, Vol. 7, No. 6, pp. 853-859, 1991.
251. G. S. Gursaran, S. Kanungo, and A. K. Sinha, "Rule-Base Content Verification Using a Digraph-Based Modelling Approach," *Artificial Intelligence in Engineering*, Vol. 13, No. 3, pp. 321-336, 1999.
252. http://www.army.mil/fcs/index.html [Online].
253. http://www.defensetech.org/archives/cat_fcs_watch.html [Online].
254. H. De Keulenaer, R. Belmans, E. Blaustein, D. Chapman, A. De Almeida, B. De Wachter, and P. Radgen, "Energy Efficient Motor Driven Systems...can save Europe 200 billion kWh of electricity consumption and 100 million tonne of greenhouse gas emissions a year," http://re.jrc.ec.europa.eu/energyefficiency/pdf/HEM_lo_all%20final.pdf [Online], 2004.
255. EU SAVEII Project, "Promotion of Energy Efficiency in Circulation Pumps, Especially in Domestic Heating Systems," Final Report, Contract No. 4.1031/-Z/99-256, June 2001.
256. J. Haataja and J. Pyrhonen, "Improving Three-Phase Induction Motor Efficiency in Europe," *IEE Power Engineering Journal*, Vol. 12, No. 2, pp. 81-86, April 1998.
257. D. Walters, "Energy-Efficient Motors-Saving Money or Costing the Earth? Part 1," *IEE Power Engineering Journal*, Vol. 13, No. 1, pp. 25-30, February 1999.
258. D. Walters, "Energy-Efficient Motors-Saving Money or Costing the Earth? Part 2," *IEE Power Engineering Journal*, Vol. 13, No. 2, pp. 17-23, April 1999.
259. B. Lu, T. G. Habetler, R. G. Harley, J. A. Gutierrezand, and D. B. Durocher, "Energy Evaluation Goes Wireless," *IEEE Industry Applications Magazine*, Vol. 13, No. 2, pp. 17-23, March 2007.
260. T. Sattia, K. Younga, and S. Groverb, "Detecting Catastrophic Failure Events in Large-Scale Milling Machines," *International Journal of Machine Tools and Manufacture*, Vol. 49, No. 14, pp. 1104-1113, November 2009.
261. M. Jamshidi, *Systems of Systems Engineering: Principles and Applications*, CRC Press, 2008.
262. M. Maier, "Architecting Principles for Systems-of-Systems," *Systems Engineering*, Vol. 1, No. 4, pp. 267-284, 1998.
263. J. Dahmann and K. J. Baldwin, "Understanding the Current State of US Defense Systems of Systems and the Implications for Systems Engineering," in *Proceedings of the IEEE Systems Conference*, Montreal, Quebec, Canada, April 7-10, 2008.

and Agent-based Techniques for Conflict Resolution in Manufacturing Systems," in *Proceedings of ACMS*, pp. [illegible]-12, Chicago, IL, USA, 1996.

[illegible]. R. A. Wysk, N. S. Yang, and S. Joshi, "Detection of Deadlocks in Flexible Manufacturing Cells," *IEEE Transactions on Robotics and Automation*, Vol. 7, No. 6, pp. 853–859, 1991.

[illegible]. G. S. [illegible], S. [illegible], and A. K. [illegible], "Rule-based Content Verification Using a [illegible] Modeling Approach," *Artificial Intelligence in Engineering*, Vol. 13, [illegible], 1999.

[illegible] [Online].

[illegible] watch.html [Online].

[illegible]

Index